Optimization in Control Applications

Optimization in Control Applications

Special Issue Editors

Guillermo Valencia-Palomo
Francisco Ronay López-Estrada

MDPI • Basel • Beijing • Wuhan • Barcelona • Belgrade

MDPI

Special Issue Editors

Guillermo Valencia-Palomo
Instituto Tecnológico de
Hermosillo
Mexico

Francisco Ronay López-Estrada
Instituto Tecnológico de Tuxtla
Gutiérrez
Mexico

Editorial Office
MDPI
St. Alban-Anlage 66
4052 Basel, Switzerland

This is a reprint of articles from the Special Issue published online in the open access journal *Mathematical and Computational Applications* (ISSN 2297-8747) in 2018 (available at: https://www. mdpi.com/journal/mca/special_issues/optimization_in_control)

For citation purposes, cite each article independently as indicated on the article page online and as indicated below:

LastName, A.A.; LastName, B.B.; LastName, C.C. Article Title. *Journal Name* **Year**, *Article Number*, Page Range.

ISBN 978-3-03897-447-5 (Pbk)
ISBN 978-3-03897-448-2 (PDF)

Contents

About the Special Issue Editors

Guillermo Valencia-Palomo was born in Merida, Yucatan, Mexico, in 1980. He received an Engineering degree in Electronics from the Instituto Tecnológico de Mérida, Mexico, in 2003; an M.Sc. in Automatic Control from the National Center of Research and Technological Development (CENIDET), Mexico, in 2006; and a Ph.D. degree in Automatic Control and Systems Engineering from The University of Sheffield, U.K., in 2010. Since 2010, Dr. Guillermo Valencia-Palomo has been a full-time professor at Tecnológico Nacional de México/Instituto Tecnológico de Hermosillo (Mexico). He is the author/co-author of more than 80 research papers published in ISI-Journals and international conferences. As a product of his research, he has one patent in commercial exploitation. He has led a number of funded research projects, and these grants' income represents a mixture of sole investigator funding, collaborative grants, and funding from industry. His research interests include predictive control, descriptor systems, linear parameter varying systems, fault detection, fault tolerant control systems, and their applications to different physical systems.

Francisco-Ronay López-Estrada received his Ph.D. in Automatic Control from the University of Lorraine, France, in 2014. He has been with Tecnológico Nacional de México/Instituto Tecnológico de Tuxtla Gutiérrez, Mexico, as a lecturer since 2008. He received his M.Sc. degree in Electronic Engineering in 2008 from the National Center of Research and Technological Development (CENIDET), Mexico. He has led several funded research projects. His research interests are descriptor systems, TS systems, fault detection, fault-tolerant control, and their applications to unmanned vehicles and pipeline leak detection systems.

Preface to "Optimization in Control Applications"

Mathematical optimization is the selection of the best element in a set with respect to a given criterion. Optimization has become one of the most used tools in modern control theory for computing the control law, adjusting the controller parameters (tuning), model fitting, finding suitable conditions in order to fulfill a given closed-loop property, etc. In the simplest case, optimization consists of maximizing or minimizing a function by systematically choosing input values from a valid input set and computing the function value. To solve optimization problems, researchers can use algorithms that end in a finite number of steps, or iterative methods that converge to a solution (in some specific class of problems), or heuristics that can provide approximate solutions to some problems (although their iterations do not necessarily converge). In practice, real-world control systems need to comply with several conditions and physical and product-quality constraints that have to be taken into account in the problem formulation. These represent challenges in the application/implementation of the optimization algorithms, particularly when the solutions of these optimization problems have to be computed in a constrained time window and/or in an embedded platform.

This Special Issue provides a forum for high-quality peer-reviewed papers that broaden the awareness and understanding of advanced optimization techniques and their applications in control engineering. This topic encompasses many algorithms and process flows and tools, including: optimal control of nonlinear systems; optimal control of complex systems; optimal observer design; numerical optimization; evolutionary optimization; and constrained optimization; among others. Specifically, this Special Issue gathers twelve papers that contribute to this topic by presenting: rapid solutions of optimal control problems by a functional spreadsheet paradigm; a novel spreadsheet direct method for optimal control problems; a fixed point method for a free isoperimetric optimal control problem to control an epidemic with restricted resources in an SIR model with a short-term controller population; optimal strategies for psoriasis treatment; an optimal control analysis of a mathematical model for breast cancer; a cost-effective analysis of control strategies to reduce the prevalence of cutaneous leishmaniasis based on a mathematical model; an optimal control and computation method for the solution of an isoperimetric problem in a discrete-time SIRS system; a solution of an optimal harvesting problem by finite difference approximations of a size-structured population model; a solution of fuzzy differential equations using fuzzy Summudu transformations; the development of a spectral observer for the reconstruction of a time signal via state estimation and its frequencies decomposition; a differential evolution algorithm for a multilevel assignment problem; and the modelling and simulation of a hydraulic network for leak diagnosis and optimal control.

We believe that the papers in this Special Issue reveal an exciting area which can be expected to continue to grow in the very near future—namely, the use of advanced optimization strategies in engineering applications. The pursuit of work in this area requires expertise in control engineering as well as in systems design and numerical analysis. We hope that this issue helps to bring these communities into closer contact with each other, as the fruitfulness of collaboration across these areas becomes clear.

Guillermo Valencia-Palomo, Francisco Ronay López-Estrada
Special Issue Editors

Mathematical and Computational Applications

MDPI

Article

Rapid Solution of Optimal Control Problems by a Functional Spreadsheet Paradigm: A Practical Method for the Non-Programmer

Chahid Kamel Ghaddar

ExcelWorks LLC, Sharon, MA 02067, USA; cghaddar@excel-works.com; Tel.: +1-781-626-0375

Received: 29 August 2018; Accepted: 26 September 2018; Published: 28 September 2018

Abstract: We devise a practical and systematic spreadsheet solution paradigm for general optimal control problems. The paradigm is based on an adaptation of a partial-parametrization direct solution method which preserves the original mathematical optimization statement, but transforms it into a simplified nonlinear programming problem (NLP) suitable for Excel NLP solver. A rapid solution strategy is implemented by a tiered arrangement of pure elementary calculus functions in conjunction with Excel NLP solver. With the aid of the calculus functions, a cost index and constraints are represented by equivalent formulas that fully encapsulate an underlining parametrized dynamical system. Excel NLP solver is then employed to minimize (or maximize) the cost index formula, by varying decision parameters, subject to the constraints formulas. The paradigm is demonstrated for several fixed and free-time nonlinear optimal control problems involving integral and implicit dynamic constraints with direct comparison to published results obtained by fundamentally different methods. Practically, applying the paradigm involves no more than defining a few formulas using basic Excel spreadsheet skills.

Keywords: optimal control; dynamic optimization; mathematical programming; differential equations; parameter estimation; Excel spreadsheet; calculus functions

1. Introduction

Many researchers and academics often need to solve optimal control problems that are frequently postulated in various engineering, social, and life sciences [1–3]. An optimal control problem is concerned with finding control functions, (or policies), that achieve optimal trajectories for a set of controlled differential state variables. The optimal trajectories are determined by solving a constrained dynamical optimization problem, such that a cost index is minimized (or maximized), subject to constraints on state variables and control functions. Mathematically, an optimal control problem may be stated generally as follows (bold symbols indicate vector-valued functions):

Find control functions $u(t) = (u_1(t), u_2(t), \ldots, u_m(t))$ and corresponding state variables $x(t) = (x_1(t), x_2(t), \ldots, x_n(t))$, $t \in [t_0, t_F]$ which minimize (or maximize) the cost index

$$J = H(x(T), T) + \int_{t_0}^{t_F} G(x(t), \dot{x}(t), \ddot{x}(t), u(t), \dot{u}(t), t) \, dt, \tag{1}$$

subject to

$$M\frac{dx}{dt} = F(x(t), u(t), t), \tag{2}$$

with initial conditions

$$x(0) = x_0, \tag{3}$$

and end conditions and bounds

$$Q(x(T), T) = 0, \qquad (4)$$

$$S\big(x(t), u(t), \dot{x}(t), \dot{u}(t)\big) \leq 0. \qquad (5)$$

In the formulation (1)–(5), the generally nonlinear H, and G are scalar functions, whereas F, Q and S are vector valued functions. Typically, either H or Q are specified but not both in the same problem. Common forms of Q and S are end conditions on the state variables, $x(T) = x_T$, and bound constraints on the controls, $u_{min} \leq u(t) \leq u_{max}$ respectively. More general forms of S considered in this paper include algebraic and integral constraints involving derivatives. The matrix M in (2) offers an optional coupling of states' temporal derivatives by a mass matrix which may be singular. If M is singular, the equation system (2) is differential algebraic, or DAE. For uncoupled derivatives, M is the identity matrix which can be omitted. Furthermore, t_F, which denotes the final time, may be fixed or free.

Numerical solution strategies for (1)–(5) can be classified into two approaches: *indirect* and *direct* methods. Indirect methods employ Pontryagin's minimum principle to transform the problem into an augmented Hamiltonian system requiring the solution of a boundary value problem which may be hard to solve [4,5]. On the other hand, direct method approaches transform the original optimal control problem into a nonlinear programming problem which can be solved by various established NLP packages. The transformation is carried out via a discretization of the control and the state functions on a time grid using some form of a collocation method [4,6,7]. Complete discretization of the state and control functions eliminate the need to iteratively solve the *inner* initial value problem (IVP) (2) but at the expense of a large numbers of decision variables for the NLP solver. Other direct approaches rely only on a partial parametrization for the control functions using piecewise constant or higher order polynomial approximations [8]. In this approach, the *inner* IVP must be solved repeatedly by the *outer* NLP algorithm while searching for the optimal parameter vector. Except for the most trivial cases, optimal control problems are inherently nontrivial to solve. They typically require a level of programming fluency, in addition to a good understanding of the general structure of the solution strategy, and the various solvers required to implement it [9].

In [10], the author introduced a practical spreadsheet method for solving a class of optimal control problems using basic spreadsheet skills. The method utilized two elementary calculus functions: an initial value problem solver and a discrete data integrator from an available Excel calculus Add-in [11] in conjunction with Excel intrinsic NLP solver to formulate a partial-parametrization direct solution strategy. With the aid of the calculus functions, a cost index was represented by an equivalent formula that fully encapsulated a control-parametrized inner IVP (2)–(3). Excel NLP solver was employed next for minimizing (or maximizing) the cost index formula, by varying a decision parameter vector, subject to bounds constraints on state and control variables. The method proved effective at solving several nonlinear optimal control problems reproduced from Elnagar and Kazemi [6] who employed a full-parametrization direct method using pseudo-spectral approximation and NLPQL optimization software.

This research paper aims at generalizing the method introduced in [10] for more general formulations of optimal control than previously considered. More specifically, this paper demonstrates a systematic solution strategy formulated by the aid of various elementary calculus functions, for optimal control problems involving one or more of the following conditions: dependence on higher order derivatives of state or control variables in the cost index and constraints; integral and algebraic dynamic constraints; as well as implicit inner IVP. In addition, this paper investigates convergence and error control of the method, and provides direct comparison of optimal trajectories with published solutions obtained by fundamentally different methods.

It should be noted that the solution strategy formulation pursued in this research, although founded on a common approach, follows closely the original mathematical problem statement, and thus implementation of the strategy varies according to the given problem. Therefore, the paper gives considerable emphasis on the application of the method using four representative problems selected from various applications. Results presented in Section 3 are remarkable, in terms of convergence, agreement with published solutions, and notably, the minimal effort required to obtain them with basic spreadsheet formulas.

In view of traditional spreadsheet applications, the devised solution strategy represents a leap in the utilization of the spreadsheet for solving general optimal control problems. The strategy departs markedly from prior spreadsheet approaches [12,13] by shifting the effort from a low-level detailed algorithmic implementation to a high-level problem modeling. Prior approaches utilized the spreadsheet explicitly as the computational grid for the discretization and solution of the inner IVP. This effectively constrained the scope to rather simple problems that can be easily discretized with an explicit differencing scheme suitable for the spreadsheet. In contrast, we employ a set of pure calculus functions for computing integrals, derivatives and solving differential equations as the building blocks for a direct solution method. The calculus functions, described in Appendix A, utilize adaptive algorithms which are independent of the spreadsheet grid and thus suitable for a general class for nonlinear stiff problems. The calculus functions are utilized in formulas just like intrinsic math functions based on a simple input/output model. In essence, the calculus functions represent a natural extension of the built-in spreadsheet math functions with the allowance that some of their input arguments are functions themselves and not just static values.

The reminder of this paper is organized as follows: In the next section, we present an outline of the general steps required to implement the direct spreadsheet solution strategy, and discuss sources of errors that impact convergence and accuracy of the solution as well as possible remedies. In Section 3, we apply the method for solving four different optimal control problems selected to demonstrate the various conditions outlined earlier. Direct comparisons of optimal trajectories obtained by the method versus published solutions obtained by fundamentally different approaches are also provided. In addition, effects of parametrization order and error control are investigated in some problems. Section 4 presents concluding remarks as well as directions for future research. Detailed descriptions of the various calculus functions utilized in this work are included in Appendix A.

2. Mechanics of Spreadsheet Direct Method

The solution strategy is based on an adaptation of the control-parametrization direct approach [4,8] by an analogous spreadsheet functional formulation. The building blocks of the functional formulation are a set of calculus spreadsheet functions [11,14] which integrate with the spreadsheet, like intrinsic pure math functions, but also accept formulas as a new type of argument for solving problems in integral, algebraic, and differential calculus. For example, an integration function accepts a formula and limits as inputs, and it outputs an accurate integral value much like an intrinsic math function accepts a number and computes its square root. Specifically, we make use of the following functions from a calculus Add-in [11]:

- Initial value problem solver, IVSOLVE, using RADAU5 an implicit 5th-order Runge-Kutta algorithm with adaptive time step [15].
- Discrete data Integrator, QUADXY, using cubic splines [16].
- Discrete data differentiator, DERIVXY, using cubic splines [16].
- Formula integrator, QUADF, using Gauss quadrature with adaptive error control [17].

The functions are utilized in combination with Excel NLP solver, which is based on the Generalized Reduced Gradient algorithm based on Lasdon and Waren [18]. A detailed description of the calculus functions usage, and respective algorithms are given in Appendix A. The critical characteristic of the calculus functions which permits their seamless utilization with the NLP solver in a functional paradigm, is the mathematical purity property. The calculus functions do not modify their inputs, and produce no side effects in the spreadsheet. They only compute and display a solution result in their allocated spreadsheet memory cells. The authority to modify the inputs to the calculus functions, via changes to the decision parameter vector, is confined to the outer NLP solver command.

3

Below, we describe the main elements of the solution strategy introduced originally in [10] but generalized in this work for solving general optimal control problem (1)–(5) with the aim of supporting the various conditions outlined earlier.

2.1. Solution Strategy

The strategy comprises three ordered steps which are implemented by the aid of calculus functions:

In the first step, we obtain an initial solution to the inner IVP (2)–(3), based on suitable parametrization for the control functions with initial guesses for the unknown parameters and a final time for free-time problems. The unknown parameters and the final time constitute the decision variables for the final optimization step by the outer NLP solver. Any prior information about the controls should be incorporated in the specified parametrization. Absent any information, a low-order polynomial is often an adequate choice. The initial IVP solution is obtained by the calculus function IVSOLVE which displays the state variables, $x(t)$, in an allocated array of the spreadsheet at uniform output time points. It should be noted that output time grid is determined by the number of rows in the allocated output array but is, otherwise, unrelated to the accuracy of the computed solution. To display a finer output time grid, a larger output array should be allocated. However, the resolution of the output time grid affects the accuracy of the computed integrals for the cost index and any integral constraints which is discussed in Section 2.2. Optional parameters to IVSOLVE could also be used to control or specify the output time points.

In the second step, we construct an analogous formula for the cost index (1) dependent on the initial solution outputted by IVSOLVE. The cost index may depend on $x(t)$, the control values, $u(t)$, as well as first and higher order derivatives of the state variables and controls. Values for $u(t)$, $\dot{u}(t)$ and higher derivatives are readily generated using the specified parametrized formula for a control $u(t)$. The spreadsheet is particularly suited for such computations using its AutoFill feature. On the other hand, values for the state variables derivatives $\dot{x}(t)$, and $\ddot{x}(t)$ are not readily available and must be approximated by differentiating $x(t)$ values obtained by IVSOLVE. We accomplish this task by the aid of a discrete data differentiator calculus function DERIVXY which computes derivatives using cubic splines to model the best function described by $x(t)$. With all the necessary values obtained, we proceed to defining an analogous formula for the cost index, which is typically defined as a continuous time integral of an algebraic integrand. The devised method is to sample the integrand expression using the obtained values for the states, controls and their derivatives, followed by employing a discrete data integrator calculus function QUADXY to integrate a cubic-spline fit function through the sampled integrand. Depending on a particular problem formulation, it may be necessary to define additional formulas to represent constraints equations (5) that may be present. Such formulas can often be constructed in a similar way to the cost index formula using appropriate calculus functions. In particular, we shall demonstrate in Section 3 using an additional formula integrator function QUADF to define an integral constraint formula.

Figure 1 illustrates the aforementioned steps applied to an optimal control problem with one control and two state variables. An initial IVP solution, which is dependent on a decision parameters vector, is obtained with IVSOLVE in an array (Figure 1a). Values for the control, $u(t)$, and any needed state derivatives such as $\ddot{x}_1(t)$, are generated in additional columns (Figure 1b,c) at the time values of the IVP solution. Next, the cost index integrand expression is sampled at the IVP solution times (Figure 1d), and the sample is then integrated to define the cost index formula (Figure 1e). The generated values interdependence hierarchy ensures that any change to the decision parameters vector, such as by an outer NLP solver, will trigger reevaluation of the cost index formula in the proper order shown in the figure. The cost index formula thus fully encapsulates the inner IVP problem.

In the last step, we configure Excel NLP solver to minimize (or maximize) the cost index formula by varying the decision parameters vector subject to bounds, end conditions and other present constraints. Bound constraints on $x(t)$, as well as end point constraints on $x(T)$, are imposed directly on the corresponding values in the IVP solution array. More general constraints are imposed on additional formulas constructed in step 2 as needed. The three steps are demonstrated on several examples in the next section.

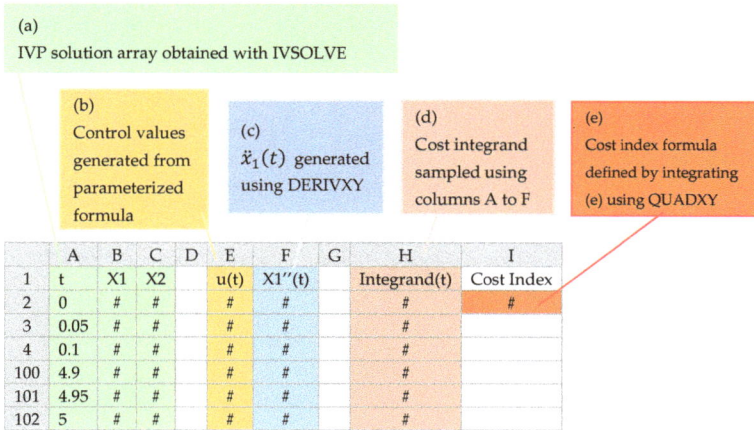

(a)
IVP solution array obtained with IVSOLVE

(b)
Control values generated from parameterized formula

(c)
$\ddot{x}_1(t)$ generated using DERIVXY

(d)
Cost integrand sampled using columns A to F

(e)
Cost index formula defined by integrating (e) using QUADXY

	A	B	C	D	E	F	G	H	I
1	t	X1	X2		u(t)	X1''(t)		Integrand(t)	Cost Index
2	0	#	#		#	#		#	#
3	0.05	#	#		#	#		#	
4	0.1	#	#		#	#		#	
100	4.9	#	#		#	#		#	
101	4.95	#	#		#	#		#	
102	5	#	#		#	#		#	

Figure 1. Illustration of the ordered steps to define an analog formula for the cost index (1) which encapsulates the inner IVP (2)–(3).

2.2. Convergence and Error Control

Two sources of errors are introduced by the spreadsheet method with respect to the original problem. The first error is introduced by restricting the space of admissible control functions to a finite-dimensional space, for example, variable-order polynomials up to a fixed degree. For some problems, it may not be possible to find a solution if the optimal control, in fact, lies outside the admissible space. The second source of error is introduced by the calculus numerical algorithms. This error can be further split into two sources. The error associated with solution of the inner IVP, and the error associated with integration (or differentiation) of discrete data sets generated from the IVP solution. The first error is bounded by the tolerances specified for IVSOLVE algorithm. The second error impacts the accuracy of the computed integral for the cost index. Under the assumption that the discrete data describe a smooth curve, the computed integral by QUADXY using cubic splines is generally quite accurate. However, it may be further improved by any of the following acts.

- Increasing the size of the data set by increasing the number of rows of the allocated IVP solution array to output a finer time grid.
- Supplying optional slopes at the end points of the curve to the calculus function when available. The slopes may be derived analytically from the integrand expression and can improve the accuracy of the spline fit near the curve edges.
- Using nonuniform output time points clustered near rapidly-varying regions of the state trajectories. This can be controlled via optional arguments to IVSOLVE including supplying exact values for the output time points.

In practice, we have found that the parametrization order and the starting guess for unknown parameters to be the most important factors influencing convergence. We have generally used polynomials up to 5th order which have performed reasonably well. On the other hand, increasing the output array for IVP solution beyond a reasonable size, on the order of 100 uniform subdivisions for the time interval, has not generally resulted in a consistent or significant improvement of the result. In the examples in the next section, we shall demonstrate the effects of both increasing the parametrization order and reducing the output time interval.

Math. Comput. Appl. **2018**, *23*, 54

3. Illustrative Optimal Control Problems

In the following subsections we apply the method to four different optimal control problems representing various engineering applications and compare the optimal trajectories with published solutions. The computations were carried out on a standard laptop computer with an Intel i7 four-core processor at 2.70 GHz running Microsoft Windows 10 and Excel 2016 with ExceLab calculus add-in [11], which enables the calculus function in Excel. A supplementary Excel workbook containing the solved examples is available for downloading from the publisher.

3.1. Minimum Energy Shape: Hanging Chain

The first example is concerned with finding the shape $u(t)$ of a chain of length L suspended between two points, such that its total energy is minimized. We state the problem as described in [19] with $L = 4$, below:

Find $u(t)$ which minimizes the total energy cost index

$$J = \int_0^1 u(t)\sqrt{1 + \dot{u}(t)^2}dt, \tag{6}$$

subject to the chain length constraint

$$\int_0^1 \sqrt{1 + \dot{u}(t)^2}dt = 4, \tag{7}$$

and the end conditions

$$u(0) = 1, \tag{8}$$

$$u(1) = 3. \tag{9}$$

Note that in this problem formulation, the inner IVP is implicitly defined by the integral constraint (7). Dolan et al. [19] reformulated the problem, via variable substitution, as a standard optimal control problem subject to a system of explicit differential equations and solved it by a direct approach. Discretization was done using a uniform time step and the trapezoidal rule for the integration. Results for the AMPL implementation were reported using several solvers including KNITRO and LOQO. The best cost index was found at 5.06852 starting from a quadratic approximation and using a grid of 800 nodes. Our spreadsheet solution below is formulated based on the original problem statement (6)–(9).

3.1.1. Solution by Direct Spreadsheet Method

Referring to Figure 2, we setup problem (6)–(9) in Excel using named variables with labels listed in column A. The shape function $u(t)$ was parametrized using a 3rd order polynomial with unknown coefficients c_0, c_1, c_2 and c_3 as shown by formula *B7*. In *B15* and *B16*, formulas for the initial and final values, $u(0)$ and $u(1)$ were defined by evaluating *B7* at time equal zero and one (these formulas are used later to impose the constraints (8)–(9)). An additional formula was defined in *B8*, (named udot), for the shape function derivative, $\dot{u}(t)$ by differentiating *B7* with respect to time. Next, we defined the cost index integral (6), by using the integration calculus function QUADF as shown in *B11*. The first parameter to QUADF is the integrand $u(t)\sqrt{1 + \dot{u}(t)^2}$ which is defined by the equivalent formula in *B10*. The 2nd parameter is the variable of integration t, and the 3rd and 4th parameters are the integration limits. Likewise, with the aid of QUADF, we defined the constraint integral (7) as shown in *B14* (named I_c). This completed the model needed to run Excel NLP solver.

3.1.2. Results and Analysis

Excel NLP solver is invoked from the Data tab on Excel Ribbon and displays a dialog to enter the problem objective, variables and constraints. Figure 3 shows the inputs for problem 3.1 in which the objective J (*B11*), was selected to be minimized, by varying the parameters c_0, c_1, c_2 and c_3, subject to the three constraints: $I\_c = 4$, corresponding to (7); $u\_0 = 1$, corresponding to (8); and $u\_1 = 3$, corresponding to (9).

	A	B
1	t	
2	**Parametrized chain shape function**	
3	c_0	0
4	c_1	0
5	c_2	0
6	c_3	0
7	u	=c_0+c_1*t+c_2*t^2+c_3*t^3
8	udot	=c_1+2*c_2*t+3*c_3*t^2
9	**Cost Index**	
10		=u*(SQRT(1+udot^2))
11	J	=QUADF(B10,t,0,1)
12	**Constraints definitions**	
13		=SQRT(1+udot^2)
14	I_c	=QUADF(B13,t,0,1)
15	u_0	=c_0
16	u_1	=c_0+c_1+c_2+c_3

Figure 2. Spreadsheet parametrized model for problem 3.1.

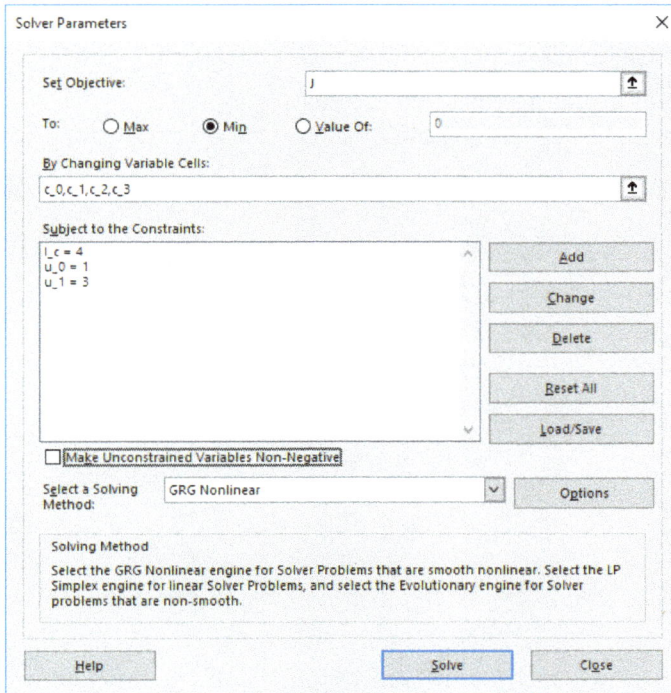

Figure 3. Input to Excel solver for problem 3.1 based on the spreadsheet model in Figure 2.

The solver converged, starting from a zero guess for the parameters in less than a second to the result shown in Figure 4 with a final cost index of 5.0751. The optimal shape function $u(t)$ is plotted in Figure 5 together with digitally-read values from the plot published in [19].

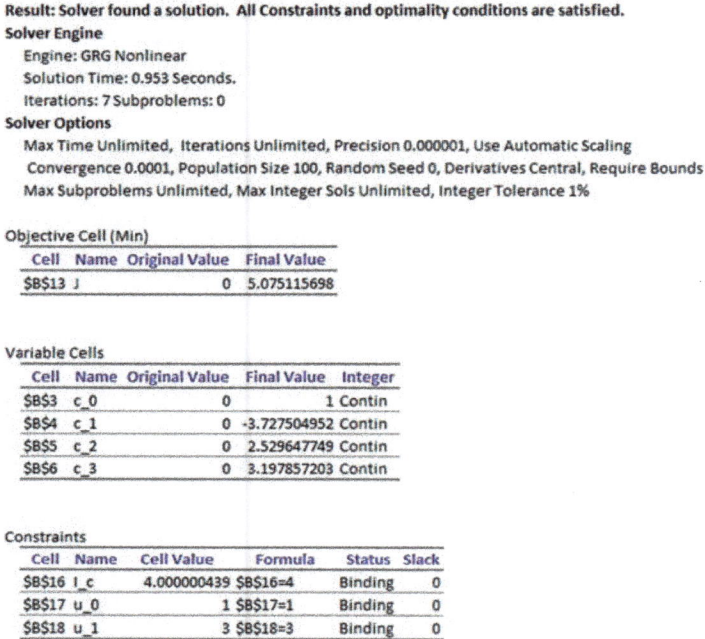

Result: Solver found a solution. All Constraints and optimality conditions are satisfied.
Solver Engine
   Engine: GRG Nonlinear
   Solution Time: 0.953 Seconds.
   Iterations: 7 Subproblems: 0
Solver Options
   Max Time Unlimited, Iterations Unlimited, Precision 0.000001, Use Automatic Scaling
   Convergence 0.0001, Population Size 100, Random Seed 0, Derivatives Central, Require Bounds
   Max Subproblems Unlimited, Max Integer Sols Unlimited, Integer Tolerance 1%

Objective Cell (Min)

Cell	Name	Original Value	Final Value
$B$13	J	0	5.075115698

Variable Cells

Cell	Name	Original Value	Final Value	Integer
$B$3	c_0	0	1	Contin
$B$4	c_1	0	-3.727504952	Contin
$B$5	c_2	0	2.529647749	Contin
$B$6	c_3	0	3.197857203	Contin

Constraints

Cell	Name	Cell Value	Formula	Status	Slack
$B$16	l_c	4.000000439	$B$16=4	Binding	0
$B$17	u_0	1	$B$17=1	Binding	0
$B$18	u_1	3	$B$18=3	Binding	0

Figure 4. Answer report generated by Excel solver using 3rd order parametrization for problem 3.1.

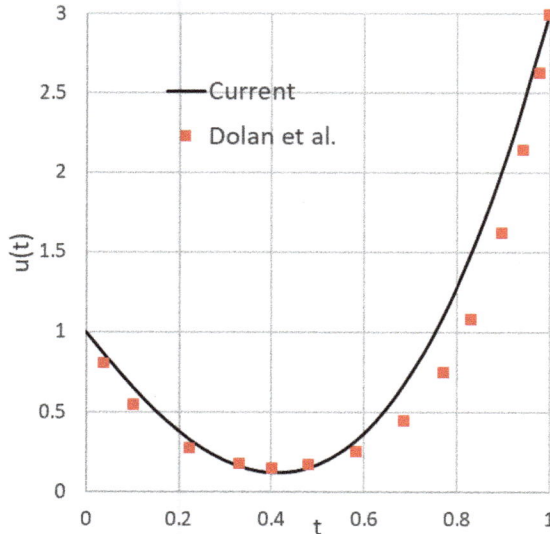

Figure 5. Optimal $u(t)$ computed using 3rd order parametrization for problem 3.1. Reported values by Dolan et al. are also shown.

The difference between the value reported by Dolan et al. [19] and our computed value using a cubic approximation for $u(t)$ is approximately 0.13%. We have tried a quadratic approximation and obtained a slightly higher cost index of 5.078412. It is likely that the small difference originated from integration error in [19] using a trapezoidal rule, whereas the integration in our solution by QUADF calculus function is based on an adaptive Gauss-quadrature scheme [17] which is accurate to machine precision for a smooth polynomial integrand.

To demonstrate the effect of control parametrization order on the result, next we tried a 5th-order polynomial approximation to the shape function $u(t)$, but also appended the problem with one additional constraint:

$$u(t) \geq 0. \tag{10}$$

Incorporating (10) into the spreadsheet model was accomplished as follows. In a new column, a vector of time values from 0 to 1 in increment of 0.1 was generated using Excel AutoFill feature, along with a corresponding vector for the parametrized shape formula as shown in Figure 6. To impose (10), it is sufficient to demand that the minimum value of the shape vector, as computed in *F13* of Figure 6, be greater than or equal to zero. Running the NLP solver with the added constraint yielded a cost index of 4.654 as shown in Figure 7 and plotted in Figure 8. The higher-order approximation to the shape function has resulted in a considerably lower cost index, by more than 8.3%, compared to that reported by Dolan et al. [19].

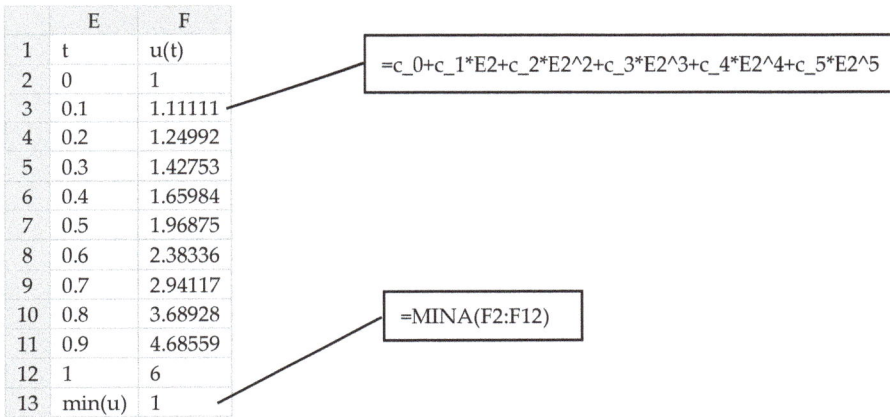

	E	F	
1	t	u(t)	=c_0+c_1*E2+c_2*E2^2+c_3*E2^3+c_4*E2^4+c_5*E2^5
2	0	1	
3	0.1	1.11111	
4	0.2	1.24992	
5	0.3	1.42753	
6	0.4	1.65984	
7	0.5	1.96875	
8	0.6	2.38336	
9	0.7	2.94117	
10	0.8	3.68928	=MINA(F2:F12)
11	0.9	4.68559	
12	1	6	
13	min(u)	1	

Figure 6. Parametrized $u(t)$ function is sampled with AutoFill to provide a handle on its minimum value for the purpose of imposing constraint (10).

Solver Engine
 Engine: GRG Nonlinear
 Solution Time: 4.266 Seconds.
 Iterations: 13 Subproblems: 0
Solver Options
 Max Time Unlimited, Iterations Unlimited, Precision 0.000001, Use Automatic Scaling
 Convergence 0.0001, Population Size 100, Random Seed 0, Derivatives Central, Require Bounds
 Max Subproblems Unlimited, Max Integer Sols Unlimited, Integer Tolerance 1%

Objective Cell (Min)

Cell	Name	Original Value	Final Value
$B$13	J	14.42885894	4.654300187

Variable Cells

Cell	Name	Original Value	Final Value	Integer
$B$3	c_0	1	1	Contin
$B$4	c_1	1	-1.941968064	Contin
$B$5	c_2	1	-1.31719562	Contin
$B$6	c_3	1	-0.28472497	Contin
$B$7	c_4	1	5.204058004	Contin
$B$8	c_5	1	0.33983065	Contin

Constraints

Cell	Name	Cell Value	Formula	Status	Slack
$B$16	l_c	4.000000057	$B$16=4	Binding	0
$B$17	u_0	1	$B$17=1	Binding	0
$B$18	u_1	3	$B$18=3	Binding	0
$F$13	Min(u) u(t)	-7.06322E-07	$F$13>=0	Binding	0

Figure 7. Answer report generated by Excel solver using a 5th order parametrization for problem 3.1 with the added constrained (10).

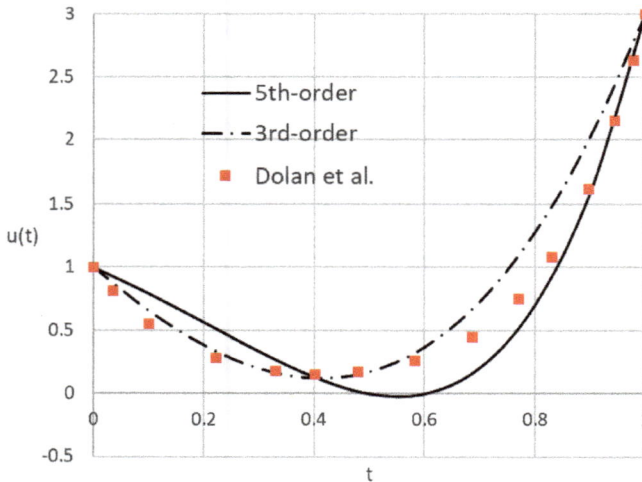

Figure 8. Optimal $u(t)$ computed by using 5th order parametrization for problem 3.1. The higher-cost solution with 3rd order parametrization and reported values by Dolan et al. are also shown.

3.2. Quadratic Control Problem with Integral Constraint

The following problem which involves an integral dynamic constraint was studied by Lim et al. [20], who showed the that the optimal control can be calculated by solving an optimal parameter selection problem together with an unconstrained LQ problem. The optimal control problem is stated as follows:

Find $u_1(t)$, $u_2(t)$, $t \in [0, 1]$ which minimize the cost index

$$J = 0.5\, x_1(1)^2 + 0.5 \int_0^1 \left(x_1{}^2 + u_1{}^2 + u_2{}^2 \right) dt, \tag{11}$$

subject to

$$\dot{x}_1 = 3x_1 + x_2 + u_1, \tag{12}$$

$$\dot{x}_2 = -x_1 + 2x_2 + u_2, \tag{13}$$

with initial conditions

$$x_1(0) = 4, x_2(0) = -4, \tag{14}$$

and integral bounds constraint (There appears to be a typographical error in [20] where (15) is stated as less than 8. The actual value appears to be 80 since 8 would clearly violate the constraint at the reported optimal solution in [20].)

$$0.5\, x_2(1)^2 + 0.5 \int_0^1 \left(x_1{}^2 + u_1{}^2 + u_2{}^2 \right) dt \leq 80. \tag{15}$$

Lim et al. [20] calculated, with aid of control software MISER 3.1, an optimal cost index J of 62.66103.

3.2.1. Solution by Direct Spreadsheet Method

Referring to Figure 9 and working with named variables shown in column A, both $u_1(t)$ and $u_2(t)$ were parametrized using 3rd-order polynomials as shown in *B10* and *B11*, and the IVP equations (12) and (13) were defined by equivalent formulas in *B13* and *B14*. The state variables x_1 and x_2 are assigned the initial conditions as shown in *B3* and *B4*. Next, an initial solution to the underlining IVP (12)–(14) was obtained by evaluating the formula

$$=IVSOLVE(B13{:}B14,\ B2{:}B4,\ \{0{,}1\}) \tag{16}$$

in an allocated array *E1:G102*. IVSOLVE was passed the IVP equations *B13:B14*, the IVP variables *B2:B4*, and the time interval [0, 1] and computed a formatted result shown partially in Figure 10. Here we have allocated 102 rows for the result array to display the solution at uniform time steps of 0.01.

To define an equivalent formula for the cost index (11), we proceeded by sampling the controls formulas, and the cost index integrand as shown in columns *I*, *J* and *K* of Figure 10 by starting from the initial formulas shown in the figure and using AutoFill to generate the values. (Note the hierarchical interdependence of the generated columns on the IVP solution). Next, we defined the cost index formula in which the discrete data integrator calculus function QUADXY was employed to integrate the sampled integrand as shown in *B16* of Figure 9. Similarly, we defined an analog formula for the integral constraint (15) as shown in *B18* of Figure 9, and thus prepared all the input needed to run Excel NLP solver next.

	A	B	C	D
1	ODE variables with initial conditions			
2	t			
3	x_1		4	
4	x_2		-4	
5	Parametrized controls with starting guess			
6	c_0		0	d_0	0
7	c_1		0	d_1	0
8	c_2		0	d_2	0
9	c_3		0	d_3	0
10	u_1	=c_0+c_1*t+c_2*t^2+c_3*t^3		
11	u_2	=d_0+d_1*t+d_2*t^2+d_3*t^3		
12	ODE rhs equations			
13	x1dot	=3*x_1+x_2+u_1		
14	x2dot	=-x_1+2*x_2+u_2		
15	Cost Index			
16	J	=0.5*F102^2+0.5*QUADXY(E2:E102,K2:K102)		
17	Constraint			
18	con	=0.5*G102^2+0.5*QUADXY(E2:E102,K2:K102)		

Figure 9. Spreadsheet parametrized model for problem 3.2. The colored ranges are inputs for IVSOLVE formula (16).

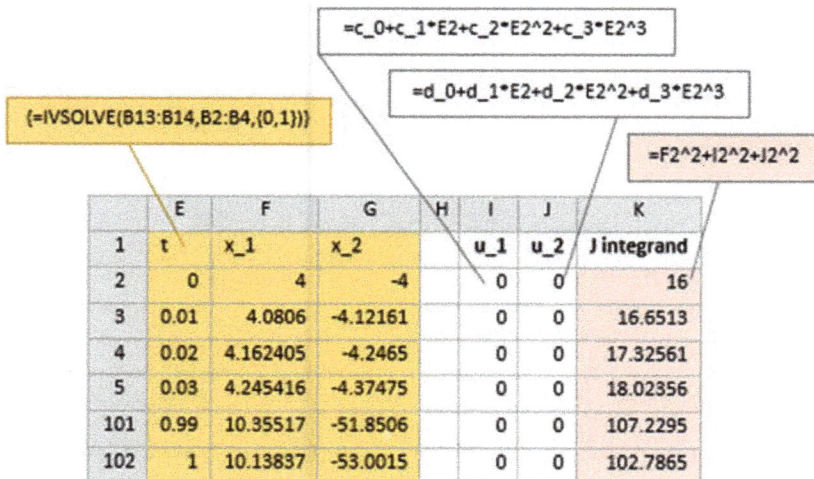

{=IVSOLVE(B13:B14,B2:B4,{0,1})}

=c_0+c_1*E2+c_2*E2^2+c_3*E2^3

=d_0+d_1*E2+d_2*E2^2+d_3*E2^3

=F2^2+I2^2+J2^2

	E	F	G	H	I	J	K
1	t	x_1	x_2		u_1	u_2	J integrand
2	0	4	-4		0	0	16
3	0.01	4.0806	-4.12161		0	0	16.6513
4	0.02	4.162405	-4.2465		0	0	17.32561
5	0.03	4.245416	-4.37475		0	0	18.02356
101	0.99	10.35517	-51.8506		0	0	107.2295
102	1	10.13837	-53.0015		0	0	102.7865

Figure 10. Partial display of IVP (12)–(14) solution obtained by IVSOLVE formula (16), and dependent generated columns for the parametrized controls formulas, and the integrand expression for the cost index (11).

3.2.2. Results and Analysis

Excel solver was configured to minimize the cost index *B16*, by varying the controls coefficients *B6:B9* and *D6:D9*, subject to the integral constraint *B18* being smaller than or equal to 80. Excel solver converged in approximately eight seconds to the solution shown in Figure 11 and plotted in Figure 12. The obtained cost index at 59.1471 was lower than reported by Lim et al. [20] at 62.66103 using an indirect approach with MISER 3.1. Figure 13 provides direct comparisons for $x_1(t)$, $u_1(t)$ and $u_2(t)$ trajectories obtained by the current method and digitized plot values from [20]. The plots show good agreement despite fundamentally different solution strategies.

Solver Engine
  Engine: GRG Nonlinear
  Solution Time: 8.75 Seconds.
  Iterations: 34 Subproblems: 0
Solver Options
  Max Time Unlimited, Iterations Unlimited, Precision 0.000001, Use Automatic Scaling
  Convergence 0.0001, Population Size 100, Random Seed 0, Derivatives Central, Require Bounds
  Max Subproblems Unlimited, Max Integer Sols Unlimited, Integer Tolerance 1%

Objective Cell (Min)

Cell	Name	Original Value	Final Value
$B$17	J	93.05906478	59.14711888

Variable Cells

Cell	Name	Original Value	Final Value	Integer
$B$6	c_0	0	-17.34628813	Contin
$B$7	c_1	0	36.41905324	Contin
$B$8	c_2	0	0.08924598	Contin
$B$9	c_3	0	-20.66984219	Contin
$D$6	d_0	0	12.93342465	Contin
$D$7	d_1	0	-10.58650725	Contin
$D$8	d_2	0	-3.190831252	Contin
$D$9	d_3	0	2.972594703	Contin

Constraints

Cell	Name	Cell Value	Formula	Status	Slack
$B$19	c1	80.00122328	$B$19<=80	Binding	0

Figure 11. Answer report generated by Excel solver for problem 3.2.

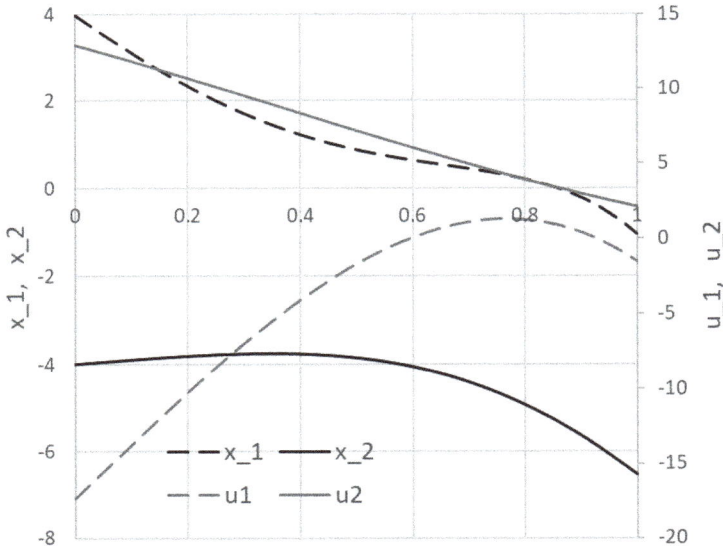

Figure 12. Optimal trajectories computed by the spreadsheet method for problem 3.2.

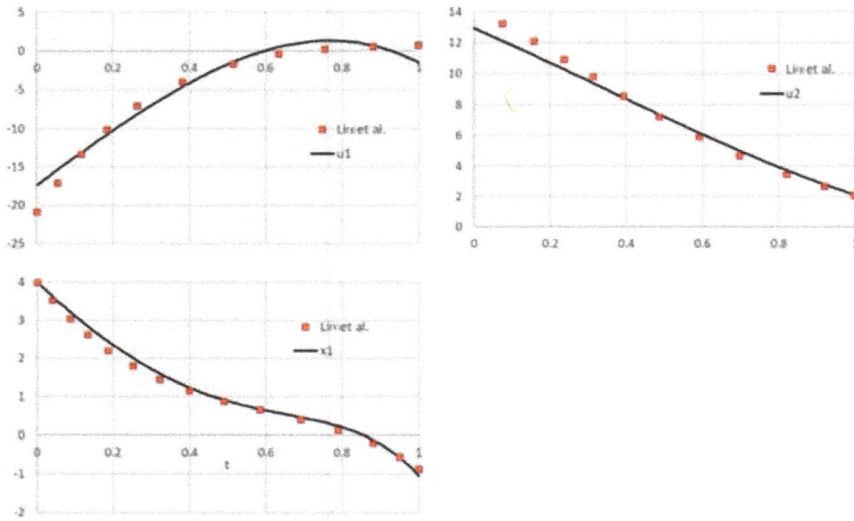

Figure 13. Direct comparison of spreadsheet solution with reported solution obtained by Lim et al. [20] for problem 3.2.

To investigate the effect of numerical integration error on the result, we increased the output array for IVSOLVE from 102 to 502 rows which reduced output time increment from 0.01 to 0.002. However, this has resulted in only minor improvement of the cost index to 59.1429, with otherwise insignificant change to the original solution which indicated the initial output time step of .01 was sufficient for accurate integration.

3.3. Robot Motion Planning: Obstacle Avoidance

The third problem is concerned with planning a 2D motion for a robot from point *A* (0, 0) to point *B* (1, 1), to avoid two circular obstacles of radius $R^2 = 0.1$, centered at (0.4, 0.5) and (0.8, 1.5), while using the least amount of energy. The two controls for the robot motion are the constant speed, *v*, and the variable angle (direction), $\theta(t)$ of the motion. The corresponding optimal control problem has the following form [21]:

Find v, $\theta(t)$, $t \in [0,1]$ which minimize the energy cost index

$$J = \int_0^1 \ddot{x}(t)^2 + \ddot{y}(t)^2 \, dt, \tag{17}$$

subject to

$$\dot{x}(t) = v * \cos(\theta), \tag{18}$$

$$\dot{y}(t) = v * \sin(\theta), \tag{19}$$

with initial conditions

$$x(0) = 0, y(0) = 0, \tag{20}$$

end conditions

$$x(1) = 1.2, y(1) = 1.6, \tag{21}$$

and trajectory constraints which model the circles to be avoided

$$(x(t) - 0.4)^2 + (y(t) - 0.5)^2 \geq 0.1, \tag{22}$$

$$(x(t) - 0.8)^2 + (y(t) - 1.5)^2 \geq 0.1. \tag{23}$$

Note that the cost index in this example depends on the second derivatives of the state variables.

3.3.1. Solution by Direct Spreadsheet Method

Referring to Figure 14, the speed was parametrized using the named variable v for *B6* with initial value of 1, and the angle (named theta in *B13*) was parametrized with a fifth order polynomial. Using the named variable t, x and y, the IVP formulas (18) and (19) were defined in *B15* and *B16*. An initial IVP solution was obtained by evaluating IVSOLVE formula (24) in array *D1:F102* shown partially in Figure 15.

$$=IVSOLVE(B15:B16, B2:B4, \{0,1\}) \tag{24}$$

The next task was to define an analog formula for the cost index (17). The integrand for the cost index depends on $\ddot{x}(t)$, and $\ddot{y}(t)$ which we needed to generate. Although $\ddot{x}(t)$, and $\ddot{y}(t)$ can be derived analytically for this particular problem by differentiating (18) and (19), we elected to compute them numerically using the discrete data differentiator calculus function DERIVXY as shown in columns H and I of Figure 15. For example, to compute $\ddot{x}(t)$, we started from the formula

$$=DERIVXY(\$D\$2:\$D\$102, \$E\$2:\$E\$102, D2, 2) \tag{25}$$

in *H2* passing in, respectively, the time and x vectors from the IVP solution array, the point of differentiation, and the order of the x derivative to compute. Next, the AutoFill was used to generate values for all the points in the time vector. Note that the first two arguments in (25) were locked using Excel $ operator to prevent these values from being incremented during the AutoFill, allowing only *D2* to be incremented. Values for the integrand expression $\ddot{x}(t)^2 + \ddot{y}(t)^2$ were then readily generated in a new column L and integrated with respect to the time vector by using the calculus function QUADXY as shown in *B18* of Figure 14.

	A	B
1	ODE system variables with initial conditions	
2	t	
3	x	0
4	y	0
5	Parametrized controls with starting guess	
6	v	1
7	c_0	1
8	c_1	0
9	c_2	0
10	c_3	0
11	c_4	0
12	c_5	0
13	theta	=c_0+c_1*t+c_2*t^2+c_3*t^3+c_4*t^4+c_5*t^5
14	ODE system equations	
15	dxdt	=v*COS(theta)
16	dydt	=v*SIN(theta)
17	Cost Index	
18	J	=QUADXY(D2:D102,J2:J102)
19	Path constraint helpers	
20	min(c1)	=MINA(L2:L102)
21	min(c2)	=MINA(M2:M102)

Figure 14. Spreadsheet parametrized model for problem 3.3. The colored ranges are inputs for IVSOLVE formula (24).

=DERIVXY($D$2:$D$102, $F$2:$F$102, D2, 2)   =(E2-0.8)^2+(F2-1.5)^2

=DERIVXY($D$2:$D$102, $E$2:$E$102, D2, 2)   =(E2-0.4)^2+(F2-0.5)^2

=IVSOLVE(B15:B16,B2:B4,{0,1})   =H2^2+I2^2

	D	E	F	G	H	I	J	K	L	M
1	t	x	y		d2xdt2	d2ydt2	J integrand		c1	c2
2	0	0	0		3.33E-14	2.66E-13	7.21E-26		0.41	2.89
3	0.01	0.005403	0.008415		-9.3E-15	0	8.56E-29		0.397363	2.856211
4	0.02	0.010806	0.016829		-5.2E-14	-2.7E-13	7.37E-26		0.384926	2.822622
101	0.99	0.534899	0.833056		-9.9E-13	-2.4E-12	6.87E-24		0.129124	0.515092
102	1	0.540302	0.841471		-7.7E-12	-9.1E-12	1.42E-22		0.136287	0.501103

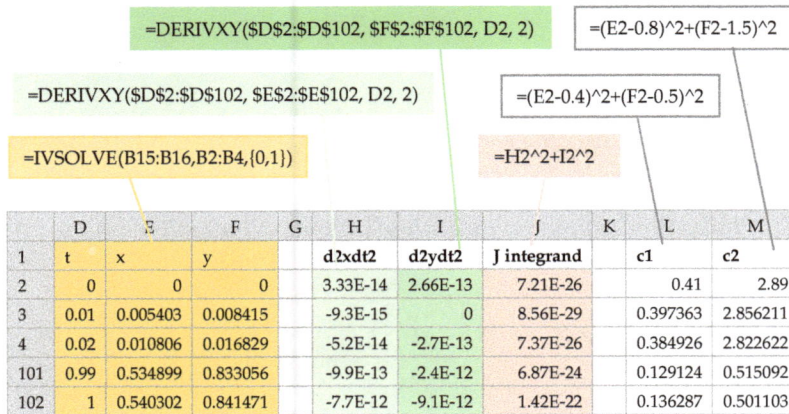

Figure 15. Partial display of the IVP (18)–(20) solution obtained by IVSOLVE formula (24), and dependent generated values needed to define the cost index and constraints formulas of problem 3.3.

The remaining task to complete the input for Excel NLP solver was to define formulas for the circle avoidance constraints. Using x and y values from the IVP solution, values for the constraints equations (22) and (23) were generated as shown in columns L and M of Figure 15. To impose the bounds, it was sufficient to require that the minimum values of columns L and M, as computed in $B20$ and $B21$ of Figure 14, be greater than or equal to the specified bound.

3.3.2. Results and Analysis

Excel solver was configured to minimize the cost index, J ($B18$), by varying the speed v ($B6$), and theta polynomial coefficients ($B7:B12$), subject to the constraints:

$$v >= 0,$$

$$E102 = 1.2, \text{ corresponding to (21)}$$

$$F102 = 1.6, \text{ corresponding to (21)}$$

$$B20 >= 0.1, \text{ corresponding to (22)}$$

$$B21 >= 0.1, \text{ corresponding to (23)}.$$

The solver converged to the expected low-energy solution shown in Figure 16b in approximately 18 s, with the result shown in Figure 17. The initial trajectory for the robot based on our starting guess for the controls is shown in Figure 16a. The cost index was found at approximately 8.02.

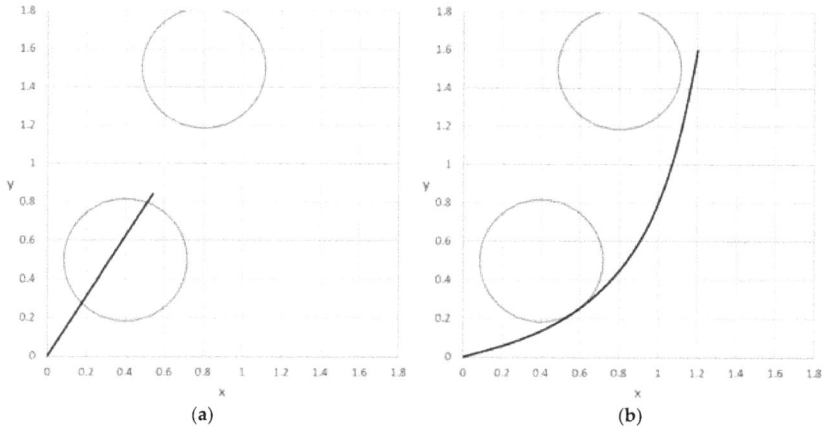

Figure 16. Initial (**a**) and optimal (**b**) trajectories for problem 3.3.

Result: Solver found a solution. All Constraints and optimality conditions are satisfied.
Solver Engine
    Engine: GRG Nonlinear
    Solution Time: 17.875 Seconds.
    Iterations: 20 Subproblems: 0
Solver Options
    Max Time Unlimited, Iterations Unlimited, Precision 0.000001
    Convergence 0.0001, Population Size 0, Random Seed 0, Derivatives Forward
    Max Subproblems Unlimited, Max Integer Sols Unlimited, Integer Tolerance 1%

Objective Cell (Min)

Cell	Name	Original Value	Final Value
$B$18	J	1.45798E-23	8.019513821

Variable Cells

Cell	Name	Original Value	Final Value	Integer
$B$6	v	1	2.170512985	Contin
$B$7	c_0	1	0.2680298	Contin
$B$8	c_1	0	-0.003745092	Contin
$B$9	c_2	0	5.602571466	Contin
$B$10	c_3	0	-5.255273392	Contin
$B$11	c_4	0	-0.842918583	Contin
$B$12	c_5	0	1.634089689	Contin

Constraints

Cell	Name	Cell Value	Formula	Status	Slack
$B$20	min_c1	0.099999978	$B$20>=0.1	Binding	0
$F$102	y	1.599999206	$F$102=1.6	Binding	0
$E$102	x	1.199999601	$E$102=1.2	Binding	0
$B$21	min_c2	0.14168052	$B$21>=0.1	Not Binding	0.04168052
$B$6	v	2.170512985	$B$6>=0	Not Binding	2.170512985

Figure 17. Answer report generated by Excel Solver for problem 3.3.

To make the problem more interesting, we added a 3rd circle obstacle by appending the additional path constraint to the problem:

$$(x(t) - 1.0)^2 + (y(t) - 0.8)^2 \geq 0.1. \tag{26}$$

The new configuration and initial trajectory are shown in Figure 18a. The incorporation of the 3rd constraint into the model setup is straight forward and the solver converged to the higher energy trajectory shown in Figure 18b at a cost index of approximately 22.69; the results are shown in Figure 19.

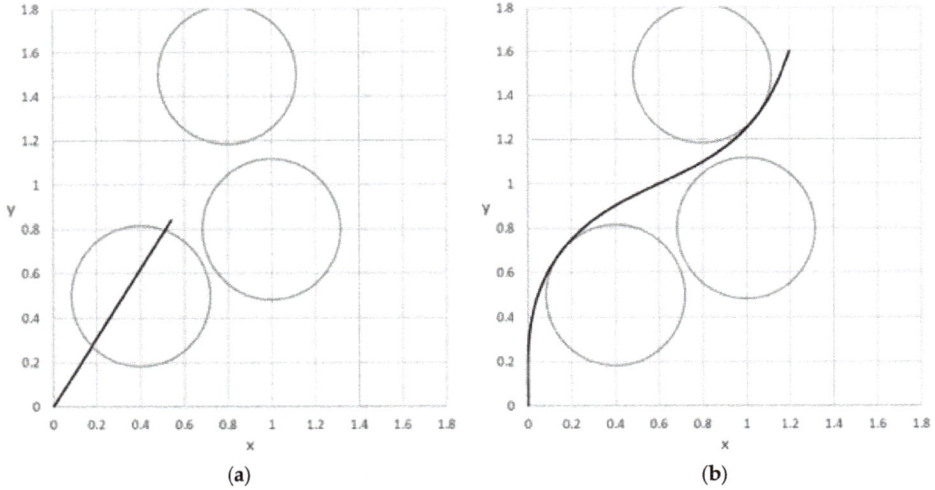

(a)   (b)

Figure 18. Initial (**a**) and optimal (**b**) trajectories for problem 3.3 with additional constraint (26).

Result: Solver found a solution. All Constraints and optimality conditions are satisfied.
Solver Engine
    Engine: GRG Nonlinear
    Solution Time: 26.219 Seconds.
    Iterations: 26 Subproblems: 0
Solver Options
    Max Time Unlimited, Iterations Unlimited, Precision 0.000001
    Convergence 0.0001, Population Size 0, Random Seed 0, Derivatives Forward
    Max Subproblems Unlimited, Max Integer Sols Unlimited, Integer Tolerance 1%

Objective Cell (Min)

Cell	Name	Original Value	Final Value
$B$18	J	1.45798E-23	22.69397908

Variable Cells

Cell	Name	Original Value	Final Value	Integer
$B$6	v	1	2.158226107	Contin
$B$7	c_0	1	1.562631814	Contin
$B$8	c_1	0	1.08802338	Contin
$B$9	c_2	0	-13.71440871	Contin
$B$10	c_3	0	10.78075362	Contin
$B$11	c_4	0	13.5293917	Contin
$B$12	c_5	0	-12.03674282	Contin

Constraints

Cell	Name	Cell Value	Formula	Status	Slack
$E$102	x	1.199999999	$E$102=1.2	Binding	0
$F$102	y	1.600000006	$F$102=1.6	Binding	0
$B$22	min_c3	0.127498013	$B$22>=0.1	Not Binding	0.027498013
$B$21	min_c2	0.099999998	$B$21>=0.1	Binding	0
$B$20	min_c1	0.100000001	$B$20>=0.1	Binding	0
$B$6	v	2.158226107	$B$6>=0	Not Binding	2.158226107

Figure 19. Answer report generated by Excel Solver for problem 3.3 with additional constraint (26).

3.4. Nonlinear Bioprocess Optimization: Batch Production

The 4th problem considers the optimal control of a fed-batch reactor for the production of ethanol [8]. The goal is to maximize the yield of ethanol using the feed rate as the control. This problem has highly nonlinear dynamic constraints and a free terminal time t_F, which is also an unknown design variable. The mathematical statement of the free end time problem is given below:

Find the flowrate $u(t)$, $t \in [0, t_F]$, and the terminal time t_F to maximize the cost index

$$J = x_3(t_F)\, x_4(t_F), \tag{27}$$

subject to

$$\dot{x}_1 = g_1 x_1 - u \frac{x_1}{x_4}, \tag{28}$$

$$\dot{x}_2 = -10 g_1 x_1 + u \frac{150 - x_2}{x_4}, \tag{29}$$

$$\dot{x}_3 = g_2 x_1 - u \frac{x_3}{x_4}, \tag{30}$$

$$\dot{x}_4 = u, \tag{31}$$

where

$$g_1 = \left(\frac{0.408}{1 + x_3/16} \right) \left(\frac{x_2}{0.22 + x_2} \right), \tag{32}$$

$$g_2 = \left(\frac{1}{1 + x_3/71.5} \right) \left(\frac{x_2}{0.44 + x_2} \right), \tag{33}$$

with Initial conditions

$$x_1(0) = 1,\; x_2(0) = 150,\, x_3(0) = 0,\; x_4(0) = 10, \tag{34}$$

and bounds constraints

$$0 \le u \le 12, \tag{35}$$

$$0 \le x_4(t_F) \le 200. \tag{36}$$

This problem was solved by Banga et al. [8] using a two-phase (stochastic-deterministic) hybrid (TPH) approach to overcome convergence difficulties reported by previous published attempts. Their best reported results found the maximum cost index J at 20839, and the terminal time t_F at 61.17 h.

3.4.1. Solution by Direct Spreadsheet Method

Following the procedure in the previous examples, the control $u(t)$ was parameterized using a 2nd order polynomial as shown in *B11* of Figure 20, and the IVP equations were defined in terms of the named variables as shown in *B18:B21*. Note that the terminal time has been assigned the variable *B14* (named tF) with initial value of 50. The IVP solution was obtained with IVSOLVE formula

$$=IVSOLVE(B18:B21, B2:B6, B13:B14) \tag{37}$$

in array *D1:H102* as shown partially in Figure 21. Note that the final time is now a variable for IVSOLVE which was passed in the 3rd parameter *B13:B14*. The cost index formula for this problem is simple and was defined by formula *B23* which references $x_3(t_F)$ and $x_4(t_F)$ of the IVSOLVE solution array. To impose the bound constraint (35), we sampled the control formula in column *J* at the solution output time points as shown in Figure 21 and demanded that the maximum and minimum values of the control vector as computed by formulas *B25* and *B26* satisfy the appropriate bounds.

	A	B
1	ODE system variables with initial conditions	
2	t	0
3	x_1	1
4	x_2	150
5	x_3	0
6	x_4	10
7	Parametrized controllers with starting guess	
8	c_0	3
9	c_1	0
10	c_2	0
11	u	=c_0+c_1*t+c_2*t^2
12	ODE time domain with final time guess	
13	ts	0
14	tF	50
15	ODE system equations	
16	g1_	=(0.408/(1+x_3/16))*(x_2/(0.22+x_2))
17	g2_	=(1/(1+x_3/71.5))*(x_2/(0.44+x_2))
18	x1dot	=g1_*x_1-u*x_1/x_4
19	x2dot	=-10*g1_*x_1+u*(150-x_2)/x_4
20	x3dot	=g2_*x_1-u*x_3/x_4
21	x4dot	=u
22	Cost Index	
23	J	=G102*H102
24	Constraints helpers	
25	max(u)	=MAXA(J2:J102)
26	min(u)	=MINA(J2:J102)

Figure 20. Spreadsheet parametrized model for problem 3.4. The colored ranges are inputs for IVSOLVE formula (37).

=c_0+c_1*D2+c_2*D2^2

=IVSOLVE(B18:B21,B2:B6,B13:B14)

	D	E	F	G	H	I	J
1	t	x_1	x_2	x_3	x_4		u
2	0	1	150	0	10		3
3	0.5	1.062808	148.0676	0.478626388	11.5		3
4	1	1.142691	146.2653	0.936035562	13		3
100	49	15.05587	0.078212	73.9061806	157		3
101	49.5	15.05539	0.077045	73.75951155	158.5		3
102	50	15.05491	0.07591	73.61434682	160		3

Figure 21. Partial display of IVP (28)–(34) solution obtained by IVSOLVE formula (37), and generated values for the parametrized control of problem 3.4.

3.4.2. Results and Analysis

Excel solver was run starting from the initial guess for the unknown control coefficients and terminal time shown in Figure 20. The cost index J (B23) was selected to be maximized by varying the terminal time t_F (B14), and the coefficients c_0, c_1 and c_2, (B8:B10) subject to the constraints:

$$B25 <= 12, \text{corresponding to (35)}$$

$$B26 >= -12, \text{corresponding to (35)}$$

$$H102 <= 200, \text{corresponding to (36)}.$$

Excel NLP solver converged in approximately 29 s to the result shown in Figure 22, and plotted in Figure 23. A partial listing of the converged control values and IVP solution reflecting the new terminal time is shown in Figure 24.

Result: Solver found a solution. All Constraints and optimality conditions are satisfied.
Solver Engine
    Engine: GRG Nonlinear
    Solution Time: 28.828 Seconds.
    Iterations: 20 Subproblems: 0
Solver Options
    Max Time Unlimited, Iterations Unlimited, Precision 0.000001, Use Automatic Scaling
    Convergence 0.0001, Population Size 100, Random Seed 0, Derivatives Central, Require Bounds
    Max Subproblems Unlimited, Max Integer Sols Unlimited, Integer Tolerance 1%

Objective Cell (Max)

Cell	Name	Original Value	Final Value
$B$24	J	11778.29549	20522.55386

Variable Cells

Cell	Name	Original Value	Final Value	Integer
$B$8	c_0	3	-0.440650315	Contin
$B$9	c_1	0	0.046376992	Contin
$B$10	c_2	0	0.001653295	Contin
$B$15	tF	50	61.63882007	Contin

Constraints

Cell	Name	Cell Value	Formula	Status	Slack
$H$102	x_4	199.999999	$H$102<=200	Not Binding	1.04472E-06
$B$26	max_u	8.699408124	$B$26<=12	Not Binding	3.300591876
$B$27	min_u	-0.440650315	$B$27>=-12	Not Binding	11.55934969

Figure 22. Answer report generated by Excel solver for problem 3.4.

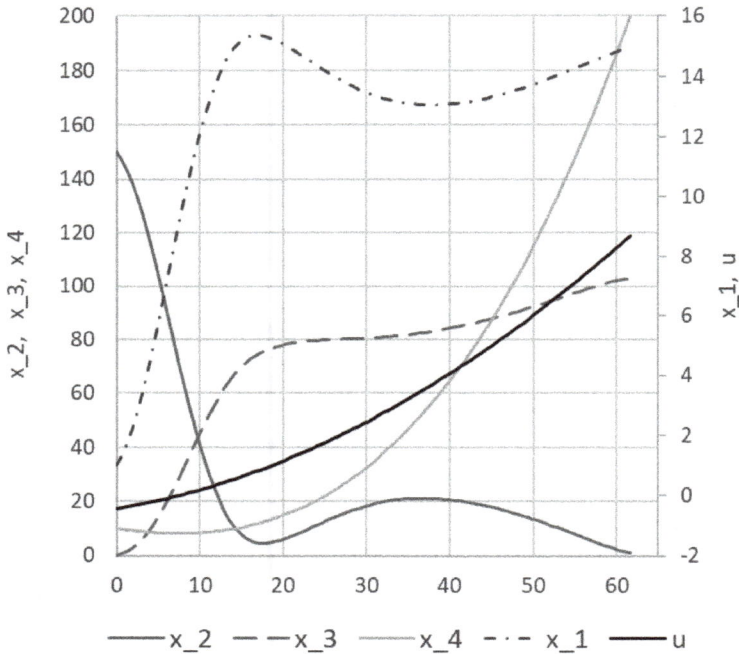

Figure 23. Optimal trajectories computed by the spreadsheet method for problem 3.4.

	D	E	F	G	H	I	J
1	t	x_1	x_2	x_3	x_4		u
2	0	1	150	0	10		-0.44065
3	0.616388	1.313316	147.1366	0.712647865	9.737328		-0.41144
4	1.232776	1.702539	143.5086	1.64960033	9.49305		-0.38097
100	60.40604	14.84059	2.12186	102.2278477	189.4647		8.393491
101	61.02243	14.88833	1.630308	102.4917196	194.6852		8.545821
102	61.63882	14.925	1.25	102.6127698	200		8.699408

Figure 24. Partial listing of the converged IVP solution and control values of problem 3.4.

The achieved maxima for the cost index was at 20522.5 and the terminal time t_F was found at approximately 61.64. These values are in very good agreement with the best results reported by Banga et al. [8] at 20839, and 61.17 h. Figure 25 shows direct comparison of the states and control trajectories with digitized plot values from Banga et al. The agreement is quite good for the most part despite fundamentally different control parametrization and algorithms employed by the two methods. In particular, the control parametrization in [8] is approximated by connected line segments, whereas our control is a continuous parabola.

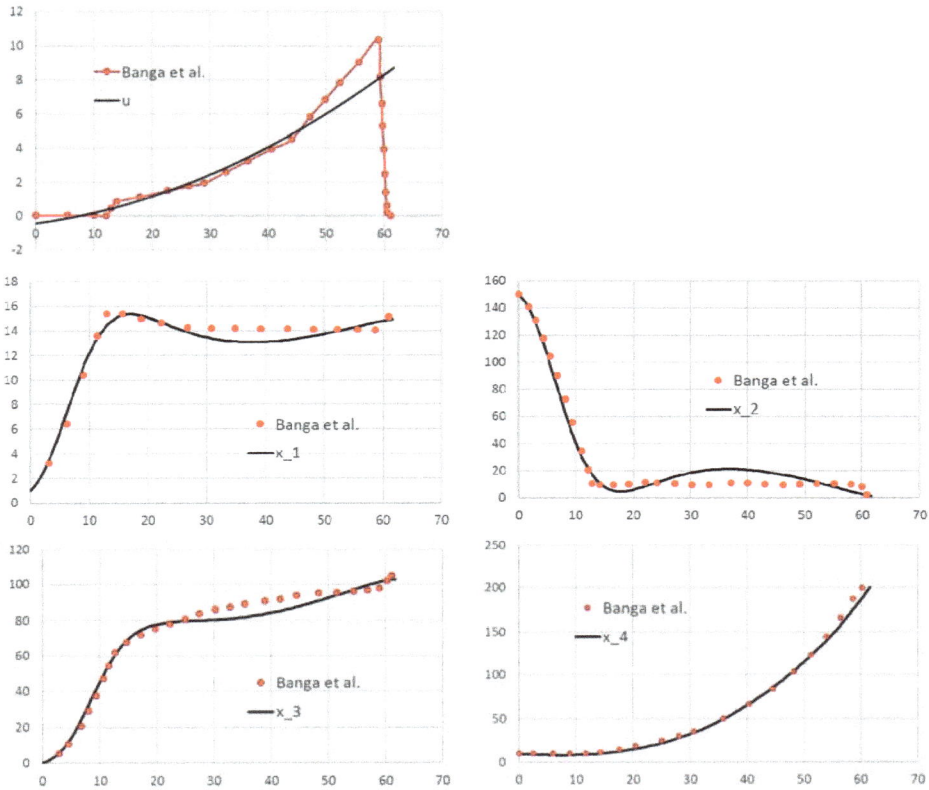

Figure 25. Direct comparison of spreadsheet solution with reported solution obtained by Banga et al. for problem 3.4.

4. Conclusions

We devised a practical and systematic spreadsheet solution strategy for solving general optimal control problems. The strategy is based on an adaptation of the partial-parametrization direct solution method which preserves the structure of the original mathematical optimization statement, but transforms it into a simplified NLP problem suitable for Excel NLP solver. The solution strategy is formulated by the aid of several elementary calculus functions from an available Excel calculus Add-in [11] for solving differential equations, computing integrals and derivatives. The calculus functions are employed as building blocks in a hierarchical functional paradigm implemented by standard Excel formulas in conjunction with Excel built-in NLP solver to carry out a dynamic optimization program.

Results were obtained for four representative problems selected to illustrate modeling of several conditions including dependence on higher order derivatives of state and controls as well as implicit IVP problems and integral constraints. The results were compared with published solutions obtained by other methods, and in some cases, were shown to be better by the measure of the cost index. The performance of the method is also notable with computing times on the order of seconds to a minute on a standard laptop. As has been illustrated by the solution procedure, applying the technique involves no more than defining a few formulas that parallel the original mathematical equations. No special programming skill is needed beyond basic familiarity with the common spreadsheet operation. The minimal problem setup effort, in combination with the ubiquity of

Excel spreadsheet, present reasons to explore the method as an alternative, simpler educational tool for rather complex problems.

The success of the devised strategy for optimal control of ordinary differential equations supports future research for extending the strategy to optimal control of partial differential equations. In [22], the author demonstrated an analogous formulation combing the NLP solver with a PDE solver from the same Add-in [11] for parameter estimation of partial differential algebraic equations, and it may be feasible that certain formulations of optimal control of partial differential equations are solvable in the spreadsheet by a similar strategy. On the other hand, it is worth investigating devising an alternative strategy based on the indirect solution method [4,5] for optimal control problems. The indirect method requires solving a boundary value problem which could be solved by the aid of a boundary value problem solver function also available in the Excel calculus Add-in [11].

Supplementary Materials: An Excel workbook containing the solved problems in this paper is available online at http://www.mdpi.com/2297-8747/23/4/54/s1.

Funding: This research received no external funding.

Conflicts of Interest: The author of the manuscript is the founder of ExcelWorks LLC of Massachusetts, USA supplying the Excel calculus Add-in [11], utilized in this research work.

Appendix A

The following subsections present brief descriptions of the calculus functions utilized in this work. The functions are enabled as an extension of Excel math functions by installing ExceLab calculus Add-in [11]. For more detailed descriptions of the functions, the reader is referred to [11].

Appendix A.1 IVSOLVE: Initial Value Problem Solver

The worksheet function IVSOLVE solves an initial value differential algebraic equation system in the interval $t \in \left[t_s, t_f\right]$

$$M\frac{dx}{dt} = F(x(t), t)$$
$$x(t_0) = x_0$$

(A1)

$x(t) = (x_1(t), x_2(t), \ldots, x_n(t))$, and M is an optional mass matrix which may be singular. IVSOLVE uses by default RADUA5 an implicit 5th-order Runge-Kutta scheme with adaptive time step [15], and at minimum, requires three input parameters to describe the ODE system:

1. Reference to the right-hand side formulas corresponding to the vector-valued function $F(x(t), t) = (f_1(x(t), t), f_2(x(t), t), \ldots, f_n(x(t), t))$.
2. Reference to the system variables in the specific order $(t, x_1, x_2, \ldots, x_n)$.
3. The integration time interval end points.

Additional optional parameters include specifying a mass matrix as well as algorithmic controls. IVSOLVE is run as an array formula in an allocated array of cells. It evaluates to an ordered tabular result where the time values are listed in the first column, and the corresponding state variables' values are listed in adjacent columns. By default, IVSOLVE reports the output at uniform intervals according to the allocated number of rows for the output array. Custom output formats can be achieved via the optional parameters including specifying custom divisions or exact points. We demonstrate IVSOLVE for the following DAE problem (reproduced from [14]):

$$\frac{dy_1}{dt} = -0.04y_1 + 10^4 y_2 y_3, t \in [0, 1000]$$
$$\frac{dy_2}{dt} = 0.04y_1 - 10^4 y_2 y_3 - 3 * 10^7 y_2^2$$
$$0 = y_1 + y_2 + y_3 - 1$$
$$y_1(0) = 1, y_2(0) = 0, y_3(0) = 0$$

(A2)

Referring to Figure A1, the system RHS formulas are defined in cells *A1:A3* using cell *T1* for the time variable and *Y1, Y2, Y3* for the state variables which are assigned the initial conditions as shown in the figure.

	A
1	=-0.04*Y1+10000*Y2*Y3
2	=0.04*Y1-10000*Y2*Y3-30000000*Y2^2
3	= Y1+Y2+Y3-1

	Y
1	1
2	0
3	0

Figure A1. Spreadsheet model for equation system (A2).

To solve the system, we execute the array formula

=IVSOLVE (A1:A3, (T1,Y1:Y3), {0,1000}, 1)

in an allocated array, (e.g., *C1:F22*) by pressing the 3 keys Control+Shift+Enter simultaneously. IVSOLVE computes and displays the solution shown Figure A2. Here we have used the 4th optional parameter to specify that the last equation, (A3) is an algebraic equation.

	C	D	E	F
1	T1	Y1	Y2	Y3
2	0	1	0	0
3	50	0.69288	8.34415E-06	0.307111
4	100	0.617245	6.15388E-06	0.382748
5	150	0.570229	5.12407E-06	0.429765
20	900	0.349743	2.13047E-06	0.650255
21	950	0.343131	2.06992E-06	0.656867
22	1000	0.336882	2.01377E-06	0.663116

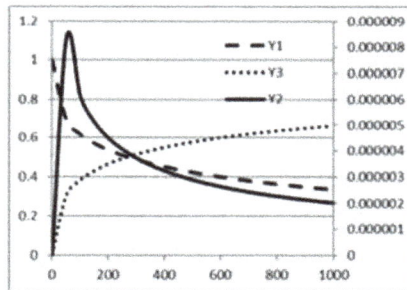

Figure A2. Partial listing of the result computed by IVSOLVE (**left**) for system (A2), and a plot of the trajectories (**right**).

Appendix A.2 QUADF: Formula Integrator Function

The spreadsheet function QUADF computes definite and improper one-dimensional integrals $\int_a^b f(x)dx$. QUADF utilizes QUADPACK numerical integration algorithms [17] suitable for smooth, irregular, and integrands with known singularities. By default, it uses QAG, an adaptive algorithm using Gauss-Kronrod 21-point integration rule. We demonstrate QUADF by computing the following integral (reproduced from [14]):

$$\int_0^1 \frac{\ln x}{\sqrt{x}}dx = -4 \tag{A3}$$

Referring to Figure A3, the integrand formula is defined in cell *A1* using *X1* as dummy variable for the integration. The QUADF integration formula

=QUADF (A1, X1, 0, 1)

is defined in *A2* passing in the integrand formula, the variable of integration, and values for the limits. Evaluating *A2* yields the result.

	A
1	=LN(X1)/SQRT(X1)
2	=QUADF(A1,X1,0,1)

⟹

	A
1	#NUM!
2	-4

Figure A3. Demonstration of QUADF for computing integral (A3) in Excel.

Using recursion, multiple integrals of any order can be computed by direct nesting of QUADF, as demonstrated by the following volume integral example:

$$\int_0^2 \int_0^{3-\frac{3}{2}x} \int_0^{6-3x-2y} 1 - x\, dz\, dy\, dx = 3 \tag{A4}$$

To compute (A4), we construct a simple program consisting of three nested calls of QUADF as shown in Figure A4. Using *X1*, *Y1* and *Z1* as dummy variables, the integrand formula is defined in *A1*, and the inner, middle, and outer integrals formulas are defined in *A2*, *A3* and *A4* respectively, with each inner QUADF formula serving as the integrand for the next outer QUADF formula. Evaluating the outer integral in *A4* computes the triple integral value.

	A
1	=1-X1
2	=QUADF(A1,Z1,0,6-3*X1-2*Y1)
3	=QUADF(A2,Y1,0,3-3*X1/2)
4	=QUADF(A3,X1,0,2)

	A
1	1
2	6
3	9
4	3

Figure A4. Demonstration of QUADF for computing multiple integral (A4) in Excel.

Appendix A.3 QUADXY: Discrete Data Integrator

The spreadsheet function QUADXY computes the integral of a curve passing through the set of data pairs $(x_i, y_i(x_i))$, $i = 0.. n$. The integration interval is defined by the endpoints $[x_0, x_n]$. QUADXY computes the integral by the aid of cubic (default) or linear splines fit to the data [16]. Start and end point slope information may be specified in the 3rd optional parameter to enhance accuracy near the end points. The slope data is defined by a reference to a key/value array as illustrated in Figure A5.

	C	D
1	ISLOPE	1
2	ESLOPE	-1

Figure A5. Optional input format to QUADXY for endpoints boundary conditions.

For example, the formula

=QUADXY(X1:X20, Y1:Y20, C1:D2)

computes an integral value of 20 data pairs in columns X1 and Y1 with supplied optional end points slope values shown in Figure A5.

Appendix A.4 DERIVXY: Discrete Data Differentiator

The spreadsheet function DERIVXY employs cubic splines to compute the dth derivative of a curve passing through a set of data pairs $(x_i, y_i(x_i))$, $i = 0..n$ at a specified point $x = p$. Similar to

QUADXY, optional start and or endpoint slope information may be specified to enhance accuracy near the end points. For example, the formula

$$=DERIVXY(X1:X20, Y1:Y20, 5, 2)$$

computes the second derivative for a curve passing through 20 points in columns *X1* and *Y1* at the point $x = 5$.

When using DERIVXY with Excel AutoFill feature to differentiate a given data set at multiple points, it is necessary to lock the parameters except for the variable point (parameter 3) so they are not incremented during the AutoFill. For example, the following formula is safe to use with AutoFill to generate derivatives at points stored in *Z1, Z2, Z3*, etc.

$$=DERIVXY(\$X\$1:\$X\$20, \$Y\$1:\$Y\$20, Z1)$$

References

1. Geering, H.P. *Optimal Control with Engineering Applications*; Springer: Berlin, Germany, 2007.
2. Sethi, S.P. *Optimal Control Theory: Applications to Management Science and Economics*, 3rd ed.; Springer: Berlin, Germany, 2019.
3. Aniţa, S.; Arnăutu, V.; Capasso, V. *An Introduction to Optimal Control Problems in Life Sciences and Economics: From Mathematical Models to Numerical Simulation with MATLAB*; Birkhäuser: Basel, Switzerland, 2011.
4. Betts, J.T. *Practical Methods for Optimal Control and Estimation Using Nonlinear Programming*, 2nd ed.; Advances in Design and Control; Society for Industrial and Applied Mathematics: Philadelphia, PA, USA, 2009.
5. Böhme, T.J.; Frank, B. Indirect Methods for Optimal Control. In *Hybrid Systems, Optimal Control and Hybrid Vehicles. Advances in Industrial Control*; Springer: Cham, Switzerland, 2017.
6. Elnagar, G.; Kazemi, M.A. Pseudospectral Chebyshev Optimal Control of Constrained Nonlinear Dynamical Systems. *Comput. Optim. Appl.* **1998**, *11*, 195–217. [CrossRef]
7. Böhme, T.J.; Frank, B. Direct Methods for Optimal Control. In *Hybrid Systems, Optimal Control and Hybrid Vehicles. Advances in Industrial Control*; Springer: Cham, Switzerland, 2017.
8. Banga, J.R.; Balsa-Canto, E.; Moles, C.G.; Alonso, A.A. Dynamic optimization of bioprocesses: Efficient and robust numerical strategies. *J. Biotechnol.* **2003**, *117*, 407–419. [CrossRef] [PubMed]
9. Rodrigues, H.S.; Torres Monteiro, M.T.; Torres, D.F.M. Optimal Control and Numerical Software: An Overview. *arXiv*, 2014; arXiv:1401.7279.
10. Ghaddar, C.K. Novel Spreadsheet Direct Method for Optimal Control Problems. *Math. Comput. Appl.* **2018**, *23*, 6. [CrossRef]
11. ExcelWorks LLC, MA, USA. ExceLab Calculus Add-in and Reference Manual. Available online: https://excel-works.com (accessed on 22 September 2018).
12. Nævdal, E. Solving Continuous Time Optimal Control Problems with a Spreadsheet. *J. Econ. Educ.* **2003**, *34*, 2. [CrossRef]
13. Weber, E.J. Optimal Control Theory for Undergraduates Using the Microsoft Excel Solver Tool. *Comput. High. Educ. Econ. Rev.* **2007**, *19*, 4–15.
14. Ghaddar, C. Unconventional Calculus Spreadsheet Functions. World Academy of Science, Engineering and Technology, International Science Index 112. *Int. J. Math. Comput. Phys. Electr. Comput. Eng.* **2016**, *10*, 194–200.
15. Hairer, E.; Wanner, G. *Solving Ordinary Differential Equations II: Stiff and Differential-Algebraic Problems*; Springer Series in Computational Mathematics; Springer: Berlin, Germany, 1996.
16. De Boor, C. *A Practical Guide to Splines (Applied Mathematical Sciences)*; Springer: Berlin, Germany, 2001.
17. Piessens, R.; de Doncker-Kapenga, E.; Ueberhuber, C.W.; Kahaner, D.K. *Quadpack: A Subroutine Package for Automatic Integration*; Springer: Berlin, Germany, 1983.
18. Lasdon, L.S.; Waren, A.D.; Jain, A.; Ratner, M. Design and Testing of a Generalized Reduced Gradient Code for Nonlinear Programming. *ACM Trans. Math. Softw.* **1978**, *4*, 34–50. [CrossRef]
19. Dolan, E.D.; More, J.J. *Benchmarking Optimization Software with Cops*; Technical Report; Argonne National Laboratory: Argonne, IL, USA, 2001.

20. Lim, A.E.B.; Liu, Y.Q.; Teo, K.L.; B, M.J. Linear-quadratic optimal control with integral quadratic constraints. *Optim. Control Appl. Methods* **1999**, *20*, 79–92. [CrossRef]
21. Bhattacharya, R. *OPTRAGEN 2.0: A MATLAB Toolbox for Optimal Trajectory Generation*; Texas A & M University: College Station, TX, USA, 2013.
22. Ghaddar, C.K. Rapid Modeling and Parameter Estimation of Partial Differential Algebraic Equations by a Functional Spreadsheet Paradigm. *Math. Comput. Appl.* **2018**, *23*, 39. [CrossRef]

Mathematical and Computational Applications

MDPI

Article

Novel Spreadsheet Direct Method for Optimal Control Problems

Chahid Kamel Ghaddar

ExcelWorks LLC, Sharon, MA 02067, USA; cghaddar@excel-works.com; Tel.: +1-781-626-0375

Received: 26 December 2017; Accepted: 23 January 2018; Published: 25 January 2018

Abstract: We devise a simple yet highly effective technique for solving general optimal control problems in Excel spreadsheets. The technique exploits Excel's native nonlinear programming (NLP) Solver Command, in conjunction with two calculus worksheet functions, namely, an initial value problem solver and a discrete data integrator, in a direct solution paradigm adapted to the spreadsheet. The technique is tested on several highly nonlinear constrained multivariable control problems with remarkable results in terms of reliability, consistency with pseudo-spectral reported answers, and computing times in the order of seconds. The technique requires no more than defining a few analogous formulas to the problem mathematical equations using basic spreadsheet operations, and no programming skills are needed. It introduces an alternative, simpler tool for solving optimal control problems in social and natural science disciplines.

Keywords: optimal control; dynamical optimization; parameter estimation; differential equations; spreadsheet; Excel Solver

1. Introduction

Optimal control problems are commonly encountered in engineering and life sciences, as well as social studies such as economics and finance [1–3]. An optimal control problem is typically concerned with finding optimal control functions (or policies) that achieve optimal trajectories for a set of controlled differential state variables. The optimal trajectories are decided by a constrained dynamical optimization problem, such that a cost functional is minimized or maximized subject to certain constraints on state variables and the control functions. Mathematically, an optimal control problem may be stated as follows:

Find the control functions $u(t) = (u_1(t), u_2(t), \ldots, u_m(t))$ and the corresponding state variables $x(t) = (x_1(t), x_2(t), \ldots, x_n(t))$, $t \in [0, T]$ which minimize (or maximize) the functional

$$J = H(x(T), T) + \int_0^T G(x(t), u(t), t) \, dt, \tag{1}$$

subject to

$$M \frac{dx}{dt} = F(x(t), u(t), t), \tag{2}$$

with initial conditions

$$x(0) = x_0, \tag{3}$$

and optional final conditions and bounds

$$Q(x(T), T) = 0, \tag{4}$$

$$S(x(t), u(t)) \leq 0. \tag{5}$$

In the formulation (1)–(5), the generally nonlinear H and G are scalar functions, whereas F, Q, and S are vector-valued functions. Typically, either H or Q are specified but not both in the same problem. Common forms of Q and S are $x(T) = x_T$ and $u_{min} \leq u(t) \leq u_{max}$, respectively. We have chosen not to include $\dot{x}(t)$ and $\ddot{x}(t)$ in the formulation because a higher-order explicit differential equation system can be restated as a first-order system via variable substitution. The matrix M in (2) offers an optional coupling of the states' temporal derivatives by a mass matrix which may be singular. If M is singular, the equation system (2) is differential algebraic, or DAE. For uncoupled derivatives, M is the identity matrix which can be omitted. Furthermore, T, which denotes the final time, may be fixed or free.

Numerical solution strategies of (1)–(5) fall into one of two approaches: an *indirect method*, where Pontryagin's minimum principle is employed to transform the problem into an augmented Hamiltonian system requiring the solution of a boundary value problem [4]; and a *direct method*, where the original system's variables are approximated by parameterized appropriate functions which, in turn, reduce the problem into a finite-dimensional nonlinear programming problem [5]. Direct methods can be further classified as full or partial parametrization methods. In the latter, only the controls are parametrized, wherein the inner initial value problem (IVP) (2)–(3) is treated as a separate dependent problem that must be solved repeatedly by the outer nonlinear programming (NLP) algorithm [6].

Except for the most trivial cases, optimal control problems can be difficult to solve, particularly for those who are not inclined towards programming and numerical methods. Despite advances in software programs, it remains a nontrivial task to utilize a standard package such as MATLAB to solve optimal control problems. The student must have sufficient programming skill, as well as a good understanding of the general structure of the solution algorithm and the various solvers required to implement it [7].

In this article, we present a systematic technique for solving optimal control problems in a spreadsheet, modeled on partial parametrization direct methods. The technique is made possible, on the one hand, by algorithmic advances [8,9] which enabled the introduction of mathematically pure calculus worksheet functions to the spreadsheet [10,11]. A pure calculus function is evaluated as a standard built-in math function; however, it accepts, via input parameters, formulas representing a problem model and outputs a formatted solution result. Specifically, we make use of two calculus functions described in Appendix A: an IVP solver, based on an implicit RADUA5 algorithm with adaptive step control [12], which we employ for solving the inner IVP (2)–(3); and a discrete data integrator, based on cubic spline approximations [13], which we employ to approximate the cost index (1). On the other hand, Excel spreadsheets include a powerful NLP Solver Command based on the Generalized Reduced Gradient Method (GRG) [14] which is compatible with the calculus functions. We devise a direct control–parametrization method based on employing the calculus functions with the NLP Solver in a dynamical optimization paradigm for the solution of (1)–(5).

Attempts to solve optimal control problems in spreadsheets are not new; however, to the best of our knowledge, no prior work has presented a practical direct spreadsheet method aimed at solving the general nonlinear multidimensional optimal control problem (1)–(5). The chief reason is that prior approaches utilized the spreadsheet explicitly as the computational grid for the discretization and solution of the underlining IVP. This limits the practical scope to rather simple problems that can be easily discretized with an explicit differencing scheme suitable for the spreadsheet. For example, Weber [15] demonstrated a direct approach to solving control problems in resource economics involving simple one-dimensional IVPs, and direct summation of discrete values for the cost index. Nævdal [16] demonstrated a basic implementation of the indirect method, with the aid of Visual Basic for Applications (VBA) programming, to solve one-dimensional optimal control problems. The method utilized Excel's Solver in conjunction with an explicit difference scheme and a shooting algorithm to solve the resulting boundary value problem. While Nævdal's work provides educational insights into the mechanics of the indirect solution method, its detail-intensive implementation makes it impractical to use or extend to higher dimensions or nonlinear stiff systems requiring adaptive implicit schemes. Our devised direct spreadsheet method, on the other hand, differs fundamentally from prior work, in that the algorithmic implementation for solving the IVP, and integrating the cost index, has been decoupled from the spreadsheet grid and encapsulated in pure spreadsheet solver functions suitable for seamless integration with the NLP Solver. The design of the solver functions, described in Section 2, permits utilization of fully implicit and adaptive algorithms which make the method applicable to a general class of nonlinear multivariable optimal control problems. Furthermore, by encapsulating the tedious implementation details in standard pure math functions with a clear divide between input and output, the method is applicable with little more than basic spreadsheet knowledge, and without any programming skills. As demonstrated in Section 3, results obtained on several highly nonlinear problems are remarkable, in terms of both the reliability and the computing time in the order of seconds. The devised method extends the utility of the spreadsheet beyond what has been practical or even feasible before.

The remainder of this paper is organized as follows: In the next section, we describe the basic steps required to model and solve an optimal control problem using the adapted direct method technique. In Section 3, we demonstrate the technique for solving four different control problems reproduced from Elnagar and Kazemi [17] who used a pseudo-spectral direct method. The problems include:

1. A bang–bang control problem.
2. A highly nonlinear and coupled system.
3. A minimum swing container transfer problem involving multiple controls and constraints.
4. A minimum time orbit transfer control problem with free end time.

Section 4 provides some practical tips for applying the technique, followed by conclusions in Section 5. Appendix A includes a description for the IVP solver, IVSOLVE, and the discrete data integrator functions, QUADXY, both of which are essential for the technique. We also remark that our main focus in this first article is to introduce and illustrate the spreadsheet direct solution method rather than formulate or study any specific optimal control problem. As such, we start from a mathematical statement of a given problem and present a feasible solution obtained by the method with relevant comparisons to the reported result in [17].

2. Spreadsheet-Adapted Direct Solution Method

The main enabling elements of the devised method are, in addition to the NLP Solver, the IVP solver function, IVSOLVE, and the discrete data integrator, QUADXY. The IVSOLVE spreadsheet function, described in Appendix A.1, is designed according to the flowchart of Figure 1, wherein a suitable highly accurate algorithm, such as RADUA5 [12], is fully shielded with a strict divide between the IVP model input and the output solution results. The model input is represented by formulas that are direct analogues to the IVP mathematical equations, and the output solution results are displayed in a formatted tabular array of elective resolution which is easily adjusted to yield an accurate integration of a dependent cost index. By design, IVSOLVE is a pure function which does not modify its input but merely computes and displays the solution in its allocated spreadsheet array. QUADXY, on the other hand, follows a standard spreadsheet User Defined Function (UDF) implementation to integrate a vector of ordered points, $(t, y(t))_i$, $i = 1, n$. QUADXY performs the integration with the aid of cubic splines fit to the data. Under the assumption that the discrete data describe a smooth curve, the computed integral is generally quite accurate and can be further improved by supplying optional slopes at the end points of the curve when they are known or can be estimated. Likewise, QUADXY is also a pure function which does not modify its inputs.

Below, we describe the general steps for employing IVSOLVE and QUADXY with the NLP Solver for solving the optimal control problem (1)–(5). Some of these steps may or may not be required for a given problem. To simplify the discussion, we shall assume a single control function, $u(t)$. Extension to multiple controls is straightforward and is demonstrated by the examples. In practice, there are three systematic tasks:

Task 1

The first step is to obtain, with the IVP solver function, IVSOLVE, an initial solution for the underlining IVP (2)–(3) using an appropriately parametrized formula for the control function and initial guesses for the unknown parameters. A continuous control function can be parametrized, for example, by a third-order polynomial with unknown coefficients, such as '=c_0+c_1*t+c_2*t^2+c_3*t^3'. On the other hand, a discontinuous control function can be modeled using the standard *IF* statement in Excel. For example, a two-stage, constant controller can be defined as follows: '=IF(t<=ts,value1,value2)'. Here ts, value1, and value2 are unknown parameters that would be assigned initial guesses.

Task 2

In the second task we define an analogous objective formula for the cost functional (1). Our strategy is to integrate, using QUADXY, a sampled vector of the integrand expression in (1) using the solution values obtained in Task 1. To accomplish this, in a new column, we generate values for the parametrized control formula evaluated at the solution's output times and, in a second column, we generate values for the integrand expression, using the solution's state variables and generated control values as needed. Both the control and integrand columns are easily generated using the AutoFill feature of Excel. To define an analogous objective formula for the cost index (1), we employ the discrete data integrator function, QUADXY, to integrate the generated integrand data column versus the solution's output times column. The ordered steps needed to define the objective formula are summarized in Figure 2.

Task 3

The last task is to configure and run Excel's NLP Solver. The NLP Solver can be configured to minimize or maximize an objective formula by changing design variables, subject to defined constraints. The design variables are the unknown parameters which are assigned initial guesses in Task 1. The constraints (4) and (5) are added directly in the Solver's dialog. Simple equality end conditions on state variables are added by referencing the corresponding cells in the solution output as illustrated in Figure 2. Bound constraints on state variables or controls are easily imposed with the aid of Excel's MAXA() and MINA() math functions which compute the maximum and minimum values of a vector. Concrete examples are presented in the next section.

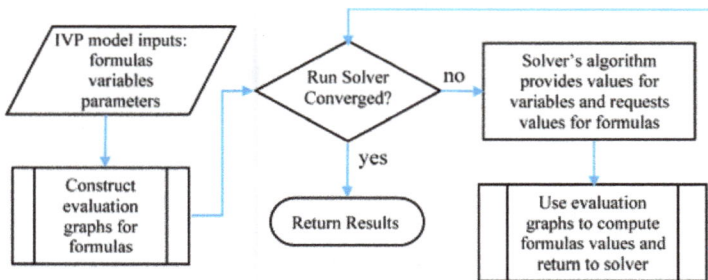

Figure 1. Flowchart for the design of a mathematically pure spreadsheet solver function which accepts formulas as input arguments. Enabling technology is described in [8,9].

How It Works

Key to the successful operation of the adapted direct solution method are two attributes: the purity of the IVP solver function, and the Automatic Calculation Mode of the spreadsheet. As described earlier, the IVSOLVE function does not modify its inputs, and the authority to modify the inputs to IVSOLVE via changes to the decision parameter vector is confined to the outer NLP Solver command. On the other hand, the spreadsheet maintains a dependency hierarchy, and updates all information whenever a change occurs. Any modification to the design parameters by the outer NLP Solver triggers reevaluation of the inner IVSOLVE solution, the dependent control and integrand columns, the objective, and any constraint formulas in the proper order. The NLP Solver always receives up-to-date values for the objective and constraints whenever it alters the design variables' values.

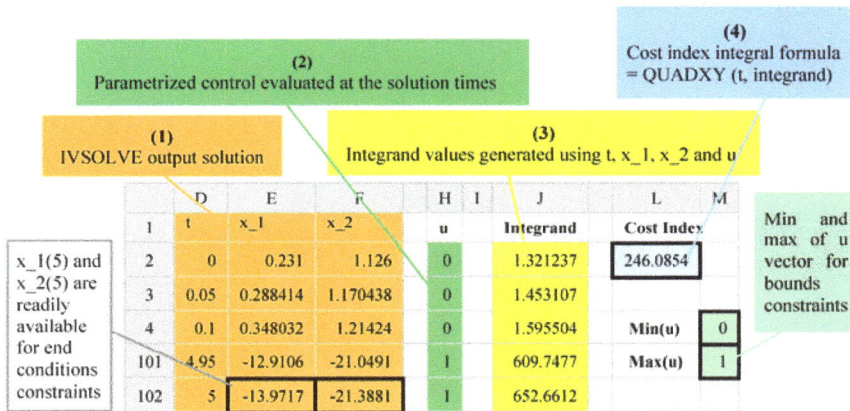

Figure 2. Illustration of the ordered steps for defining an analogous objective formula to the cost index functional (1). The illustration assumes a problem with two state variables and one control.

3. Illustrative Examples

In this section, we apply the spreadsheet method to solve four optimal control problems reproduced from Elnagar and Kazemi [17], who used a pseudo-spectral high order Chebyshev approximation scheme in conjunction with the general-purpose sequential programming software package NLPQL. Relevant comparisons with reported results in [17] are included. The examples are representative of various types of optimal control problems and intended to serve as a template as well as validation for the effectiveness of the devised spreadsheet method. We recommend that the reader review Appendix A prior to reading the examples. Also, some basic familiarity with the spreadsheet operation is assumed, including naming variables, defining formulas, and running the NLP Solver command.

3.1. A Bang–Bang Control Problem

The first example describes a bang–bang (two-stage) control problem. The mathematical problem is stated as follows:

Minimize

$$J = \frac{1}{2} \int_0^5 \left[x(t)^2 + y(t)^2 \right] dt \tag{6}$$

subject to

$$\dot{x}(t) = y(t), \quad t \in [0, 5], \tag{7}$$

$$\dot{y}(t) = y(t) - x(t) + u(t), \tag{8}$$

$$x(0) = 0.231, \quad y(0) = 1.126, \tag{9}$$

$$-0.8 \le u(t) \le 0.8. \tag{10}$$

3.1.1. Spreadsheet Model

Working with named variables with adjacent labels shown in Column A of Figure 3, we parametrized the two-stage control function, $u(t)$, using a standard $IF()$ statement, as shown in B9. The unknown parameters switchT, stage1, and stage2 were assigned the initial guess values 0.1, 0, and 1. The right-hand sides of the IVP differential Equations (7) and (8) were represented by the equivalent formulas B11 and B12, and the initial conditions (9) were assigned to the variables x and y in B3 and B4. The colored ranges in Figure 3 represent the model input required to obtain the initial solution for the IVP (7)–(9) using the IVSOLVE function. The initial solution was obtained by evaluating the formula

$$\text{=IVSOLVE(B11:B12,B2:B4,\{0,5\})} \tag{11}$$

in an allocated array D2:F103. The result is shown partially in Figure 4, and the initial trajectories of $x(t)$, $y(t)$, and $u(t)$ are plotted in Figure 7a.

	A	B
1	**ODE variables**	
2	t	0
3	x	0.231
4	y	1.126
5	**Parametrized control formula**	
6	switchT	0.1
7	stage1	0
8	stage2	1
9	u	=IF(t<=switchT,stage1,stage2)
10	**ODE rhs formulas**	
11	xdot	=y
12	ydot	=u−x+y

Figure 3. Spreadsheet model for IVP (7)–(9) with parametrized control function. The colored ranges are input parameters for the IVSOLVE Formula (11).

	D	E	F	H	J	L	M
1	**IVP Solution**						
2	t	x	y	u	Integrand	Cost functional	
3	0	0.231	1.126	0	1.321237	Objective	246.0854
4	0.05	0.288414	1.170438	0	1.453107		
5	0.1	0.348032	1.21424	0	1.595504	**Constraint formulas**	
6	0.15	0.411089	1.308339	1	1.880744	Max(u)	1
7	0.2	0.478891	1.403911	1	2.200301	Min(u)	0
102	4.95	−12.9106	−21.0491	1	609.7477		
103	5	−13.9717	−21.3881	1	652.6612		

Figure 4. Partial listing of computed results by Formula (11). Also shown are generated control and integrand columns, and initial objective formula value. The associated formulas are listed in Table 1.

In order to define the objective formula for the cost functional (6) as described in Task 2 of Section 2, we first generated, based on the obtained initial solution array, two new columns labeled **u** and **Integrand** (see Figure 4) for the control function and the integrand expression. The control column, **u**, was generated with the AutoFill feature of Excel, using the formula H3 shown in Table 1. The integrand column was generated in a similar way using the formula J3 in Table 1. Here, we simply evaluated the expression $x(t)^2 + y(t)^2$ using the corresponding output solution values for t, x, and y from the IVSOLVE solution.

Table 1. Formula definitions used for solving optimal control problem (6)–(10).

Purpose	Cell	Formula
Initial value problem solution	D2:F103	=IVSOLVE(B11:B12,B2:B4,{0,5})
AutoFill formula for control values	H3	=IF(D3<=switchT,stage1,stage2)
AutoFill formula for integrand values	J3	=E3^2+F3^2
Objective formula	M3	=0.5*QUADXY(D3:D103,J3:J103)
Maximum value of control column	M6	=MAXA(H3:H103)
Minimum value of control column	M7	=MINA(H3:H103)

Next, we employed the discrete data integrator function QUADXY to integrate the generated integrand column versus the solution output times column, as shown by formula M3 of Table 1. The initial value of the objective formula was 246.0854, as shown in Figure 4. To impose the bound constraint (10) on $u(t)$, we defined two aid formulas in M6 and M7 (see Table 1) which computed the maximum and minimum of the generated control column values. We made use of these aid formulas during the configuration of the NLP Solver.

3.1.2. Results and Discussion

We invoked Excel's Solver from the Data Tab which brings up a dialog as shown in Figure 5. We configured the Solver to minimize the objective formula M3 by varying the control parameters B6:B8 (corresponding to switchT, stage1 and stage2) subject to the constraints

$$M6 \leq 0.8 \text{ corresponds to } \max(u) \leq 0.8, \tag{}$$

$$M7 \geq -0.8 \text{ corresponds to } \min(u) \geq -0.8, \tag{}$$

which are needed impose (10). We unchecked the box which reads 'Make Unconstrained Variables Non-Negative' to allow the variables to take on negative values as well. In the options for the GRG Nonlinear solver, we switched the derivative scheme from the default Forward to Central, and then ran the Solver, which reports a feasible solution in less than 3 s. By accepting the Solver's solution, all the values and plots in the spreadsheet were automatically updated to reflect the optimal result. The NLP Solver also generates an optional Answer Report, as shown in Figure 6. The optimal trajectories are plotted in Figure 7b. As shown in the Answer Report, the optimal switching time was found at approximately 1.26 which is within 1% of the 1.25 value reported by Elnagar and Kazemi [17] using a pseudo-spectral Chebyshev approximation of order 15. (In [17], the time domain was transformed to $[-1, 1]$, and the switching time was found at negative 0.5 which maps to 1.25 in the original $[0, 5]$ time domain.)

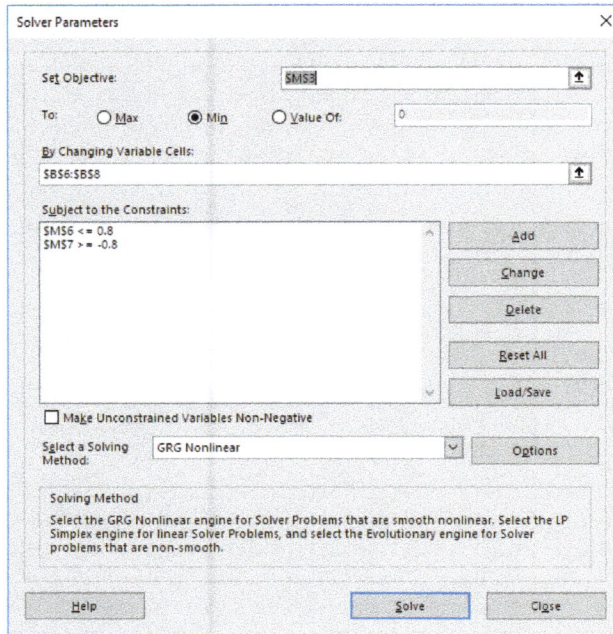

Figure 5. Excel's Solver dialog configured for optimal control problem (6)–(10).

Microsoft Excel 16.0 Answer Report
Worksheet: [Examples.xlsx]Example 1
Report Created: 11/21/2017 8:05:18 PM
Result: Solver found a solution. All Constraints and optimality conditions are satisfied.
Solver Engine
   Engine: GRG Nonlinear
   Solution Time: 2.594 Seconds.
   Iterations: 15 Subproblems: 0
Solver Options
   Max Time Unlimited, Iterations Unlimited, Precision 0.000001, Use Automatic Scaling
   Convergence 0.0001, Population Size 100, Random Seed 0, Derivatives Central, Require Bounds
   Max Subproblems Unlimited, Max Integer Sols Unlimited, Integer Tolerance 1%

Objective Cell (Min)

Cell	Name	Original Value	Final Value
$M$3	Objective	246.0853608	5.649829193

Variable Cells

Cell	Name	Original Value	Final Value	Integer
$B$6	switchT	0.1	1.263702828	Contin
$B$7	stage1	0	-0.799999875	Contin
$B$8	stage2	1	0.8	Contin

Constraints

Cell	Name	Cell Value	Formula	Status	Slack
$M$6	Max u	0.8	$M$6<=0.8	Binding	0
$M$7	Min u	-0.799999875	$M$7>=-0.8	Binding	0

Figure 6. Answer Report generated by Excel's Solver for optimal control problem (6)–(10).

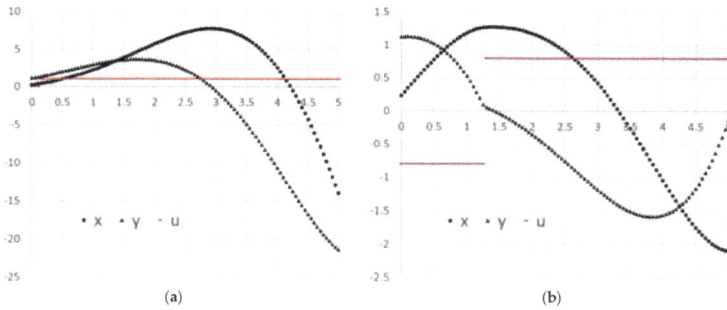

Figure 7. (**a**) Initial trajectories for optimal control problem (6)–(10) based on the default values shown in Figure 3; (**b**) Optimal trajectories found by Excel's nonlinear programming (NLP) Solver.

3.2. Unconstrained Nonlinear Optimal Control Problem

The second example represents an unconstrained optimal control problem in the fixed interval $t \in [-1, 1]$, but with highly nonlinear equations. The mathematical problem is stated as follows:
Minimize

$$J = \frac{0.78}{2} \int_{-1}^{1} \left[x_1(t)^2 + x_2(t)^2 + 0.1u(t)^2 \right] dt \tag{12}$$

subject to

$$\dot{x}_1(t) = \frac{0.78}{2} \left(-2[x_1(t) + 0.25] + [x_2(t) + 0.5]exp\left[\frac{25x_1(t)}{x_1(t) + 2} \right] - [x_1(t) + 0.25]u(t) \right), \tag{13}$$

$$\dot{x}_2(t) = \frac{0.78}{2} \left(0.5 - x_2(t) - [x_2(t) + 0.25]exp\left[\frac{25x_1(t)}{x_1(t) + 2} \right] - [x_1(t) + 0.25] \right), \tag{14}$$

$$x(-1) = 0.05, \quad x_2(-1) = 0.0. \tag{15}$$

3.2.1. Spreadsheet Model

The spreadsheet model for the IVP (13)–(15) with a parametrized control function using a third-order polynomial is shown in Figure 8. The initial solution to the IVP was obtained by evaluating the formula

$$=IVSOLVE(B12:B13,B2:B4,\{-1,1\}) \tag{16}$$

in an allocated array E2:G103, which is shown partially in Figure 9 and plotted in Figure 10a. Clearly, our initial guess for the control coefficients B6:B9 was not good, since the solution exhibits instabilities at larger time values. The control and integrand vectors, needed to construct the objective formula for the cost index (12), were generated based on the obtained initial solution using formulas I3 and K3, listed in Table 2. The objective formula was defined using the data integrator QUADXY as shown in N3 of Table 2, with an initial value of 1.92×10^{18}, as shown in Figure 9.

Table 2. Formula definitions used for solving optimal control problem (12)–(15).

Purpose	Cell	Formula
Initial value problem solution	E2:G103	=IVSOLVE(B12:B13,B2:B4,{−1,1})
AutoFill formula for control values	I3	=c_0+c_1*E3+c_2*E3^2+c_3*E3^3
AutoFill formula for integrand values	K3	=F3^2+G3^2+0.1*I3^2
Objective	N3	=0.78*QUADXY(E3:E103,K3:K103)/2

	A	B
1	**ODE variables**	
2	t	-1
3	x_1	0.05
4	x_2	0
5	**Parametrized control formula**	
6	c_0	1
7	c_1	0
8	c_2	0
9	c_3	0
10	u	=c_0+c_1*t+c_2*t^2+c_3*t^3
11	**ODE rhs equations**	
12	x1dot	=0.78*(-2*(x_1+0.25)+(x_2+0.5)*EXP(25*x_1/(x_1+2))-(x_1+0.25)*u)/2
13	x2dot	=0.78*(0.5-x_2-(x_2+0.25)*EXP(25*x_1/(x_1+2)))/2

Figure 8. Spreadsheet model for the IVP (13)–(15) with parametrized control function. The colored ranges are input parameters for IVSOLVE Formula (16).

	E	F	G	H I J	K	L	M	N
1	**IVP Solution**							
2	t	x_1	x_2	u	Integrand		Cost functional	
3	−1.00	0.05	0	1	0.1025		Objective	1.92396E+18
4	−0.98	0.050163	0.000305	1	0.102516			
5	−0.96	0.050341	0.000596	1	0.102535			
100	0.94	3.59E+09	−0.25	1	1.29E+19			
101	0.96	3.64E+09	−0.25	1	1.33E+19			
102	0.98	3.7E+09	−0.25	1	1.37E+19			
103	1.00	3.75E+09	−0.25	1	1.41E+19			

Figure 9. Partial listing of computed results by Formula (16). Also shown are generated control and integrand columns, and the initial objective formula value. The associated formulas are listed in Table 2.

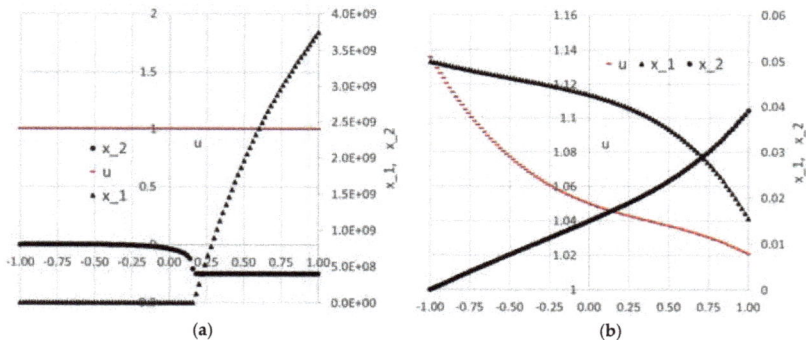

Figure 10. (**a**) Initial trajectories for optimal control problem (12)–(15) based on the default values shown in Figure 8; (**b**) Optimal trajectories found by Excel's NLP Solver.

3.2.2. Results and Discussion

We ran Excel's Solver to minimize the objective formula N3 by varying the control parameters in B6:B9 with no added constraints. Despite the bad initial values for the control parameters, the Solver reported a feasible solution in about 2 seconds with the Answer Report shown in Figure 11. The optimal trajectories for the system variables are plotted in Figure 10b.

Microsoft Excel 16.0 Answer Report
Worksheet: [Examples.xlsx]Example 2
Report Created: 9/27/2017 7:33:46 PM
Result: Solver found a solution. All Constraints and optimality conditions are satisfied.
Solver Engine
Engine: GRG Nonlinear
Solution Time: 2.359 Seconds.
Iterations: 2 Subproblems: 0
Solver Options
Max Time Unlimited, Iterations Unlimited, Precision 0.000001, Use Automatic Scaling
Convergence 0.0001, Population Size 100, Random Seed 0, Derivatives Central, Require Bounds
Max Subproblems Unlimited, Max Integer Sols Unlimited, Integer Tolerance 1%

Objective Cell (Min)

Cell	Name	Original Value	Final Value
$N$3	Objective	1.92396E+18	0.089191915

Variable Cells

Cell	Name	Original Value	Final Value	Integer
$B$6	c_0	1	1.05	Contin
$B$7	c_1	0	-0.034540768	Contin
$B$8	c_2	0	0.028300278	Contin
$B$9	c_3	0	-0.022981318	Contin

Constraints

NONE

Figure 11. Answer Report generated by Excel's Solver for optimal control problem (12)–(15) based on the initial guess values in Figure 8.

The reported objective in [17] is 0.026621417, which is better than the achieved objective value of 0.08919 found by Excel. To improve the result, we tried a different initial guess for the parameters (c_0, c_1, c_2, c_3) by changing their values in Figure 8 to (1, 0, 1, 0). A second run of the Solver reported the feasible solution shown in Figure 12. The new objective value was reduced by more than 50% to 0.040245. The new solution is plotted in Figure 13 and shows noticeably different trajectories for x_1 and x_2 than those obtained initially in Figure 10. This is expected, given the highly nonlinear and unconstrained problem.

Solution Time: 6.094 Seconds.
Iterations: 18 Subproblems: 0
Solver Options
Max Time Unlimited, Iterations Unlimited, Precision 0.000001, Use Automatic Scaling
Convergence 0.0001, Population Size 100, Random Seed 0, Derivatives Central, Require Bounds
Max Subproblems Unlimited, Max Integer Sols Unlimited, Integer Tolerance 1%

Objective Cell (Min)

Cell	Name	Original Value	Final Value
$N$3	Objective	0.160849029	0.042025578

Variable Cells

Cell	Name	Original Value	Final Value	Integer
$B$6	c_0	1	0.287624409	Contin
$B$7	c_1	0	-0.454314103	Contin
$B$8	c_2	1	0.649993818	Contin
$B$9	c_3	0	-0.546843336	Contin

Constraints

NONE

Figure 12. Answer Report for optimal control problem (12)–(15) using a different initial guess and yielding improved minimum.

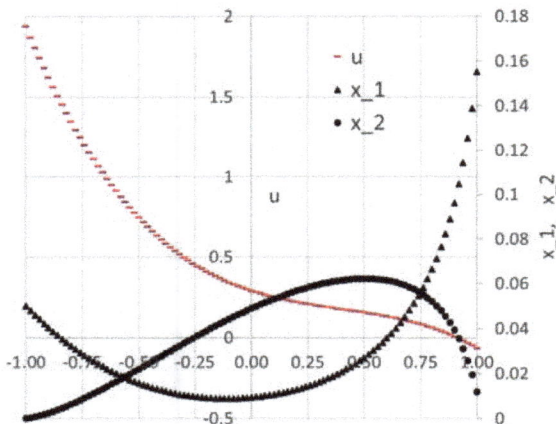

Figure 13. Optimal trajectories for optimal control problem (12)–(15) found by Excel's Solver starting from a different initial guess, leading to a lower objective value.

3.3. Minimal Swing Container Transfer Problem

The third example represents the problem of transferring containers, driven by a hoist motor and a trolley drive motor, from a ship to a cargo truck. The goal is to minimize the swing during and at the end of the transfer. The mathematical optimal control problem is described by (17)–(29). The problem is nonlinear with six state variables and two controllers subject to multiple final and bound constraints.

Minimize

$$J = 4.5 \int_0^1 [x_3(t)^2 + x_6(t)^2]\, dt \tag{17}$$

subject to

$$\dot{x}_1(t) = 9x_4(t), \tag{18}$$

$$\dot{x}_2(t) = 9x_5(t), \tag{19}$$

$$\dot{x}_3(t) = 9x_6(t), \tag{20}$$

$$\dot{x}_4(t) = 9[u_1(t) + x_3(t)], \tag{21}$$

$$\dot{x}_5(t) = 9u_2(t), \tag{22}$$

$$\dot{x}_6(t) = \frac{9(u_1(t) + 27.0756x_3(t) + 2x_5(t)x_6(t))}{x_2(t)}, \quad t \in [0,1]. \tag{23}$$

Initial conditions:

$$x_1(0) = 0,\ x_2(0) = 22,\ x_3(0) = 0,\ x_4(0) = 0,\ x_5(0) = -1,\ x_6(0) = 0. \tag{24}$$

Final conditions:

$$x_1(1) = 10,\ x_2(1) = 14,\ x_3(1) = 0,\ x_4(1) = 2.5,\ x_5(1) = 0,\ x_6(1) = 0. \tag{25}$$

Bounds:

$$|u_1(t)| \leq 2.83374, \tag{26}$$

$$-0.80865 \leq u_2(t) \leq 0.71265, \tag{27}$$

$$|x_4(t)| \leq 2.5, \tag{28}$$

$$|x_5(t)| \leq 1.0. \tag{29}$$

3.3.1. Spreadsheet Model

Following the same procedure as that in the previous examples, we prepared the spreadsheet model for the IVP (18)–(24) using third-order parametrized polynomial control functions $u_1(t)$ and $u_2(t)$, as shown in Figure 14. Initial values and guesses were assigned to the state variables and unknown parametrization coefficients as shown in the figure. Figure 15 shows a partial listing of the initial solution obtained by evaluating the formula

$$=\text{IVSOLVE(B17:B22,B2:B8,\{0,1\})} \tag{30}$$

in array F2:L103, and the generated control columns, u_1, u_2, and the integrand expression column using the corresponding formulas listed in Table 3. The initial system trajectories are plotted in Figure 16.

	A	B	C	D
1	**ODE variables**			
2	t	0		
3	x_1	0		
4	x_2	22		
5	x_3	0		
6	x_4	0		
7	x_5	−1		
8	x_6	0		
9	**Parametrized controls formulas**			
10	c_0		1	d_0 1
11	c_1		1	d_1 1
12	c_2		−5	d_2 -5
13	c_3		−5	d_3 -5
14	u_1	=c_0+c_1*t+c_2*t^2+c_3*t^3		
15	u_2	=d_0+d_1*t+d_2*t^2+d_3*t^3		
16	**ODE rhs equations**			
17	x1dot	=9*x_4		
18	x2dot	=9*x_5		
19	x3dot	=9*x_6		
20	x4dot	=9*(u_1+x_3)		
21	x5dot	=9*u_2		
22	x6dot	=9*(u_1+27.0756*x_3+2*x_5*x_6)/x_2		

Figure 14. Spreadsheet model for the IVP (18)–(24) with parametrized control functions. The colored ranges are input parameters for IVSOLVE Formula (30).

Next, we defined the objective formula, S3, corresponding to the cost index (17) as shown in Table 3, in which the data integrator QUADXY was used to integrate the generated integrand expression values. The objective formula evaluated to an initial value of 24,229.22793. Table 3 also lists a number of aid formulas which compute the minimum and maximum values for the state variables x_4 and x_5 and the generated control columns. These aid formulas were used to define the bound constraints for the NLP Solver.

	F	G	H	I	J	K	L	M	N	O	P
1	IVP Solution										
2	t	x_1	x_2	x_3	x_4	x_5	x_6		u_1	u_2	Integrand
3	0	0	22	0	0	−1	0		1	1	0
4	0.01	0.004063	21.91406	0.000185	0.09044	−0.90957	0.00411		1.009495	1.009495	1.69E-05
5	0.02	0.016305	21.8363	0.000742	0.181723	−0.81832	0.008285		1.01796	1.01796	6.92E-05
6	0.03	0.036797	21.76679	0.001679	0.273786	−0.72636	0.012569		1.025365	1.025365	0.000161
101	0.98	230.6731	15.34798	189.5472	205.5287	−12.3527	147.8948		−7.52796	−7.52796	57801.01
102	0.99	249.9267	14.20542	203.2503	222.511	−13.0407	156.6717		−7.762	−7.762	65856.72
103	1	270.7634	13	217.7568	240.7409	−13.75	165.7417		-8	-8	74888.35

Figure 15. Partial listing of computed result by Formula (30). Also shown are generated control and integrand columns. The associated formulas are listed in Table 3.

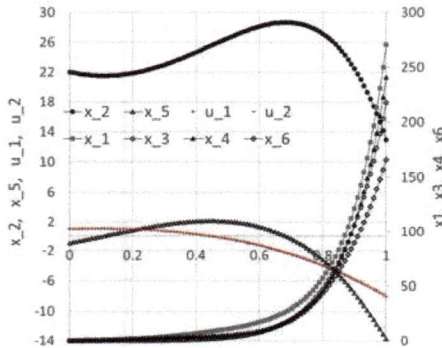

Figure 16. Initial trajectories for optimal control problem (17)–(29) based on default values shown in Figure 14.

Table 3. Formulas definitions used for solving optimal control problem (17)–(29).

Purpose	Cell	Formula
Initial value problem solution	F2:L103	=IVSOLVE(B17:B22,B2:B8,{0,1})
AutoFill formula for u_1 control values	N3	=c_0+c_1*F3+c_2*F3^2+c_3*F3^3
AutoFill formula for u_2 control values	O3	=d_0+d_1*F3+d_2*F3^2+d_3*F3^3
AutoFill formula for integrand values	P3	=I3^2+L3^2
Objective Formula	S3	=4.5*QUADXY(F3:F103,P3:P103)
u_1 column max value	S6	=MAXA(N3:N103)
u_1 column min value	S7	=MINA(N3:N103)
u_2 column max value	S8	=MAXA(O3:O103)
u_2 column min value	S9	=MINA(O3:O103)
x_4 column max value	S10	=MAXA(J3:J103)
x_4 column min value	S11	=MINA(J3:J103)
x_5 column max value	S12	=MAXA(K3:K103)
x_5 column min value	S13	=MINA(K3:K103)

3.3.2. Results and Discussion

We configured Excel's Solver to minimize the objective formula S3 by varying the controls' coefficients B10:B14 and D10:D14 subject to the added constraints listed in Table 4. Here, we made use of the aid formulas listed in Table 3 to define the inequality bound constraints (26)–(29). The end point equality constraints on the state variables (25) were imposed directly onto the corresponding cells at t = 1 (last row) of the IVSOLVE solution array (see Figure 15). The Solver spun for a few seconds, then reported that it did not find a feasible solution when, in fact, it already had, judging by the best-found solution results shown partially in Figure 17 and plotted in Figure 18. The solution indicates that all constraints were satisfied within a reasonable tolerance of 1×10^{-5}, except for x_6(1), which was satisfied within a tolerance of 1×10^{-3}. This is verified by the feasibility report generated by the Solver, and shown in Figure 19. The report indicates that the Solver had difficulty satisfying end point constraints for x_4 and x_6 at the Solver's default tolerances, while all other constraints were satisfied.

Table 4. List of constraints added to the NLP Solver for optimal control problem (17)–(29) and their corresponding equations. The bound constraints were defined using aid formulas listed in Table 3.

Added Constraints	Purpose
G103 = 10	$x_1(1) = 10$
H103 = 14	$x_2(1) = 14$
I103 = 0	$x_3(1) = 0$
J103 = 2.5	$x_4(1) = 2.5$
K103 = 0	$x_5(1) = 0$
L103 = 2.5	$x_6(1) = 0$
S10 ≤ 2.5 S11 ≥ −2.5	(4.12)
S12 ≤ 1 S13 ≥ −1	(4.13)
S6 ≤ 2.83374 S7 ≥ −2.83374	(4.10)
S8 ≤ 0.71265 S9 ≥ −0.80865	(4.11)

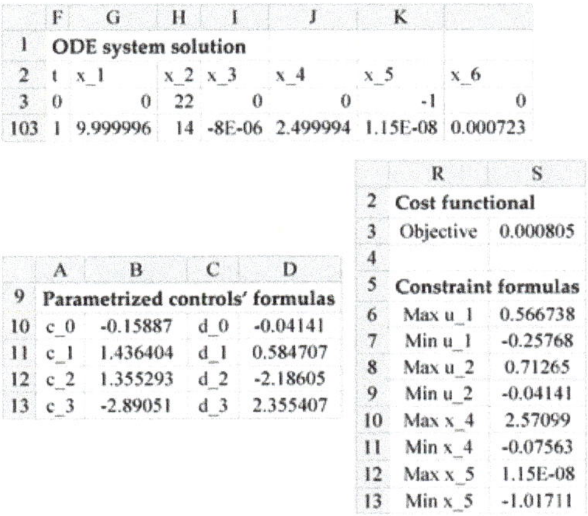

	F	G	H	I	J	K	
1	ODE system solution						
2	t	x_1	x_2	x_3	x_4	x_5	x_6
3	0	0	22	0	0	-1	0
103	1	9.999996	14	-8E-06	2.499994	1.15E-08	0.000723

	R	S
2	Cost functional	
3	Objective	0.000805
4		
5	Constraint formulas	
6	Max u_1	0.566738
7	Min u_1	-0.25768
8	Max u_2	0.71265
9	Min u_2	-0.04141
10	Max x_4	2.57099
11	Min x_4	-0.07563
12	Max x_5	1.15E-08
13	Min x_5	-1.01711

	A	B	C	D
9	Parametrized controls' formulas			
10	c_0	-0.15887	d_0	-0.04141
11	c_1	1.436404	d_1	0.584707
12	c_2	1.355293	d_2	-2.18605
13	c_3	-2.89051	d_3	2.355407

Figure 17. Best-found solution obtained by Excel's Solver for optimal control problem (17)–(29).

The reported objective in [17] is 0.005361, but the exact tolerances used for the NLPQL solver are unknown. We note that in this problem, Excel's Solver best-found solution has a much lower minimum for the objective at 0.000805, as shown in Figure 17.

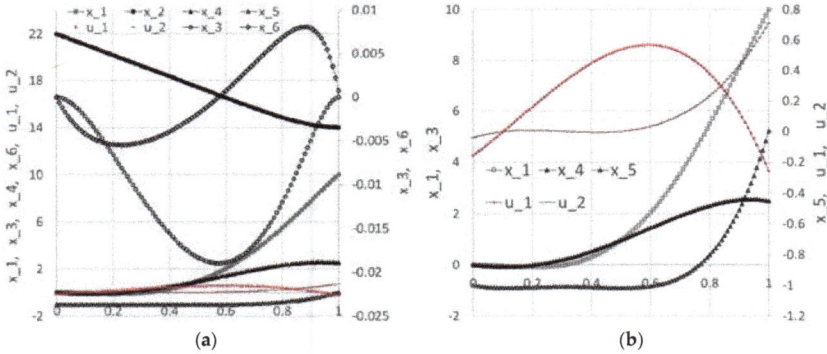

Figure 18. (a) Optimal trajectories for all variables of optimal control problem (17)–(29); (b) Selected optimal trajectories.

Microsoft Excel 16.0 Feasibility Report
Worksheet: [Examples.xlsx]Example 3
Report Created: 9/26/2017 6:43:45 PM
Constraints Which Make the Problem Infeasible

Cell	Name	Cell Value	Formula	Status	Slack
$H$103	x_2	14.00000001	$H$103=14	Binding	0
$J$103	x_4	2.499994248	$J$103=2.5	Violated	−5.75165E-06
$L$103	x_6	0.00072322	$L$103=0	Violated	−0.00072322

Figure 19. Feasibility Report generated by Excel's Solver for optimal control problem (17)–(29).

3.4. Minimum Time Orbit Transfer Problem

The fourth example describes a minimum time orbit transfer problem. The goal is to minimize the transfer time of a constant thrust rocket between the orbits of Earth and Mars, and to determine the optimal thrust angle control. The original free-time mathematical problem is stated as follows:
   Minimize

$$J = t_F \tag{31}$$

subject to

$$\dot{x}_1(t) = x_2(t), \tag{32}$$

$$\dot{x}_2(t) = \frac{x_3(t)^2}{x_1(t)} - \frac{\gamma}{x_1(t)^2} + \frac{R_0 \sin u(t)}{m_0 + \dot{m}\,t}, \tag{33}$$

$$\dot{x}_3(t) = -\frac{x_2(t)x_3(t)}{x_1(t)} + \frac{R_0 \cos u(t)}{m_0 + \dot{m}\,t}, \quad t \in [0, t_F], \tag{34}$$

with initial conditions

$$x_1(0) = 1, \; x_2(0) = 0, x_3(0) = 1, \tag{35}$$

and final conditions

$$x_1(t_F) = 1.525, \; x_2(t_F) = 0, x_3(t_F) = 0.8098. \tag{36}$$

In [17], the original problem was transformed into a fixed time domain [−1, 1]. The corresponding value for the transfer time, t_F, was reported at 3.31873 using ninth-degree Chebyshev polynomial approximations. Here, we solve the original free-time problem as stated in (31)–(36).

3.4.1. Spreadsheet Model

Referring to Figure 20, the IVP (32)–(35) was modeled using a third-order parametrized polynomial approximation for $u(t)$, with an initial guess of zero for each of the unknown coefficients. The differential equations in B21:B23 are defined in terms of the system variables in B2:B5, the control u in B19, and the constants γ, m_0, $\dot{m}$, and R_0 which are assigned corresponding names in the figure. In this problem, the end time, t_F, is a design variable and is therefore assigned its own variable tF in B13 with an initial guess of 10. Figure 21 shows a partial listing of the initial solution obtained by evaluating the IVSOLVE formula

$$=\text{IVSOLVE(B21:B23,B2:B5,B12:B13)} \tag{37}$$

in array E2:H103, along with the generated control column and the initial objective value. The corresponding formulas are listed in Table 5, and the initial trajectories for the system states and control are plotted in Figure 23a. Note that the third argument to the IVSOLVE formula is the variable time domain $[0, t_F]$ which is represented by the range B12:B13.

	A	B
1	**ODE variables**	
2	t	0
3	x_1	1
4	x_2	0
5	x_3	1
6	**Parameters**	
7	gamma	1
8	R0	0.1405
9	m0	1
10	mdot	−0.07487
11	**Time Domain**	
12	t0	0
13	tF	10
14	**Parametrized control formula**	
15	c_0	0
16	c_1	0
17	c_2	0
18	c_3	0
19	u	=c_0+c_1*t+c_2*t^2+c_3*t^3
20	**ODE rhs equations**	
21	x1dot	=x_2
22	x2dot	=x_3^2/x_1−gamma/x_1^2+R0*SIN(u)/(m0+mdot*t)
23	x3dot	=−x_2*x_3/x_1+R0*COS(u)/(m0+mdot*t)

Figure 20. Spreadsheet model for the IVP (32)–(35) with parametrized control function. The colored ranges are input parameters for IVSOLVE Formula (37).

Table 5. Formula definitions used for solving optimal control problem (31)–(36).

Purpose	Cell	Formula
Initial value problem solution	E2:H103	=IVSOLVE(B21:B23,B2:B5,B12:B13)
AutoFill formula for control values	J3	=c_0+c_1*E3+c_2*E3^2+c_3*E3^3
Objective formula	L3	=t_F

	E	F	G	H	I	J	K	L
1	**IVP Solution**							
2	t	x_1	x_2	x_3		u	**Cost functional**	
3	0.00	1	0	1		0	Objective	10
4	0.10	1.000047	0.001413	1.014055		0		
5	0.20	1.000378	0.005681	1.027927		0		
6	0.30	1.001278	0.012807	1.041313		0		
7	0.40	1.003034	0.02276	1.053905		0		
100	9.70	8.305837	1.499819	1.391005		0		
101	9.80	8.456923	1.521927	1.417734		0		
102	9.90	8.610249	1.544573	1.445546		0		
103	10.00	8.765865	1.567783	1.474497		0		

Figure 21. Partial listing of computed result by Formula (37). Also shown are generated control column and the initial objective formula value. The associated formulas are listed in Table 5.

3.4.2. Results and Discussion

We configured Excel's Solver to minimize the objective formula L3 by varying the end time t_F, B13, and the control coefficients B15:B18, subject to the end point equality constraints on the state variables (36). The constraints were added directly into the Solver's dialog by referencing the corresponding cells in the last row of the IVSOLVE solution array in Figure 21. The Solver reported, in under 20 s, the feasible solution shown in the Answer Report of Figure 22. The minimum orbit time, t_F, was found to be 3.58656. This compares reasonably well to the value reported in [17] at 3.31873 using ninth-degree Chebyshev polynomial approximations. The optimal trajectories are plotted in Figure 23b. In Figure 24, we show a partial listing of the updated IVSOLVE solution result reflecting the new end time, and the decreased output time increment in comparison to the initial result shown in Figure 21.

Microsoft Excel 16.0 Answer Report
Worksheet: [Examples.xlsx]Example 4
Report Created: 11/22/2017 7:42:37 AM
Result: Solver found a solution. All Constraints and optimality conditions are satisfied.
Solver Engine
Engine: GRG Nonlinear
Solution Time: 19.437 Seconds.
Iterations: 41 Subproblems: 0
Solver Options
Max Time Unlimited, Iterations Unlimited, Precision 0.000001, Use Automatic Scaling
Convergence 0.0001, Population Size 100, Random Seed 0, Derivatives Central, Require Bounds
Max Subproblems Unlimited, Max Integer Sols Unlimited, Integer Tolerance 1%

Objective Cell (Min)

Cell	Name	Original Value	Final Value
$M$3	Objective	10	3.586555548

Variable Cells

Cell	Name	Original Value	Final Value	Integer
$B$14	c_0	0	0.126729662	Contin
$B$15	c_1	0	4.214481179	Contin
$B$16	c_2	0	-3.781677247	Contin
$B$17	c_3	0	0.718794874	Contin
$B$12	tF	10	3.586555548	Contin

Constraints

Cell	Name	Cell Value	Formula	Status	Slack
$F$103	x_1	1.525000209	$F$103=1.525	Binding	0
$G$103	x_2	1.1803E-06	$G$103=0	Binding	0
$H$103	x_3	0.80979919	$H$103=0.8098	Binding	0

Figure 22. Answer Report generated by Excel's Solver for optimal control problem (31)–(36).

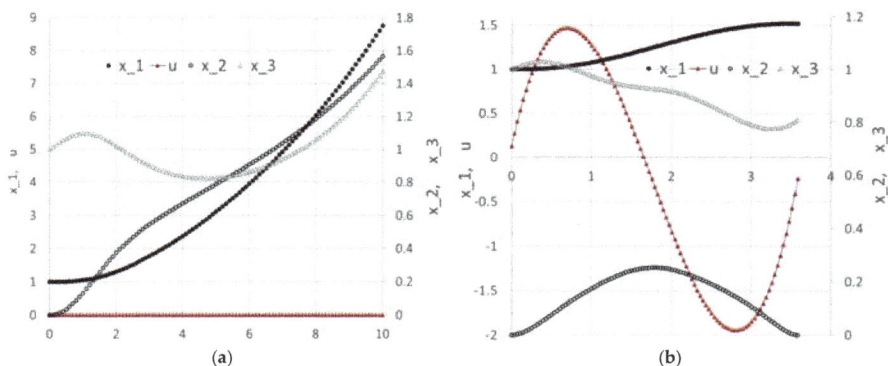

(a)   (b)

Figure 23. (**a**) Initial trajectories for optimal control problem (31)–(36) based on default values shown in Figure 20; (**b**) Optimal trajectories found by Excel's NLP Solver.

	E	F	G	H	I	J	K	L
1	**IVP solution**							
2	t	x_1	x_2	x_3		u	**Cost functional**	
3	0	1	0	1		0.12673	Objective	3.586556
4	0.03586	1.000018	0.001183	1.004922		0.273053		
5	0.07173	1.000097	0.003406	1.009605		0.409846		
6	0.10759	1.000274	0.00658	1.013935		0.537309		
7	0.14346	1.000579	0.010605	1.017831		0.655639		
100	3.47899	1.524526	0.01056	0.792036		−0.71564		
101	3.51482	1.524818	0.005901	0.79735		−0.56733		
102	3.55069	1.524962	0.002325	0.803313		−0.40924		
103	3.58555	1.525	1.18E-06	0.809799		−0.24119		

Figure 24. Partial listing of the updated initial result of Figure 21 which reflects the optimal final time and adjusted output time values in Column E.

4. Practical Tips

Successful nonlinear optimization is often the result of numerical experimentation. In this section, we share a few practical tips for effective use of the presented spreadsheet optimization method.

4.1. Excel's NLP Solver and Settings

The standard NLP Solver shipped with Excel uses the Generalized Reduced Gradient algorithm [14], which has proved effective for smooth nonlinear problems. The standard Solver also offers a simplex and evolutionary genetic algorithm options that may be suitable for linear or nonsmooth problems. Nonetheless, it is possible to expand the available pool of NLP algorithms, including sequential programming, interior point, and active set methods, by upgrading to a premium version of the NLP Solver [18].

Perhaps the most important factor is the starting guess for the decision parameter vector which may require a nonzero initial value for some problems. The author has found it generally quick to find a good initial guess interactively by trial and error in just a few attempts given the fast response of the Solver. Excel Solver's dialog offers a few settings, two of which have proved influential in aiding the convergence for some problems. In particular, the 'Derivatives' scheme is recommended to be switched to Central from the default Forward, and the 'Use Automatic Scaling' option is recommended to be left enabled (default setting).

4.2. Spreadsheet Tips

- Naming spreadsheet variables (e.g., naming B1 as *t*) makes the formulas easier to read and spot errors. However, it is also recommended to restrict the scope of a named variable to the specific sheet it will be used on, and not the whole workbook. This prevents accidental interdependence between multiple problems on different sheets sharing variables with the same name.
- The shown layouts for the model setup with labels ensures that the Answer Report generated by Excel's Solver has proper descriptive names for the variables and constraints.
- Excel gives precedence to the unary negation operator which may be confused with the binary minus operator since they both use the same symbol. This can lead to hard-to-find errors in formulas. For example, Excel evaluates the formula '$=-X1\hat{}2$' as '$=(-X1)\hat{}2$'. The intention may have been to do '$-(X1\hat{}2)$' instead. A simple fix is to either use parentheses when needed, or to use the intrinsic POWER(X1,2) function instead of the operator ^. Also, when using the IF statement in a formula, it is important to verify that the formula evaluates to a numeric value for all possible conditions. Otherwise, the formula may evaluate to a nonnumeric Boolean condition, leading to a solver error.
- The calculus functions are designed to operate in two modes: a silent mode, where only standard spreadsheet errors are returned like #VALUE!, and a verbose mode, where the function may display an informative error or warning message alert in a popup window. It is recommended to work in the verbose mode when setting up the problem, but switch to the silent mode before running the NLP Solver. Switching between the two modes is triggered by evaluating the formula '=VERBOSE(TRUE)' or '=VERBOSE(FALSE)' in any cell in the workbook. For some problems, the Solver may wander into illegal input space before it recovers and adjusts its search. The silent mode blocks any occasional error alerts from the calculus functions.

4.3. Generalization to a Special Class of Control Problems

By restricting the space of admissible control functions to, for example, variable-order polynomials up to a fixed degree, it may not be always possible to find a solution to a certain class of control problems for which the optimal control, in fact, lies outside the admissible space. It is difficult to know a priori when an algorithm may or may not work, but this may arise in problems with a particularly long time horizon. When all fails, it may well be that the only alternative is to use an indirect method and maximize the Hamiltonian for a number of points, then use interpolation to construct the complete control function(s). Nonetheless, we propose below a general idea that we have not tested but which fits within our presented direct method, and may offer a potential solution in certain cases.

The idea, effectively, is to enlarge the admissible space by stitching together different parametrized control functions defined over nonoverlapping subintervals of the time horizon. Both the interior end points of the subintervals (stitching points), and the control functions' parameters are unknown design variables for the NLP Solver. The stitching is enforced by imposing additional constraints for the NLP Solver that demand continuity of the control functions at the stitching points. A continuity condition is easily derived algebraically by matching two parametrized functions' values at a stitching point. Excel NLP Solver poses no restriction on the number of design variables, although performance and convergence will be impacted as the problem dimension grows.

5. Conclusions

We have demonstrated a practical technique which adapts the control parametrization direct method for optimal control problems to the spreadsheet. The technique combines Excel's NLP Solver command with two calculus worksheet functions, an initial value problem solver, and a discrete data integrator in a simple systematic procedure. The employed calculus functions are utilized from an

Excel calculus Add-in library [19]. The technique has proved very effective on several highly nonlinear, multivariable, constrained control problems, and produced results consistent with reported answers obtained with highly accurate pseudo-spectral approximation and NLPQL optimization software package using a full parametrization direct method. Excel's Solver's computing time has been in the order of seconds to a minute on a laptop with an Intel 4-Core i7 CPU, which makes the technique highly interactive for experimenting with different initial guesses and variations. As demonstrated by the examples, using the devised method requires no more than defining a few formulas with basic knowledge of the spreadsheet and requires no programming skills, offering a simpler solution approach to optimal control problems.

A slightly modified version of the technique can also be applied for parameter estimation of initial value problems where the parameters may include initial conditions or coefficients. In future work, it may also be worth investigating a dual technique based on indirect methods for optimal control problems. The indirect method requires the modeler to recast the problem in a different form by applying Pontryagin's maximum principle. However, the ensuing boundary value problem could be solved with the aid of a boundary value problem solver function, BVSOLVE, also included in the Excel calculus Add-in [19].

Supplementary Materials: An Excel workbook containing all the solved examples presented in this article is available at www.mdpi.com/2297-8747/23/1/6/s1, (file Location to be determined.)

Acknowledgments: No funding has been received in support of this research work.

Conflicts of Interest: The author of the manuscript is the founder of ExcelWorks LLC of Massachusetts, USA supplying the Excel calculus Add-in, ExceLab, utilized in this research work.

Appendix

Appendix A.1 Initial Value Problem Solver Spreadsheet Function

The spreadsheet function

=IVSOLVE(equations, variables, time_interval, mass_matrix, options)

is utilized from the calculus Add-in [19] to solve an initial value ordinary differential algebraic equation system in the interval $t \in [t_s, t_e]$:

$$M\frac{dx}{dt} = F(x(t), t),$$
$$x(t_s) = x_s, \tag{A1}$$

$x(t) = (x_1(t), x_2(t), \ldots, x_n(t))$, and M is an optional mass matrix. If M is singular, the system is differential algebraic. IVSOLVE implements several adaptive integration schemes [12,20], suitable for stiff and smooth problems. By default, it uses the RADUA5 algorithm [12]. Algorithm selection and control parameters, as well as the system analytic Jacobian can be supplied via optional arguments [19]. At minimum, IVSOLVE requires three arguments:

1. Reference to the right-hand side formulas corresponding to the vector-valued function $F(x(t), t) = (f_1(x(t), t), f_2(x(t), t), \ldots, f_n(x(t), t))$. Any algebraic equations should be ordered last.
2. Reference to the system variables corresponding to t and $x(t)$ in the specific order $(t, x_1, x_2, \ldots, x_n)$.
3. The integration time interval end points.

IVSOLVE is run as a standard array formula in an allocated array of cells. It evaluates to an ordered tabular result where the time values are listed in the first column and the corresponding state variables' values are listed in adjacent columns. By default, IVSOLVE reports the output at uniform intervals according to the allocated number of rows for the output array. Custom output formats can be achieved via the optional parameters, including specifying custom divisions or output points [19]. We demonstrate IVSOLVE for the following DAE example:

$$\frac{dy_1}{dt} = -0.04y_1 + 10^4 y_2 y_3, \quad t \in [0, 1000],$$
$$\frac{dy_2}{dt} = 0.04y_1 - 10^4 y_2 y_3 - 3*10^7 y_2^2,$$
$$0 = y_1 + y_2 + y_3 - 1,$$
$$y_1(0) = 1, \ y_2(0) = 0, \ y_3(0) = 0.$$

(A2)

The system RHS formulas are defined in cells A1:A3 using cell T1 for the time variable and Y1, Y2, Y3 for the state variables with the specified initial conditions shown in Figure A1.

	A		Y
1	=-0.04*Y1+10000*Y2*Y3	1	1
2	=0.04*Y1-10000*Y2*Y3-30000000*Y2^2	2	0
3	= Y1+Y2+Y3-1	3	0

Figure A1. Spreadsheet input model for equation system A2.

To solve the system, we evaluate in the array C1:F22 the formula

$$\text{=IVSOLVE (A1:A3,(T1,Y1:Y3),\{0,1000\},1)}$$

(A3)

which computes and displays the solution shown in Figure A2. Note that the fourth argument, 1, instructs the solver that the last equation in A1:A3 is an algebraic equation.

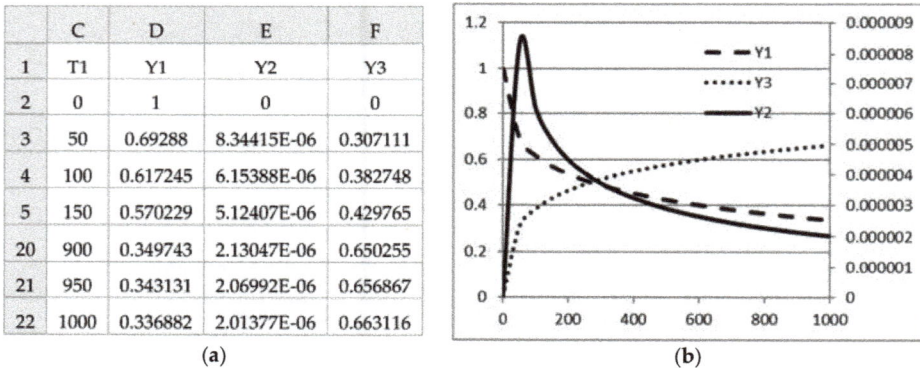

	C	D	E	F
	T1	Y1	Y2	Y3
1	T1	Y1	Y2	Y3
2	0	1	0	0
3	50	0.69288	8.34415E-06	0.307111
4	100	0.617245	6.15388E-06	0.382748
5	150	0.570229	5.12407E-06	0.429765
20	900	0.349743	2.13047E-06	0.650255
21	950	0.343131	2.06992E-06	0.656867
22	1000	0.336882	2.01377E-06	0.663116

(a)

(b)

Figure A2. (a) Partial listing of the computed solution by formula A3; (b) Plots of the trajectories.

Appendix A.2 Discrete Data Integrator Spreadsheet Function

The spreadsheet function

$$\text{=QUADXY(x, y, options)}$$

is utilized from the calculus Add-in [19] to integrate a set of discrete $(x, y(x))$ data points. The integration limits are determined from the endpoints of the x vector. QUADXY performs the integration with the aid of cubic (default) or linear splines [13]. The third optional argument allows the specification of boundary conditions for the data, including starting and end slopes. Options are specified by (key, value) pairs as detailed in [19].

References

1. La Torre, D.; Kunze, H.; Ruiz-Galan, M.; Malik, T.; Marsiglio, S. Optimal Control: Theory and Application to Science, Engineering, and Social Sciences. *Abstr. Appl. Anal.* **2015**, *2015*, 890527. [CrossRef]
2. Geering, H.P. *Optimal Control with Engineering Applications*; Springer: Berlin/Heidelberg, Germany, 2007.
3. Anița, S.; Arnăutu, V.; Capasso, V. *An Introduction to Optimal Control Problems in Life Sciences and Economics: From Mathematical Models to Numerical Simulation with MATLAB*; Birkhäuser: Basel, Switzerland, 2011.
4. Böhme, T.J.; Frank, B. Indirect Methods for Optimal Control. In *Hybrid Systems, Optimal Control and Hybrid Vehicles. Advances in Industrial Control*; Springer: Cham, Switzerland, 2017.
5. Böhme, T.J.; Frank, B. Direct Methods for Optimal Control. In *Hybrid Systems, Optimal Control and Hybrid Vehicles. Advances in Industrial Control*; Springer: Cham, Switzerland, 2017.
6. Banga, J.R.; Balsa-Canto, E.; Moles, C.G.; Alonso, A.A. Dynamic optimization of bioprocesses: Efficient and robust numerical strategies. *J. Biotechnol.* **2003**, *117*, 407–419. [CrossRef] [PubMed]
7. Rodrigues, H.S.; Monteiro, M.T.T.; Torres, D.F.M. Optimal Control and Numerical Software: An Overview. This is a preprint of a paper whose final and definite form will appear in the book. In *Systems Theory: Perspectives, Applications and Developments*; Miranda, F., Ed.; Nova Science Publishers: Hauppauge, NY, USA, 2014. Available online: https://arxiv.org/abs/1401.7279 (accessed on 23 January 2018).
8. Ghaddar, C. Method, Apparatus, and Computer Program Product for Optimizing Parameterized Models Using Functional Paradigm of Spreadsheet Software. U.S. Patent 9,286,286, 15 March 2016.
9. Ghaddar, C. Method, Apparatus, and Computer Program Product for Solving Equation System Models Using Spreadsheet Software. U.S. Patent 15,003,848, 2018. in Press.
10. Ghaddar, C. Unconventional Calculus Spreadsheet Functions. World Academy of Science, Engineering and Technology, International Science Index 112. *Int. J. Math. Comput. Phys. Electr. Comput. Eng.* **2016**, *10*, 194–200. Available online: http://waset.org/publications/10004374 (accessed on 23 January 2018).
11. Ghaddar, C. Unlocking the Spreadsheet Utility for Calculus: A Pure Worksheet Solver for Differential Equations. *Spreadsheets Educ.* **2016**, *9*, 5. Available online: http://epublications.bond.edu.au/ejsie/vol9/iss1/5 (accessed on 23 January 2018).
12. Hairer, E.; Wanner, G. *Solving Ordinary Differential Equations II: Stiff and Differential-Algebraic Problems*; Springer Series in Computational Mathematics; Springer: Berlin/Heidelberg, Germany, 1996.
13. De Boor, C. *A Practical Guide to Splines (Applied Mathematical Sciences)*; Springer: Berlin/Heidelberg, Germany, 2001.
14. Lasdon, L.S.; Waren, A.D.; Jain, A.; Ratner, M. Design and Testing of a Generalized Reduced Gradient Code for Nonlinear Programming. ACM Trans. *Math. Softw.* **1978**, *4*, 34–50. [CrossRef]
15. Weber, E.J. Optimal Control Theory for Undergraduates Using the Microsoft Excel Solver Tool. *Comput. High. Educ. Econ. Rev.* **2007**, *19*, 4–15.
16. Nævdal, E. Solving Continuous Time Optimal Control Problems with a Spreadsheet. *J. Econ. Educ.* **2003**, *34*, 99–122. [CrossRef]
17. Elnagar, G.; Kazemi, M.A. Pseudospectral Chebyshev Optimal Control of Constrained Nonlinear Dynamical Systems. *Comput. Optim. Appl.* **1998**, *11*, 195–217. [CrossRef]
18. FrontlineSolvers. Available online: https://www.solver.com/ (accessed on 23 January 2018).
19. ExcelWorks LLC; MA, USA. ExceLab Calculus Add-in for Excel and Reference Manual. Available online: https://excel-works.com (accessed on 23 January 2018).
20. Hindmarsh, A.C. ODEPACK, A Systematized Collection of ODE Solvers. In *Scientific Computing*; Stepleman, R.S., Carver, M., Peskin, R., Ames, W.F., Vichnevetsky, R., Eds.; North-Holland Publishing: Amsterdam, The Netherlands, 1983; pp. 55–64.

Mathematical and Computational Applications

MDPI

Article

Time Needed to Control an Epidemic with Restricted Resources in SIR Model with Short-Term Controlled Population: A Fixed Point Method for a Free Isoperimetric Optimal Control Problem

Imane Abouelkheir, Fadwa El Kihal[ID]**, Mostafa Rachik and Ilias Elmouki** *[ID]

Department of Mathematics and Computer Sciences, Faculty of Sciences Ben M'Sik, Hassan II University of Casablanca, Casablanca 20000, Morocco; abouelkheir88@gmail.com (I.A.); fadwa.elkihal@gmail.com (F.E.K.); m_rachik@yahoo.fr (M.R.)
* Correspondence: i.elmouki@gmail.com

Received: 10 October 2018; Accepted: 20 October 2018; Published: 22 October 2018

Abstract: In this paper, we attempt to determine the optimal duration of an anti-epidemic control strategy which targets susceptible people, under the isoperimetric condition that we could not control all individuals of this category due to restricted health resources. We state and prove the local and global stability conditions of free and endemic equilibria of a simple epidemic compartmental model devised in the form of four ordinary differential equations which describe the dynamics of susceptible-controlled-infected-removed populations and where it is taken into account that the controlled people cannot acquire long-lived immunity to move towards the removed compartment due to the temporary effect of the control parameter. Thereafter, we characterize the sought optimal control and we show the effectiveness of this limited control policy along with the research of the optimal duration that is needed to reduce the size of the infected population. The isoperimetric constraint is defined over a fixed horizon, while the objective function is defined over a free horizon present under a quadratic form in the payoff term. The complexity of this optimal control problem requires the execution of three numerical methods all combined together at the same time, namely, the forward–backward sweep method to generate the optimal state and control functions, the secant method adapted to the isoperimetric restriction, and, finally, the fixed point method to obtain the optimal final time.

Keywords: epidemic model; optimal control; isoperimetric constraint; free horizon; fixed point method

---

1. Introduction

1.1. Background

Many epidemiological models have been interested in the study of the dynamics of susceptible, infected and removed individuals who belong to a sample of a population threatened by an infection. Many theoretical models in epidemiology have been devised to show the effect of different anti-epidemic control strategies when they are followed to prevent transmission of a particular type of infection to the susceptible population. As examples of these control approaches, we can cite Refs. [1,2] where the authors introduced an awareness control function in their models and which aimed to prevent the susceptible people from Human Immunodeficiency Virus infection and Acquired Immune Deficiency Syndrome (HIV/AIDS) epidemic. Roy et al. [3] treated the idea of awareness control programs in HIV/AIDS prevention after the addition of a new variable in their models and which defines the number of individuals in the aware class. Other examples of control models have used vaccination of the susceptible individuals as a control policy, while considering the number of vaccinated people as additional compartment in their systems (see [3–7]). Most optimal control strategies suggested for preventing an infection to spread consider a fixed final time. However, health policy-makers need very often to know when it is appropriate to stop their anti-epidemic measures as this information is important for managing their medical resources [8], and then, it becomes not reasonable in such situations to study dynamics of an epidemic under control without an estimation of the final time. The present paper tries to find the optimal value of this mentioned variable through a free horizon optimal control approach applied to an epidemic model with four compartments, namely classes of susceptible, controlled, infected and removed people.

1.2. Formulation of the Problem of Interest for this Investigation

Here, we devise a simple generalized model where the control utilized has only a temporary effect on the immunity of the targeted population so the susceptible people under control do not acquire long-lived immunity to move to the removed class, while taking into account the presence of an equation which describes the evolution of the number of the controlled people. First, we study the local and global stability of our epidemic model, which is devised in the form of four ordinary differential equations, and wherein a control is introduced as a constant parameter; and, second, we seek the optimal duration needed for reaching the goal of our strategy while determining the optimal value of this control when it is changed to a function of time under the hypothesis that this anti-epidemic preventive measure can reach only a specific fraction of susceptible people due to the limited health resources. We should note that authors of [8,9] tried to find the optimal final time in a first case through their plots and checked the values where this optimal function verified the obtained additional necessary condition, and in two other cases via discrete numerical schemes. We believe this work is more interesting, as it provides a more precise numerical method. In fact, as the necessary condition on final time also represents here a fixed point equation, the incorporation of the fixed point method better facilitates our task, as this technique seems more accurate and convincing since it meets the theoretical aspect of the found condition.

1.3. Literature Survey

Zhou and Fan [10] discussed different forms of functions introduced in epidemic models to explore the impact of limited medical resources in the transmission of infectious diseases. Abdelrazec et al. [11] introduced, in a mathematical model of dengue fever, a function in place of the recovery rate for similar purpose. More recently, Yu et al. proposed an optimal control approach to investigate the optimal distribution policy of the limited vaccination resources based on the research of a parameter introduced in their model and which minimizes the basic reproduction number [12].

1.4. Scope and Contribution of this Study

We present here an optimal control approach which treats the problem of limited resources differently to the three above-mentioned references. In fact, our method considers a constraint, so-called "isoperimetric", which is used on the control function as done in [13] for the resolution of a dosage problem, on the fraction of controlled variables as done in [14] in the case of an epidemic model, or when adapted to a discrete-time SIRS epidemic model as in [15], and it supposes also that the final time or horizon of the objective function is free (non-fixed) as used in many applications (see, for example, Ref. [8], which discusses the problem of optimal duration needed for reaching the intravesical therapy goal, and Ref. [9], where it is explained why such considerations are very important to health-policy makers and managers in the health sector when there is an epidemic that is controlled through awareness of the susceptible class).

1.5. Organization of the Paper

Based on the theory of mathematical epidemiology in [16], the spread of an epidemic can be described mathematically by SIR models which in turn have been developed later to extended forms such as SEIR [17], SIRS [18], or SIS [19], where each letter refers to a class of individuals. A class of controlled people can also be considered, as done in [20] where authors added a vaccination compartment in a model of pertussis and tuberculosis, and in [21] where they studied nonfatal diseases, and [22,23] in case of influenza. Sharomi and Malik [24] represented an other form of SIR model with an additional equation corresponding to the vaccinated category in the case the vaccination is not 100% effective; such considerations can also be found in [4]. Based on similar assumptions as in the two last mentioned references, we devise our present model.

In the following parts of the paper, we start with the presentation of our mathematical model and study its stability in cases of free and endemic equilibria. Furthermore, we seek the optimal value of the free horizon considered in the objective function, along with the determination of the optimal value of the control function. Finally, we discuss our numerical results.

2. The Mathematical Model and Stability

In this section, we consider a mathematical model with the four following main compartments:

- S is the number of susceptible people to infection or who are not yet infected.
- C_S is the number of susceptible people who are temporarily controlled so they cannot move to the removed class due to the limited effect of control.
- I is the number of infected people who are capable of spreading the epidemic to those in the susceptible and temporary controlled categories.
- R is the number of removed people from the epidemic.

In our modeling approach, we choose to describe dynamics of variables S, C_S, I and R at time t, based on the following differential system

$$\begin{cases} \dot{S}(t) &= \Pi(t) - \beta S(t) I(t) - a\theta S(t) - \mu S(t) \\ \dot{C}_S(t) &= a\theta S(t) - b\beta C_S(t) I(t) - \mu C_S(t) \\ \dot{I}(t) &= \beta(S(t) + bC_S(t))I(t) - \gamma I(t) - \mu I(t) \\ \dot{R}(t) &= \gamma I(t) - \mu R(t) \end{cases} \tag{1}$$

with initial conditions $S(0) > 0$, $C_S(0) \geq 0$, $I(0) \geq 0$ and $R(0) \geq 0$, and where $\Pi(t) = \mu N(t)$, with $N(t) = S(t) + C_S(t) + I(t) + R(t)$ as the total population size, gives the newborn people at time t; $a\theta$ ($0 \leq a \leq 1$) is the recruitment rate of susceptibles to the controlled class with θ defining the control parameter as a constant between 0 and 1 and "a" modeling the reduced chances of a susceptible individuals to be controlled; $\beta = \dfrac{\delta}{N(t)}$ with δ the infection transmission rate, μ the natural death rate, $b\theta$ ($0 \leq b \leq 1$) the recruitment rate of controlled people to the infected class even in the presence of θ and "b" modeling the reduced chances of a temporarily controlled individual to be infected; and γ is the recovery rate. We note that the population size is constant because $\dot{N}(t) = \dot{S}(t) + \dot{C}_S(t) + \dot{I}(t) + \dot{R}(t) = 0$. Hence, $N(t) = N = $ *a constant*, and then, $\Pi(t) = \Pi = $ *a constant.*

For the sake of readability, hereafter, we use S, C_S, I and R as notations of the time functions $S(t)$, $C_S(t)$, $I(t)$ and $R(t)$.

Recalling that $R_0 = \dfrac{\beta}{\mu + \gamma}$ is the basic reproduction number of the standard SIR model (see [25] where it is concluded that the disease free equilibrium E^0 is global asymptotically stable if $R_0 \leq 1$, and there exists a global asymptotically stable and unique endemic equilibrium E^+ if $R_0 > 1$).

Since the two first equations are independent of the last equation, we only study the stability of the following differential system

$$\begin{cases} \dot{S} &= \Pi - \beta SI - a\theta S - \mu S \\ \dot{C}_S &= a\theta S - b\beta C_S I - \mu C_S \\ \dot{I} &= \beta(S + bC_S)I - \gamma I - \mu I \end{cases} \tag{2}$$

A disease free equilibrium in our case can be defined as $E^0 = (S^0, C_S^0, 0)$ where S^0 and C_S^0 are obtained based on the assumptions $\dot{S} = 0$ and $\dot{I} = 0$ when there is no infection.

Explicitly, we have $\dot{S} = 0$ when $I = 0$, gives $S^0 = \dfrac{\Pi}{\mu + a\theta}$. In addition, we have $\dot{C}_S = 0$ when $I = 0$, gives $C_S^0 = \dfrac{a\theta\Pi}{\mu(\mu + a\theta)}$.

If $\lim_{t \to +\infty} C_S(t) = 0$ as a consequence of the case when $\theta = 0$, we define the basic reproduction number for our case by R_0^C which is the average new infections produced by one infected individual during his life cycle when the population is at E^0.

Since I is the only infected compartment, then $R_0^C = \beta(S^0 + bC_S^0) \times \dfrac{1}{\mu + \gamma}$. Thus, we have

$$R_0^C = \frac{\beta\Pi}{(\mu + a\theta)(\mu + \gamma)} + \frac{ab\beta\theta\Pi}{\mu(\mu + a\theta)(\mu + \gamma)} = \frac{\beta\Pi(\mu + ab\theta)}{\mu(\mu + a\theta)(\mu + \gamma)} \tag{3}$$

Now, we try to find the components of the endemic equilibrium $E^+ = (S^+, C_S^+, I^+)$ where S^+ and C_S^+ are obtained based on the assumptions $\dot{S} = 0$ and $\dot{C}_S = 0$ when there is an infection.

Explicitly, we have $\dot{S} = 0$ when $I > 0$, which gives $S^+ = \dfrac{\Pi}{\mu + a\theta + \beta I^+}$. In addition, we have $\dot{C}_S = 0$ when $I > 0$, which gives $C_S^+ = \dfrac{a\theta S^+}{\mu + b\beta I^+}$.

On the other part, we have $\dot{I} = 0$ when $I > 0$, which gives

$$\beta S^+ I^+ + b\beta C_S^+ I^+ - \gamma I^+ - \mu I^+ = 0$$

$$\Rightarrow \beta \frac{\Pi}{\mu + a\theta + \beta I^+} + \beta \frac{ab\theta S^+}{\mu + b\beta I^+} = \gamma + \mu$$

$$\Rightarrow \beta\Pi(\mu + b\beta I^+) + ab\beta\theta S^+(\mu + a\theta + \beta I^+) = (\gamma + \mu)(\mu + a\theta + \beta I^+)(\mu + b\beta I^+)$$

$$\Rightarrow \beta\Pi(\mu + b\beta I^+) + ab\beta\theta\Pi = (\gamma + \mu)(\mu + a\theta + \beta I^+)(\mu + b\beta I^+)$$

$$\Rightarrow b\beta^2\Pi I^+ + \beta\Pi(\mu + ab\theta) = (\gamma + \mu)(\mu + a\theta + \beta I^+)(\mu + b\beta I^+)$$

$$\Rightarrow (\gamma + \mu)(\mu + a\theta)R_0^C \mu - (\gamma + \mu)(a\theta\mu + \mu^2)$$

$$= b\beta^2(\gamma + \mu)I^{+^2} + [(\gamma + \mu)(\beta\mu(1 + b) + ab\beta\theta) - b\beta^2\Pi]I^+$$

Thus, we find that I^+ is the root of the function $f(I^+) = \alpha_1 I^{+^2} + \alpha_2 I^+ + (1 - R_0^C)\alpha_3$ where α_1, α_2 and α_3 are constants.

In the following three theorems, we state and prove stability results on free and endemic equilibria.

Theorem 1. *E^0 always exists and is locally asymptotically stable if $R_0^C < 1$ (respectively, E^0 is unstable if $R_0^C > 1$).*

Proof. The existence of E^0 is trivial.

For the stability of E^0, we define the Jacobian Matrix associated to the system in Equation (2) by

$$
\begin{pmatrix}
-\beta I - a\theta - \mu & 0 & -\beta S \\
a\theta & -b\beta I - \mu & -b\beta C_S \\
\beta I & b\beta I & \beta(S + bC_S) - \gamma - \mu
\end{pmatrix}
\tag{4}
$$

At E^0, (4) becomes

$$
\begin{pmatrix}
-a\theta - \mu & 0 & -\beta S^0 \\
a\theta & -\mu & -b\beta C_S^0 \\
0 & 0 & \beta(S^0 + bC_S^0) - \gamma - \mu
\end{pmatrix}
$$

whose eigenvalues are

$$
\lambda_1 = -\mu < 0,
$$

$$
\lambda_2 = -(\mu + a\theta) < 0
$$

$$
\lambda_3 = \beta(S^0 + b\beta C_S^0) - \gamma - \mu = (\gamma + \mu)(R_0^C - 1),
$$

which imply the local asymptotic stability of E^0 when $R_0^C < 1$, and its instability when $R_0^C > 1$. $\square$

Theorem 2. *The differential system in Equation* (2) *admits* $E^+ = \left(\dfrac{\Pi}{\mu + a\theta + \beta I^+}, \dfrac{a\theta S^+}{\mu + b\beta I^+}, I^+ \right)$ *as the unique positive equilibrium and which is asymptotically stable when it exists, if and only if $R_0^C > 1$.*

Proof. First, we have

$$
\alpha_1 = b\beta^2(\gamma + \mu) > 0,
$$

$$
\alpha_2 = [(\gamma + \mu)(\beta\mu(1 + b) + ab\beta\theta) - b\beta^2\Pi],
$$

$$
\alpha_3 = \mu(\gamma + \mu)(\mu + a\theta) > 0
$$

For the sufficiency of the existence and uniqueness of E^+, so we have $\alpha_1 > 0$ and since $f(0) = (1 - R_0^C)\alpha_3 < 0$ if $R_0^C > 1$, then $f(I^+)$ has two real roots, one is positive and the other is negative. For the necessity, let us assume that $R_0^C \leq 1$ and prove that $f(I^+)$ has no positive roots. In this case, the first fraction in Equation (3) verifies

$$
\beta\Pi \leq (\mu + \gamma)(\mu + a\theta)
$$

$$
\Rightarrow \alpha_2 = (b\beta(\mu + a\theta) + \mu\beta)(\gamma + \mu) - b\beta^2\Pi
$$

$$
\geq (b\beta(\mu + a\theta) + \mu\beta)(\gamma + \mu) - b\beta(\mu + a\theta)(\mu + \gamma) = \mu\beta(\gamma + \mu) > 0.
$$

Thus, we have $\alpha_1 > 0$ and since $f(0) = (1 - R_0^C)\alpha_3 \geq 0$, $f(I^+)$ is increasing and $f(I^+) > f(0) \geq 0$, then we reach the non-positivity of the roots.

For the stability of E^+, at E^+, Equation (4) is defined as

$$
\begin{pmatrix}
-\beta I^+ - a\theta - \mu & 0 & -\beta S^+ \\
a\theta & -b\beta I^+ - \mu & -b\beta C_S^+ \\
\beta I^+ & b\beta I^+ & \beta(S^+ + bC_S^+) - \gamma - \mu
\end{pmatrix}
$$

$$
= \begin{pmatrix} -\dfrac{\Pi}{S^+} & 0 & -\beta S^+ \\[2mm] a\theta & -\dfrac{a\theta S^+}{C_S^+} & -b\beta C_S^+ \\[2mm] \beta I^+ & b\beta I^+ & 0 \end{pmatrix}
$$

whose characteristic equation is $\lambda^3 + \sigma_1 \lambda^2 + \sigma_2 \lambda + \sigma_3$ and where

$$
\sigma_1 = \frac{\Pi}{S^+} + \frac{a\theta S^+}{C_S^+}
$$

$$
\sigma_2 = \frac{a\theta \Pi}{C_S^+} + b^2 \beta^2 C_S^+ I^+ + \beta^2 S^+ I^+
$$

$$
\sigma_3 = ab\beta^2 \theta S^+ I^+ + \frac{a\theta \beta^2 S^{+^2} I^+}{C_S^+} + \frac{b^2 \beta^2 \Pi C_S^+ I^+}{S^+}.
$$

Hence, we have

$$
\sigma_1 \sigma_2 - \sigma_3
$$

$$
= \frac{\Pi}{S^+} \left(\frac{a\theta \Pi}{C_S^+} + \beta^2 S^+ I^+ \right)
$$

$$
+ \frac{a\theta S^+}{C_S^+} \left(\frac{a\theta \Pi}{C_S^+} + b^2 \beta^2 C_S^+ I^+ \right) - ab\beta^2 \theta S^+ I^+
$$

$$
= \frac{a\theta \Pi^2}{S^+ C_S^+} + (\Pi + a\theta + \beta I^+) \beta^2 S^+ I^+
$$

$$
+ \frac{a\theta S^+}{C_S^+} \left(\frac{a\theta \Pi}{C_S^+} + b^2 \beta^2 C_S^+ I^+ \right) - ab\beta^2 \theta S^+ I^+
$$

$$
= \frac{a\theta \Pi^2}{S^+ C_S^+} + (\Pi + \beta I^+) \beta^2 S^+ I^+ + \frac{a^2 \theta^2 \Pi S^+}{C_S^{+^2}}
$$

$$
+ a\theta S^+ I^+ (\beta - b\beta)^2 + ab\beta^2 \theta S^+ I^+
$$

$$
> 0
$$

Finally, based on the Routh–Hurwitz Criterion, we deduce the local asymptotic stability of E^+. $\quad\square$

Theorem 3. *If $R_0^C \leq 1$, then E^0 is globally asymptotically stable. If $R_0^C > 1$, then E^+ is globally asymptotically stable.*

Proof. We suppose that $R_0^C \leq 1$ and we prove that E^0 is globally asymptotically stable. Let us define the Lyapunov function by

$$
L_0 = S - S^0 - S^0 \ln \frac{S}{S^0} + C_S - C_S^0 - C_S^0 \ln \frac{C_S}{C_S^0} + I.
$$

Its derivative is then defined by

$$
\dot{L}_0 = \dot{S} + \dot{C}_S + \dot{I} - S^0 \frac{\dot{S}}{S} - C_S^0 \frac{\dot{C}_S}{C_S}
$$

$$
= \Pi - \mu S - \mu C_S - \mu I - \gamma I - S^0 \frac{\Pi}{S} + \mu S^0 + \beta S^0 I + a\theta S^0 - C_S^0 \frac{a\theta S}{C_S} + b\beta C_S^0 I + \mu C_S^0.
$$

Since $\mu = \dfrac{a\theta S^0}{C_S^0}$ and $\Pi = \mu S^0 + a\theta S^0$, then this derivative becomes

$$\dot{L}_0 = -\mu S + 2\mu S^0 + 3a\theta S^0 - (\mu + \gamma - \beta S^0 - b\beta C_S^0)I$$

$$- \frac{a\theta C_S^0 S}{C_S} - \frac{S^0}{S}(\mu S^0 + a\theta S^0) - \frac{a\theta S^0 C_S}{C_S^0}$$

$$= -\mu S^0 \left(\frac{S}{S^0} + \frac{S^0}{S} - 2 \right)$$

$$- a\theta S^0 \left(\frac{C_S}{C_S^0} + \frac{S^0}{S} + \frac{C_S^0 S}{C_S S^0} - 3 \right)$$

$$- (\mu + \gamma)(1 - R_0^C)I$$

Now, we have $\dfrac{S}{S^0} + \dfrac{S^0}{S} - 2 \geq 0$ and $\dfrac{C_S}{C_S^0} + \dfrac{S^0}{S} + \dfrac{C_S^0 S}{C_S S^0} - 3 \geq 0$ due to the fact that arithmetic mean is larger than or equals to the geometric mean, and the equalities hold if $S = S^0$ and $C_S = C_S^0$. Thus, $\dot{L}_0 \leq 0$ which implies the global asymptotic stability of E^0 based on Lyapunov–LaSalle's invariance principle.

Similarly, we study the global asymptotic stability of E^+ by considering the following Lyapunov function

$$L_+ = S - S^+ - S^+ \ln \frac{S}{S^+} + C_S - C_S^+ - C_S^+ \ln \frac{C_S}{C_S^+} + I - I^+ - I^+ \ln \frac{I}{I^+}.$$

The derivative is then defined as

$$\dot{L}_+ = \dot{S} + \dot{C}_S + \dot{I} - S^+ \frac{\dot{S}}{S} - C_S^+ \frac{\dot{C}_S}{C_S} - I^+ \frac{\dot{I}}{I}$$

$$= \Pi - \mu S - \mu C_S - (\mu + \gamma)I - \Pi \frac{S^+}{S} + \mu S^+ + \beta S^+ I + a\theta S^+ - a\theta S \frac{C_S^+}{C_S} + b\beta C_S^+ I + \mu C_S^+$$

$$- \beta S I^+ - b\beta C_S I^+ + (\mu + \gamma)I^+$$

Since $\mu + \gamma = \beta S^+ + b\beta C_S^+$, $\mu = \dfrac{a\theta S^+ - b\beta C_S^+ I^+}{C_S^+}$ and $\Pi = \mu S^+ + \beta S^+ I^+ + a\theta S^+ = \mu S^+ + \mu C_S^+ + (\mu + \gamma)I^+$, this derivative becomes

$$\dot{L}_+ = 2\Pi - \mu S - \frac{a\theta S^+ - b\beta C_S^+ I^+}{C_S^+} C_S - (\mu S^+ + \beta I^+ S^+ + a\theta S^+)\frac{S^+}{S} + a\theta S^+$$

$$- a\theta S \frac{C_S^+}{C_S} - \beta S I^+ - b\beta C_S I^+$$

$$= 2(\mu S^+ + \beta S^+ I^+ + a\theta S^+) - \mu S - \frac{a\theta S^+ C_S}{C_S^+} + a\theta S^+ - \frac{\mu S^{+2}}{S} - \frac{\beta S^{+2} I^+}{S}$$

$$- \frac{a\theta S^{+2}}{S} - \frac{a\theta S C_S^+}{C_S} - \beta S I^+$$

$$= -(\mu S^+ + \beta S^+ I^+) \left(\frac{S}{S^+} + \frac{S^+}{S} - 2 \right)$$

$$- a\theta S^+ \left(\frac{C_S}{C_S^+} + \frac{S^+}{S} + \frac{C_S^+ S}{C_S S^+} - 3 \right)$$

$$\leq 0$$

which is the final result, sought to prove for deducing that E^+ is globally asymptotically stable. □

3. Free Horizon Isoperimetric Optimal Control Approach

Now, we consider the mathematical model in Equation (1) with θ as a control function of time t.

Motivated by the desire to find the optimal time needed to reduce the number of infected people as much as possible while minimizing the value of the control $\theta(t)$ over a free (non-fixed) horizon t_f, our objective is to seek a couple $(\theta^*(t), t_f^*)$ such that

$$J(\theta^*(t), t_f^*) = \min_{(\theta(t), t_f) \in U \times R^+} J(\theta(t), t_f) \tag{5}$$

where J is the functional defined by

$$J(\theta(t), t_f) = t_f^2 + \int_0^{t_f} \left(a'I(t) + \frac{b'}{2}\theta^2(t) \right) dt \tag{6}$$

and where the control space U is defined by the set

$$U = \{\theta(t) | 0 \le \theta(t) \le 1, \ \theta(t) \ measurable, \ t \in [0, t_f], \ t_f \ free\}$$

where a' and b' represent constant severity weights associated to functions I and θ, respectively. Alkama et al. treated three cases of the form of the free horizon t_f^* in the final gain function of their objective function when applying a free final time optimal control approach to a cancer model [9]. Here, we suppose that t_f^* takes the quadratic form as formulated in Equation (6) to obtain a direct formula which characterizes t_f^*. In fact, if t_f^* is taken linear or the final gain function is zero, t_f^* would just be approximated numerically due to the nature of necessary conditions in these two cases (see [9] for explanation).

Since managers of the anti-epidemic resources cannot well-predict whether their control strategy will reach the entire susceptible population over a fixed horizon T, we treat here an example where the number of targeted people in the susceptible class is equal for example to only a constant $C = 3026$ for $T = 50$ months. Hence, we try to find $(\theta^*(t), t_f^*)$ under the definition of the following isoperimetric restriction

$$\int_0^T a\theta(t)S(t)dt = C \tag{7}$$

In [13], the authors defined an isoperimetric constraint on the control variable only to model the total tolerable dosage amount of a therapy along the treatment period. In their conferences talks [26,27], Kornienko et al. and De Pinho et al. introduced state constraints in an optimal control problem that is subject to an S-Exposed-I-R differential system to model the situation of limited supply of vaccine based on work in [14] and where the isoperimetric constraint is defined on the product of the control and state variables.

In our case, to take into account the constraint in Equation (7) for the resolution of the optimal control problem in Equation (5), we consider a new variable Z defined as

$$Z(t) = \int_0^t a\theta(v)S(v)dv \tag{8}$$

Then, we have $\dot{Z}(t) = a\theta(t)S(t)$. Using notations of the state variables in the previous section and keeping θ as a notation of $\theta(t)$ and Z in place of $Z(t)$, we study the differential system defined as follows

$$\begin{cases} \dot{S} &= \Pi - \beta SI - a\theta S - \mu S \\ \dot{C_S} &= a\theta S - b\beta C_S I - \mu C_S \\ \dot{I} &= \beta(S + bC_S)I - \gamma I - \mu I \\ \dot{R} &= \gamma I - \mu R \\ \dot{Z} &= a\theta S \end{cases} \tag{9}$$

instead of the model in Equation (1). We also note that, when the minimization problem in Equation (5) is under the constraint in Equation (7), the application of Pontryagin's Maximum Principle would not be appropriate for this case, but the new variable Z has the advantage to convert Equations (5)–(7) to a classical optimal control problem under one restriction which is the system in Equation (9) only [28]. If we follow most optimal control approaches in the literature, the objective function in Equation (6) will be defined over a fixed time interval. However, t_f^* is free here, and, to find the optimal duration needed to control an epidemic, it would be advantageous to managers of medical or health resources to control an epidemic before reaching the fixed time T for lesser costs. For this purpose, we need to assume that $0 \leq t_f^* \leq T$ to guarantee the sufficient condition for an optimal θ^* in the case of a free horizon. This is because θ^* exists for the minimization problem in Equation (5) when Equation (6) is defined over T based on the verified properties of the sufficient conditions as stated in details in Theorem 4.1, pp. 68–69 of [29] and that can easily be verified for many examples as ours, and this implies in our case that the existence of an optimal control θ^* and associated optimal trajectories S^*, C_S^*, I^*, R^* and Z^* comes directly from the convexity of the integrand term in Equation (6) with respect to the control θ and the Lipschitz properties of the state system with respect to state variables S, C_S, I, R and Z. Then, it exists for any time in the interval $[0, T]$ including t_f^*. As regard the necessary conditions, we state and prove the following theorem.

Theorem 4. *If there exist optimal control u^* and optimal horizon t_f^* which minimize Equation (6) along with the optimal solutions S^*, C_S^*, I^* and R^* associated to the differential system in Equation (9), then there exist adjoint variables λ_k, $k = 1, 2, 3, 4, 5$ as notations of $\lambda_k(t)$ and which satisfy the following adjoint differential system*

$$\begin{cases} \dot{\lambda_1} &= \lambda_1(\beta I^* + \mu + a\theta^*) - a\lambda_2\theta^* - \beta\lambda_3 I^* - a\theta\lambda_5 \\ \dot{\lambda_2} &= \lambda_2(b\beta I^* + \mu) - b\lambda_3\beta I^* \\ \dot{\lambda_3} &= -a' + \lambda_1\beta S^* + b\lambda_2\beta C_S^* - \lambda_3(\beta(S^* + bC_S^*) - \mu - \gamma) - \lambda_4\gamma \\ \dot{\lambda_4} &= \lambda_4\mu \\ \dot{\lambda_5} &= 0 \end{cases} \tag{10}$$

with the transversality conditions $\lambda_k(t_f^*) = 0$, $k = 1, 2, 3, 4$ and $\lambda_5(t_f^*) = $ constant which should be determined. Furthermore, the sought optimal control is characterized by

$$\theta^* = \min\left(\max\left(0, \frac{aS^*(\lambda_1 - \lambda_2 - \lambda_5)}{b'}\right), 1\right) \tag{11}$$

while the sought optimal horizon is characterized by

$$t_f^* = -\frac{H(t_f^*, S(t_f^*), C_S(t_f^*), I(t_f^*), R(t_f^*), Z(t_f^*), \lambda_1(t_f^*), \lambda_2(t_f^*), \lambda_3(t_f^*), \lambda_4(t_f^*), \lambda_5(t_f^*), \theta(t_f^*))}{2} \tag{12}$$

where $H(t_f^*, S(t_f^*), C_S(t_f^*), I(t_f^*), R(t_f^*), Z(t_f^*), \lambda_1(t_f^*), \lambda_2(t_f^*), \lambda_3(t_f^*), \lambda_4(t_f^*), \lambda_5(t_f^*), \theta(t_f^*))$ defines the Hamiltonian function as the sum of the integrand term of Equation (6) and the term $\lambda_1\dot{S} + \lambda_2\dot{C_S} + \lambda_3\dot{I} + \lambda_4\dot{R} + \lambda_5\dot{Z}$ at t_f^*.

Moreover, t_f^* is positive only when $H(t_f^*, S(t_f^*), C_S(t_f^*), I(t_f^*), R(t_f^*), Z(t_f^*), \lambda_1(t_f^*), \lambda_2(t_f^*), \lambda_3(t_f^*),$
$\lambda_4(t_f^*), \lambda_5(t_f^*), \theta(t_f^*))$ is negative.

Proof. Let H be a notation of the Hamiltonian function $H(t, S(t), C_S(t), I(t), R(t), Z(t), \lambda_1(t), \lambda_2(t),$
$\lambda_3(t), \lambda_4(t), \lambda_5(t), \theta(t))$ in all time t. Then, we have

$$
\begin{aligned}
H &= a'I + \frac{b'}{2}\theta^2 + \lambda_1 \dot{S} + \lambda_2 \dot{C_S} + \lambda_3 \dot{I} + \lambda_4 \dot{R} + \lambda_5 \dot{Z} \\
&= a'I + \frac{b'}{2}\theta^2 + \lambda_1(\Pi - \beta SI - a\theta S - \mu S) + \lambda_2(a\theta S - b\beta C_S I - \mu C_S) \\
&\quad + \lambda_3(\beta(S + bC_S)I - \gamma I - \mu I) + \lambda_4(\gamma I - \mu R) + a\theta S \lambda_5
\end{aligned}
$$

Using Pontryagin's maximum principle [30], we have

$$
\begin{aligned}
\dot{\lambda_1} &= -\frac{\partial H}{\partial S} = \lambda_1(\beta I^* + \mu + a\theta^*) - a\lambda_2\theta^* - \beta\lambda_3 I^* - a\theta\lambda_5 \\
\dot{\lambda_2} &= -\frac{\partial H}{\partial C_S} = \lambda_2(b\beta I^* + \mu) - b\lambda_3 \beta I^* \\
\dot{\lambda_3} &= -\frac{\partial H}{\partial I} = -a' + \lambda_1 \beta S^* + b\lambda_2 \beta C_S^* - \lambda_3(\beta(S^* + bC_S^*) - \mu - \gamma) - \lambda_4\gamma \\
\dot{\lambda_4} &= -\frac{\partial H}{\partial R} = \lambda_4\mu \\
\dot{\lambda_5} &= -\frac{\partial H}{\partial R} = 0
\end{aligned}
$$

while the transversality conditions defined as minus the derivative of the final gain function with respect to the state variables S, C_S, I and R. Since the final gain function in Equation (6) does not contain any term of these variables, then $\lambda_k(t_f^*) = 0$, $k = 1, 2, 3, 4$ and $\lambda_5(t_f^*)$ is unknown but we are sure it is a constant since $\dot{\lambda_5}(t) = 0 \ \forall t \in [0, t_f^*]$. The solution of this problem is treated in the next section.

The optimality condition at $\theta = \theta^*$ implies that $\frac{\partial H}{\partial \theta} = 0$. Then, after setting $S = S^*$, we have

$$
b'\theta - aS\lambda_1 + aS\lambda_2 + aS\lambda_5 = 0 \Rightarrow \theta = \frac{aS(\lambda_1 - \lambda_2 - \lambda_5)}{b'}
$$

Taking into account the bounds of the control, we obtain,

$$
\theta^* = \min\left(\max\left(0, \frac{aS^*(\lambda_1 - \lambda_2 - \lambda_5)}{b'}\right), 1\right)
$$

Now, let us prove the necessary conditions on t_f^*. As $J(\theta, t_f)$ reaches its minimum at θ^* and t_f^*, we have

$$
\lim_{h \to 0} \frac{J\left(\theta^*, t_f^* + h\right) - J\left(\theta^*, t_f^*\right)}{h} = 0
$$

with the consideration of the final gain function ϕ that we deduce it is defined in Equation (6) by $\phi(t_f, S(t_f), C_S(t_f), I(t_f), R(t_f), Z(t_f)) = t_f^2$, while setting $\theta = \theta^*$ and $t_f = t_f^*$, we obtain

$$\lim_{h \to 0} \frac{1}{h} \left[\phi(t_f + h, S(t_f + h), C_S(t_f + h), I(t_f + h), R(t_f + h), Z(t_f + h)) + \int_0^{t_f+h} \left(aI(t) + \frac{b}{2}\theta^2(t) \right) dt \right.$$

$$\left. -\phi(t_f, S(t_f), C_S(t_f), I(t_f), R(t_f), Z(t_f)) - \int_0^{t_f} k_1 \left(aI(t) + \frac{b}{2}\theta^2(t) \right) dt \right] = 0$$

$$\Rightarrow \lim_{h \to 0} \left[\frac{\phi(t_f + h, S(t_f + h), C_S(t_f + h), I(t_f + h), R(t_f + h), Z(t_f + h)) - \phi(t_f, S(t_f), C_S(t_f), I(t_f), R(t_f), Z(t_f))}{h} \right.$$

$$\left. + \frac{1}{h} \int_{t_f}^{t_f+h} \left(aI(t) + \frac{b}{2}\theta^2(t) \right) dt \right] = 0$$

$$\Rightarrow \frac{\partial \phi}{\partial t}(t_f) + \frac{\partial \phi}{\partial S}(t_f)\dot{S}(t_f) + \frac{\partial \phi}{\partial C_S}(t_f)\dot{C}_S(t_f) + \frac{\partial \phi}{\partial I}(t_f)\dot{I}(t_f) + \frac{\partial \phi}{\partial R}(t_f)\dot{R}(t_f) + \frac{\partial \phi}{\partial Z}(t_f)\dot{Z}(t_f) + aI(t) + \frac{b}{2}\theta^2(t) = 0$$

$$\Rightarrow 2t_f + H(t_f, S(t_f), C_S(t_f), I(t_f), R(t_f), Z(t_f), \lambda_1(t_f), \lambda_2(t_f), \lambda_3(t_f), \lambda_4(t_f), \lambda_5(t_f), \theta(t_f)) = 0$$

$$\Rightarrow t_f + \frac{H(t_f, S(t_f), C_S(t_f), I(t_f), R(t_f), Z(t_f), \lambda_1(t_f), \lambda_2(t_f), \lambda_3(t_f), \lambda_4(t_f), \lambda_5(t_f), \theta(t_f))}{2} = 0$$

Finally, we have

$$t_f = -\frac{H(t_f, S(t_f), C_S(t_f), I(t_f), R(t_f), Z(t_f), \lambda_1(t_f), \lambda_2(t_f), \lambda_3(t_f), \lambda_4(t_f), \lambda_5(t_f), \theta(t_f))}{2}$$

Otherwise, the positivity of t_f^* under the condition of negativity of

$$H(t_f^*, S(t_f^*), C_S(t_f^*), I(t_f^*), R(t_f^*), Z(t_f^*), \lambda_1(t_f^*), \lambda_2(t_f^*), \lambda_3(t_f^*), \lambda_4(t_f^*), \lambda_5(t_f^*), \theta(t_f^*))$$

is trivial, but this is not a condition we should have necessarily for θ^* since the Hamiltonian could change signs any time along the interval of study. $\square$

4. Numerical Simulations

Based on the formulation of Equation (8), we have $Z(0) = 0$ and $Z(t_f) = C$. Since the optimal control problem consists to resolve the two-point boundary value problem defined by the two systems in Equations (2) and (10), the differential system in Equation (2) will be numerically resolved forward in time because of its initial conditions and the value of $Z(0)$ does not change, while the differential system in Equation (10) will be numerically resolved backward in time because of its final or transversality conditions but with the condition that $Z(t_f)$ varies depending on the value of k. Based on the numerical approach in [13], we propose also here to define a real function g such that $k \to g(k) = \tilde{Z}_f - Z_f$ and where $\tilde{Z}_f$ is the value of Z at t_f for various values of k and Z_f is the value fixed by C. This leads to the combination of the Forward–Backward–Sweep Method (FBSM) which resolves the two-point boundary value problem in Equations (2)–(10), with the secant-method to find the value of the root $'k'$ of the function g [31]. The necessary condition on t_f^* defined by the characterization in Equation (12), which leads to seek a fixed point of a real function F such that $F(t_f^*) = t_f^*$. We choose to solve this numerical problem differently to the method used in [8,9] using the fixed point method. In brief, the four steps of numerical calculus associated to the resolution of our free optimal control problem (5) under isoperimetric constraint (7), are described in Algorithm 1.

Algorithm 1: Resolution steps of the two-point boundary value optimal control problem (9) and (10).

Step 0:
    Guess an initial estimation to θ and t_{final}.

Step 1:
    Use the initial condition $S(0)$, $C_S(0)$, $I(0)$, $R(0)$ and $Z(0)$ and the stocked values by θ and t_f.

    Find the optimal states S^*, C_S^*, I^*, R^* and Z^* which iterate forward in the two-point boundary value problem (2)–(10).

Step 2:
    Use the stocked values by θ and the transversality conditions $\lambda_k(t_f)$ for $k = 1, 2, 3, 4$ while searching the constant $\lambda_5(t_f)$ using the secant-method.

    Find the adjoint variables λ_k for $k = 1, 2, 3, 4, 5$ which iterate backward in the two-point boundary value problem (2)–(10).

Step 3:
    Update the control utilizing new S, C_S, I, R, Z and λ_k for $k = 1, 2, 3, 4, 5$ in the characterization of θ^* as presented in (11) while searching the optimal time t_f^* characterized by (12) using the fixed point method.

Step 4:
    Test the convergence. If the values of the sought variables in this iteration and the final iteration are sufficiently small, check out the recent values as solutions. If the values are not small, go back to Step 1.

Figure 1 depicts the SC_SIRZ dynamics in the absence and presence of the control and we can see that the number of susceptible people has increased linearly from its initial condition to a number higher than 92.5 individuals when we choose $\theta = 0$, while the optimal state S^* increases during the first months of the optimal control strategy and it decreases when we work with the characterization of Equation (11). Simultaneously, the number of removed people increases to only a value close to eight people while it reaches a value higher than this number with a maximal peak equaling to 17 when $\theta \neq 0$. As regards to the number of infected people, it decreases from its initial condition to a value close to an important value of 30 individuals because of the natural death and recovery only, while it decreases towards a value very close to zero after the introduction of the control θ. We can see the relationship between the number of controlled people and the optimal values taken by θ so when this is increasing, the optimal state C_S^* is also increasing. In fact, we can deduce that, with only small values of θ, we reach our goal by minimizing I function, and maximizing R function while the total number of the susceptible who received the control along T and which is represented by the function Z has not exceeded the imposed constant C. The dashed lines introduced in this figure show the highest fixed point value of the sought final time, and we can understand that, at this point, we have already reached our goal which concerns the minimization of the number of infected people and maximization of the number of removed people. The next figure gives more information about the obtained value of t_f^*.

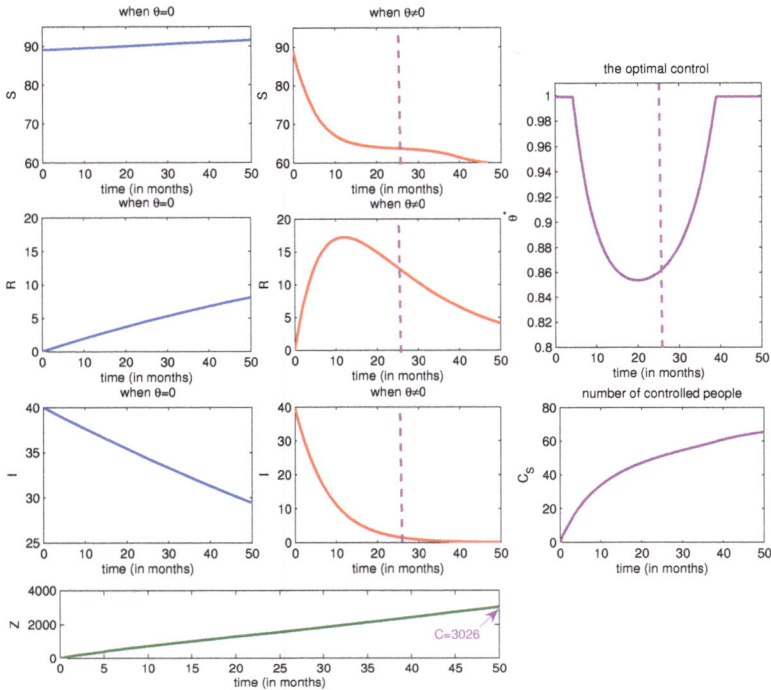

Figure 1. $SC_S IRZ$ dynamics in the absence and presence of the control in the two cases $\theta = 0$ and $\theta \neq 0$. Parameters values: $\Pi = 6.45$, $a = 0.06$, $b = 0.001$, $\beta = 0.0003$, and $\mu = 0.05$, $\gamma = 0.1$. Initial conditions: $S(0) = 89$, $C_S(0) = 0$, $I(0) = 40$, and $R(0) = 0$. Severity weights constants: $a' = 1$ and $b' = 50$.

In Figure 2, we present dynamics of the functions S, I and R, and we can see the fixed points t_f^* in the first plot above. The solution of the equation $F(t_f^*) = t_f^*$ starts from an initial guess which equals zero, and increases to values that are very close or sometimes equal to 26 months (we note that, even if they appeared taking the value 26, this is not the case at all iterations but just because all values are very close to 26 with a small precision of about 10^{-4}). As noted in this figure, for instance, the highest value of $t_f^* = 26.4081$ found at iteration 292 among 1000. In the same figure, in the plot below, we observe that, at t_f^* indicated by the dashed purple line, the number of infected people has already taken the direction towards zero values, while the number of removed people has already reached its positive peak and started to decrease because of the decrease of the optimal control function θ^*, as shown in the previous figure. This means that there is no need in this case to extend the optimal control approach for other months since, at t_f^*, Equation (5) has been almost realized.

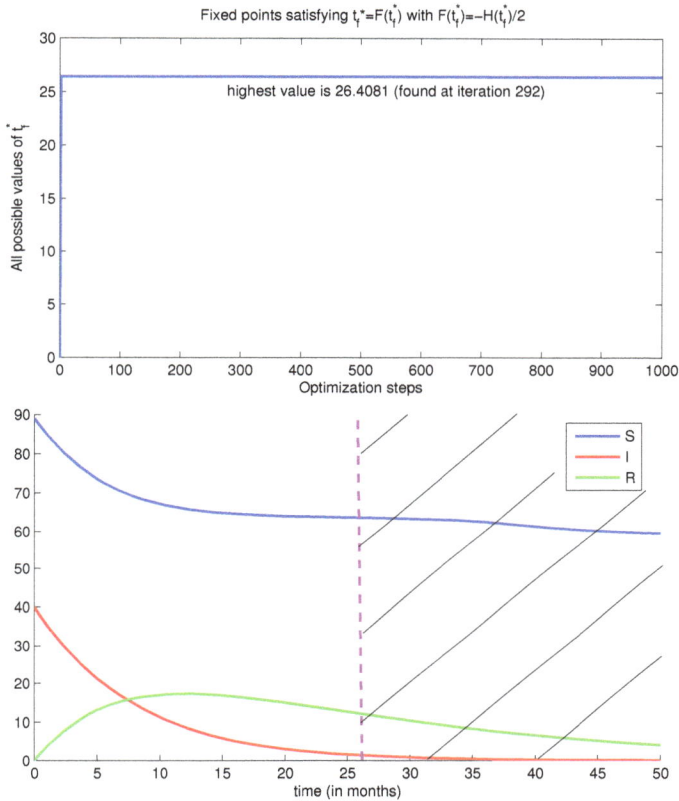

Figure 2. SIR dynamics with the precision of the optimal horizon t_f^* with the same parameters, initial conditions and severity weights constants as in Figure 1.

In Figure 3, the fixed points t_f^* for different values of the control severity weight b' suggest that, as the value of b' increases, t_f^* increases. In fact, the bigger is b', the lesser is the optimal control θ^*, which is important, as we can deduce from the formulation in Equation (11), and this is reasonable since, when θ^* is small, we need more time to control the epidemic. The obtained results in this figure can be summarized as follows:

- When $b' = 60$: $I(t_f^*) = 6.734$ with $\theta(t_f^*) = 0.8658$ (iteration 278), which implies that 83.165% of infected people have left the I compartment.
- When $b' = 70$: $I(t_f^*) = 6.0627$ with $\theta(t_f^*) = 0.8624$ (iteration 295), which implies that 84.84325% of infected people have left the I compartment.
- When $b' = 80$: $I(t_f^*) = 4.7619$ with $\theta(t_f^*) = 0.8573$ (iteration 332), which implies that 88.09525% of infected people have left the I compartment.

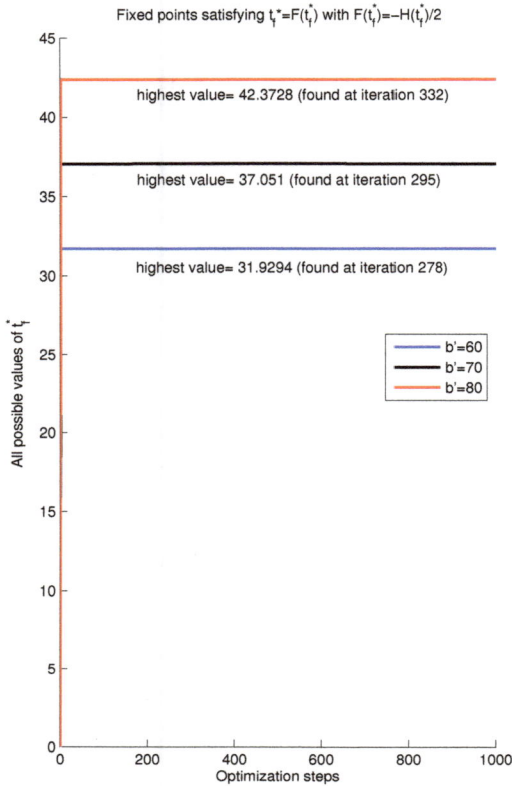

Figure 3. t_f^* for different values of b' with the same parameters, initial conditions and severity weights constants as in Figure 1.

In Figure 4, we show the impact of the initial condition of I function, namely I_0, on fixed points t_f^*, and we can deduce from the obtained optimal horizons that, as I_0 increases, t_f^* increases, and this is reasonable since, when the number of infected people is important, the anti-epidemic measures need longer time for controlling the situation. The obtained results in this figure can be summarized as follows:

- When $I(0) = 50$: $I(t_f^*) = 8.4416$ with $\theta(t_f^*) = 0.8758$ (iteration 278), which implies that 83.1168% of infected people have left the I compartment.
- When $I(0) = 60$: $I(t_f^*) = 5.7018$ with $\theta(t_f^*) = 0.8662$ (iteration 338), which implies that 90.497% of infected people have left the I compartment.
- When $I(0) = 70$: $I(t_f^*) = 8.1909$ with $\theta(t_f^*) = 0.8802$ (iteration 310), which implies that 88.2987% of infected people have left the I compartment.

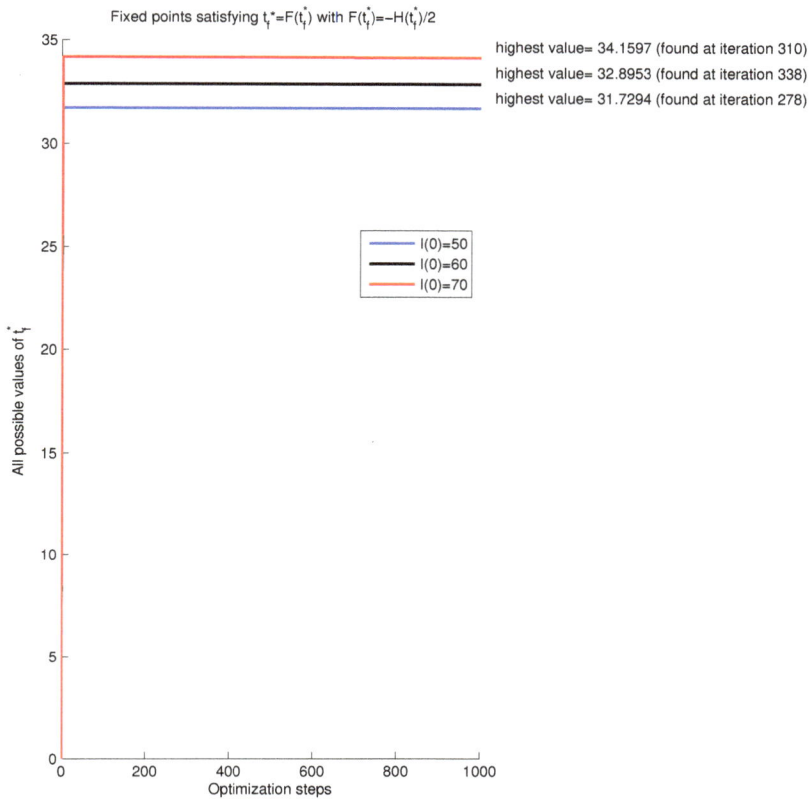

Figure 4. t_f^* for different values of the initial condition $I(0)$ with the same parameters, severity weights constants and , initial conditions for $S(0)$, $C_S(0)$, and $R(0)$ as in Figure 1.

5. Conclusions

In this paper, we have determined the optimal duration needed for controlling an epidemic based on a free horizon optimal control approach with an isoperimetric constraint and which has been applied to a four-compartmental epidemic model where it is supposed that the controlled population does not reach the removed class due to the temporary effect of the control. The isoperimetric restriction which has been proposed to define the number of susceptible people who receive the control along the anti-epidemic measures period, allowed us to find the optimal horizon of the optimal control strategy when there are limited resources devised to fight against a disease. In the numerical simulations, we used the fixed point method since the necessary condition on the free horizon led to a fixed point equation. Our results prove their usefulness, since, at the obtained optimal horizons for different values of parameters and initial conditions on infection, the infected population size has been reduced and this presents an advantage of the followed control approach to managers of the health resources even when these are limited.

Math. Comput. Appl. **2018**, *23*, 64

Author Contributions: All authors contributed equally to this work. All authors read and approved the final manuscript.

Acknowledgments: The authors would like to thank all Managing Editors and members of the Editorial Board who were responsible for dealing with this paper, and the anonymous referees for their valuable comments and suggestions, improving the content of this paper.

Conflicts of Interest: The authors declare no conflict of interest.

References

1. Zakary, O.; Rachik, M.; Elmouki, I. On the impact of awareness programs in HIV/AIDS prevention: An SIR model with optimal control. *Int. J. Comput. Appl.* **2016**, *133*, 1–6. [CrossRef]
2. Zakary, O.; Larrache, A.; Rachik, M.; Elmouki, I. Effect of awareness programs and travel-blocking operations in the control of HIV/AIDS outbreaks: A multi-domains SIR model. *Adv. Differ. Equ.* **2016**, *2016*, 169. [CrossRef]
3. Roy, P.K.; Saha, S.; Al Basir, F. Effect of awareness programs in controlling the disease HIV/AIDS: An optimal control theoretic approach. *Adv. Differ. Equ.* **2015**, *2015*, 217. [CrossRef]
4. Rodrigues, H.S.; Monteiro, M.T.T.; Torres, D.F. Vaccination models and optimal control strategies to dengue. *Math. Biosci.* **2014**, *247*, 1–12. [CrossRef] [PubMed]
5. Kumar, A.; Srivastava, P.K. Vaccination and treatment as control interventions in an infectious disease model with their cost optimization. *Commun. Nonlinear Sci. Numer. Simul.* **2017**, *44*, 334–343. [CrossRef]
6. Liu, X.; Takeuchi, Y.; Iwami, S. SVIR epidemic models with vaccination strategies. *J. Theor. Biol.* **2008**, *253*, 1–11. [CrossRef] [PubMed]
7. Nainggolan, J.; Supian, S.; Supriatna, A.K.; Anggriani, N. Mathematical model of tuberculosis transmission with reccurent infection and vaccination. *J. Phys. Conf. Ser.* **2013**, *423*, 012059. [CrossRef]
8. Zakary, O.; Rachik, M.; Elmouki, I. How much time is sufficient for benefiting of awareness programs in epidemics prevention? A free final time optimal control approach. *Int. J. Adv. Appl. Math. Mech.* **2017**, *4*, 26–40.
9. Alkama, M.; Larrache, A.; Rachik, M.; Elmouki, I. Optimal duration and dosage of BCG intravesical immunotherapy: A free final time optimal control approach. *Math. Methods Appl. Sci.* **2018**, *41*, 2209–2219. [CrossRef]
10. Zhou, L.; Fan, M. Dynamics of an SIR epidemic model with limited medical resources revisited. *Nonlinear Anal. Real World Appl.* **2012**, *13*, 312–324. [CrossRef]
11. Abdelrazec, A.; Bélair, J.; Shan, C.; Zhu, H. Modeling the spread and control of dengue with limited public health resources. *Math. Biosci.* **2016**, *271*, 136–145. [CrossRef] [PubMed]
12. Yu, T.; Cao, D.; Liu, S. Epidemic model with group mixing: Stability and optimal control based on limited vaccination resources. *Commun. Nonlinear Sci. Numer. Simul.* **2018**, *61*, 54–70. [CrossRef]
13. Elmouki, I.; Saadi, S. BCG immunotherapy optimization on an isoperimetric optimal control problem for the treatment of superficial bladder cancer. *Int. J. Dyn. Control* **2016**, *4*, 339–345. [CrossRef]
14. Neilan, R.M.; Lenhart, S. An Introduction to Optimal Control with an Application in Disease Modeling. In *Modeling Paradigms and Analysis of Disease Trasmission Models*; Gumel, A.B., Lenhart, S., Eds.; American Mathematical Society: Providence, RI, USA, 2010; pp. 67–82.
15. El Kihal, F.; Abouelkheir, I.; Rachik, M.; Elmouki, I. Optimal Control and Computational Method for the Resolution of Isoperimetric Problem in a Discrete-Time SIRS System. *Math. Comput. Appl.* **2018**, *23*, 52. [CrossRef]
16. Kermack, W.O.; McKendrick, A.G. A Contribution to the Mathematical Theory of Epidemics. *Proc. R. Soc. A* **1927**, *115*, 700–721. [CrossRef]
17. Korobeinikov, A. Lyapunov functions and global properties for SEIR and SEIS epidemic models. *Math. Med. Biol. J. IMA* **2004**, *21*, 75–83. [CrossRef]
18. Gonçalves, S.; Abramson, G.; Gomes, M.F. Oscillations in SIRS model with distributed delays. *Eur. Phys. J. B* **2011**, *81*, 363. [CrossRef]
19. Gray, A.; Greenhalgh, D.; Mao, X.; Pan, J. The SIS epidemic model with Markovian switching. *J. Math. Anal. Appl.* **2012**, *394*, 496–516. [CrossRef]

20. Kribs-Zaleta, C.M.; Velasco-Hernández, J.X. A simple vaccination model with multiple endemic states. *Math. Biosci.* **2000**, *164*, 183–201. [CrossRef]
21. Kribs-Zaleta, C.M.; Martcheva, M. Vaccination strategies and backward bifurcation in an age-since-infection structured model. *Math. Biosci.* **2002**, *177*, 317–332. [CrossRef]
22. Alexander, M.E.; Bowman, C.; Moghadas, S.M.; Summers, R.; Gumel, A.B.; Sahai, B.M. A vaccination model for transmission dynamics of influenza. *SIAM J. Appl. Dyn. Syst.* **2004**, *3*, 503–524. [CrossRef]
23. Shim, E. A note on epidemic models with infective immigrants and vaccination. *Math. Biosci. Eng.* **2006**, *3*, 557. [CrossRef] [PubMed]
24. Sharomi, O.; Malik, T. Optimal control in epidemiology. *Ann. Oper. Res.* **2017**, *251*, 55–71. [CrossRef]
25. Hethcote, H.W. The mathematics of infectious diseases. *SIAM Rev.* **2000**, *42*, 599–653. [CrossRef]
26. Kornienko, I.; Paiva, L.T.; De Pinho, M.D.R. Introducing state constraints in optimal control for health problems. *Procedia Technol.* **2014**, *17*, 415–422. [CrossRef]
27. De Pinho, M.D.R.; Kornienko, I.; Maurer, H. Optimal Control of a SEIR Model with Mixed Constraints and L^1 Cost. In Proceedings of the 11th Portuguese Conference on Automatic Control, Porto, Portugal, 21–23 July 2014; pp. 135–145.
28. Lenhart, S.; Workman, J.T. *Optimal Control Applied to Biological Models*; CRC Press: New York, NY, USA, 2007.
29. Fleming, W.H.; Rishel, R.W. *Deterministic and Stochastic Optimal Control*; Springer Science & Business Media: New York, NY, USA, 2012.
30. Pontryagin, L.S. *Mathematical Theory of Optimal Processes*; CRC Press: New York, NY, USA, 1987.
31. Gumel, A.B.; Lenhart, S. *Modeling Paradigms and Analysis of Disease Transmission Models*; American Mathematical Society: Providence, RI, USA, 2010.

Mathematical and Computational Applications
MDPI

Article

Optimal Strategies for Psoriasis Treatment

Ellina Grigorieva [1],* and Evgenii Khailov [2]

[1] TWU Department of Mathematics and Computer Science, Texas Woman's University, Denton, TX 76204, USA

[2] MSU Faculty of Computational Mathematics and Cybernetics, Lomonosov Moscow State University, Moscow 119992, Russia; khailov@cs.msu.su

* Correspondence: egrigorieva@mail.twu.edu or egrigorieva@twu.edu; Tel.: +1-940-898-2452

Received: 1 August 2018; Accepted: 31 August 2018; Published: 4 September 2018

Abstract: Within a given time interval we consider a nonlinear system of differential equations describing psoriasis treatment. Its phase variables define the concentrations of T-lymphocytes, keratinocytes and dendritic cells. Two scalar bounded controls are introduced into this system to reflect medication dosages aimed at suppressing interactions between T-lymphocytes and keratinocytes, and between T-lymphocytes and dendritic cells. For such a controlled system, a minimization problem of the concentration of keratinocytes at the terminal time is considered. For its analysis, the Pontryagin maximum principle is applied. As a result of this analysis, the properties of the optimal controls and their possible types are established. It is shown that each of these controls is either a bang-bang type on the entire time interval or (in addition to bang-bang type) contains a singular arc. The obtained analytical results are confirmed by numerical calculations using the software "BOCOP-2.0.5". Their detailed analysis and the corresponding conclusions are presented.

Keywords: psoriasis; nonlinear control system; optimal control; Pontryagin maximum principle; switching function; Lie brackets; singular arc; chattering control

---

1. Introduction

Psoriasis is an immune-mediated inflammatory skin disease that affects 2–3% of the population around the world [1]. The most characteristic features of the pathology are hyperproliferation and disrupted epidermal differentiation, altered immunological and vascular skin profiles. Manifestations of psoriasis can vary according to severity: from weak, when only a few characteristic psoriatic plaques are present on the body of patients, to extremely severe ones, when the lesion affects almost the entire surface of the body and joints. Disease can significantly reduce the standard of living and performance of patients [2]. There are established links between psoriasis and other diseases: obesity, diabetes, cardiovascular diseases, metabolic syndrome, and depression [3]. Molecular-genetic causes of the disease are still not fully established. At the same time, various polymorphisms associated with psoriasis as well as several environmental factors that can lead to the manifestation of this disease are identified [4]. To date, about 150 million people around the world are suffering from psoriasis. In the USA the annual cost of treatment is about $12 billion, while the existing therapies can only relieve symptoms and increase remission time. In addition, recent studies have revealed the division of psoriasis into subtypes, depending on which patients may respond differently to therapy. In some cases, despite costly and prolonged treatment, the patient's condition may not improve [5]. In recent years, studies have greatly expanded the understanding of the pathogenesis of psoriasis and allowed the development of numerous therapies.

In psoriasis, mathematical models are effectively used to predict cellular behavior of the skin in both normal and pathological conditions. Among all possible models [6], we highlight the models that are described by systems of differential equations [6–15]. In turn, controlled mathematical models are used to simulate the use of medication and dosage regimen, compare the effects of various drugs on the affected areas of the skin, and to determine the most effective methods of treatment. Within the framework of a specific model, optimal control theory is used to find the best strategies in one sense or another for psoriasis treatment.

For mathematical models of psoriasis treatment, optimal control problems were considered in [15–17], where the optimal treatment strategies minimizing the weighted sum of the total concentration of keratinocytes and the total cost of the treatment were found numerically. This cost of psoriasis treatment was expressed by an integral of the square of control, and these models were described using systems of differential equations linear in control. After applying the Pontryagin maximum principle as a necessary optimality condition, the corresponding optimal control problems were reduced to two-point boundary value problems for the maximum principle, which were then solved numerically, applying standard mathematical software. This was because the right-hand sides of the systems of differential equations of such boundary value problems were Lipschitz functions of the phase and adjoint variables.

In the optimal control problems for mathematical models of diseases and the spread of epidemics, the total cost of treatment can also be expressed by an integral of control [18–21]. In this case, the models are still described by systems of differential equations that are linear in control. Then, as shown in [21,22], after applying the Pontryagin maximum principle to such problems, the corresponding optimal controls can contain singular arcs on which these controls are not uniquely determined from the maximum condition. After the existence of such singular arcs are established on appropriate singular intervals, the corresponding optimality conditions are checked, and the concatenations of singular and nonsingular intervals are found, the finding of specific optimal solutions in optimal control problems is still performed only numerically. This paper shows that optimal controls can contain singular arcs in the minimization problem for the mathematical model of psoriasis treatment even in the absence of the controls in the integral terms of the functional to be minimized responsible for the cost of this treatment.

This paper is organized as follows. Section 2 is devoted to the description of the mathematical model of psoriasis treatment and the formulation of the corresponding optimal control problem for it. Namely, we consider a nonlinear controlled system of three differential equations that presents the process of treating this disease. Its phase variables determine the concentrations of T-lymphocytes, keratinocytes, and dendritic cells. The interaction of these types of cells leads to the appearance and development of psoriasis. There are two scalar bounded controls in the system that reflect all types of psoriasis treatment: skin creams that prevent or inhibit the formation, development and spread of psoriatic skin lesions, as well as pills and injections, also aimed at achieving similar results. At the same time, skin creams suppress the interaction between T-lymphocytes and keratinocytes; pills and injections weaken the interaction between T-lymphocytes and dendritic cells. All such medications are aimed at achieving the main goal, which is to reduce the number of keratinocytes. Therefore, for such a controlled system, at a given time interval the problem of minimizing the concentration of keratinocytes at the end of the time interval is stated. An optimal solution for such a problem, consisting of the optimal controls and the corresponding optimal solutions of the system, is analyzed using the Pontryagin maximum principle. Its application to the considered minimization problem is discussed in Section 3. Here, also possible types of the optimal controls are discussed: whether they are only bang-bang functions, or they can contain singular arcs in addition to the bang-bang intervals. A detailed analysis of the behavior of the optimal controls is given in Section 4. Here, we study the properties of the corresponding switching functions, which are crucial in such analysis. Using the Lie brackets of the geometric control theory allows us to obtain the Cauchy problem for the switching functions. Such a Cauchy problem makes it possible to draw the conclusions about possible singular arcs of the optimal controls. The first one shows that these controls do not simultaneously have singular arcs on the same singular intervals. The second of them is that when one optimal control has a singular arc on some singular interval, the second optimal control is constant on this interval and takes one of its boundary values. The next two sections, Sections 5 and 6, are devoted to a detailed analysis of singular arcs of each optimal control. In these sections it is shown that one of them can have a singular arc of the order of two, and the other can contain a singular arc of order one. Also, for each of these singular arcs the corresponding necessary optimality conditions are checked (Kelly-Cope-Moyer condition and Kelly condition). Finally, the forms of concatenations of the singular arcs of the optimal controls with nonsingular intervals on which such controls are bang-bang functions are considered. Since, as previously mentioned, after discussing all such issues, finding the specific optimal solutions is carried out numerically, then in Section 7 the results of the corresponding numerical calculations and their discussion are presented. Section 8 contains our conclusions.

2. Mathematical Model and Optimal Control Problem

Let a time interval $[0, T]$ be given, which is the period of psoriasis treatment. We consider on this interval a mathematical model that establishes the links between the concentrations $l(t)$, $k(t)$, $m(t)$ of T-lymphocytes, keratinocytes, and dendritic cells, respectively, because interactions between these types of cells cause a disease such as psoriasis. Such a model is the nonlinear system of differential equations:

$$\begin{cases} l'(t) = \sigma - \delta v(t)l(t)m(t) - \gamma_1 u(t)l(t)k(t) - \mu l(t), \\ k'(t) = (\beta + \delta)v(t)l(t)m(t) + \gamma_2 u(t)l(t)k(t) - \lambda k(t), \\ m'(t) = \rho - \beta v(t)l(t)m(t) - v m(t), \end{cases} \tag{1}$$

with given initial values:

$$l(0) = l_0, \ k(0) = k_0, \ m(0) = m_0; \ l_0, m_0, k_0 > 0. \tag{2}$$

In System (1) we suppose that σ is the constant rate of accumulation for T-lymphocytes and ρ is the constant accumulation rate of dendritic cells. The rate of activation of T-lymphocytes by dendritic cells is δ and β is the activation rate of dendritic cells by T-lymphocytes. Also, we consider that the removal rates of T-lymphocytes, keratinocytes and dendritic cells are denoted by μ, λ and v, respectively. Finally, we suppose that γ_1 is the rate of activation of keratinocytes by T-lymphocytes and growth of keratinocytes due to T-lymphocytes occurs at the rate γ_2.

Next, in System (1) two control functions $u(t)$ and $v(t)$ are introduced. The control $u(t)$ is at the places of interaction between T-lymphocytes and keratinocytes and reflects the medication dosage to restrict the excessive growth of keratinocytes. In a similar way, the control $v(t)$ is at the places of interaction between T-lymphocytes and dendritic cells and reflects the medication dosage for the restriction of the excessive growth of keratinocytes as well. The controls $u(t)$ and $v(t)$ subject to the following restrictions:

$$0 < u_{\min} \leq u(t) \leq 1, \ \ 0 < v_{\min} \leq v(t) \leq 1. \tag{3}$$

We consider that the set of all admissible controls $\Omega(T)$ is formed by all possible pairs of Lebesgue measurable functions $(u(t), v(t))$, which for almost all $t \in [0, T]$ satisfy Inequality (3).

Let us introduce a region:

$$\Lambda = \left\{ (l, k, m)^\top \in \mathbb{R}^3 : l > 0, \ k > 0, \ m > 0, \ l + k + m < M \right\},$$

where M is a positive constant that depends on the parameters σ, ρ, β, δ, μ, λ, v, γ_1, γ_2 of System (1) and its initial values l_0, k_0, m_0 from (2). Here $\mathbb{R}^3$ is the Euclidean space consisting of all column vectors and the sign $\top$ means transposition.

Then, the boundedness, positiveness, and continuation of the solutions for System (1) is stated by the following lemma.

Lemma 1. *Let the inclusion* $(l_0, k_0, m_0)^\top \in \Lambda$ *be valid. Then, for any pair of admissible controls* $(u(t), v(t))$ *the corresponding absolutely continuous solutions* $l(t)$, $k(t)$, $m(t)$ *for System* (1) *are defined on the entire interval* $[0, T]$ *and satisfy the inclusion:*

$$(l(t), k(t), m(t))^\top \in \Lambda, \quad t \in (0, T]. \tag{4}$$

Remark 1. *Relationship* (4) *implies that the region* Λ *is a positive invariant set for System* (1)*. The proof of Lemma 1 is fairly straightforward, and we omit it. Proofs of such statements are given, for example, in [16,23,24].*

Now, let us consider for System (1) on the set of admissible controls $\Omega(T)$ the following minimization problem:

$$J(u(\cdot), v(\cdot)) = k(T) \to \min_{(u(\cdot), v(\cdot)) \in \Omega(T)}, \tag{5}$$

which consists in minimizing the concentration of keratinocytes at the final moment T of psoriasis treatment. As already noted in [25], the optimal control problem (5) differs from problems that are typically considered in the literature on the control of psoriasis models [15–17,26] in that the functional from (5) does not include an integral of the weighted sum of the squares of the controls $u(t)$ and $v(t)$, which is responsible for the total cost of drug dosages. In psoriasis treatment, in most cases, either a skin cream, or an oral medication are used. Both prescribed medications have regular daily dosage and are not as harmful for patients as the drugs used in chemotherapy for cancer treatment [21]. Therefore, the total cost of psoriasis treatment in the meaning "harm" to a patient and that usually mathematically is described by an integral of the weighted sum of the squares of the controls, can be ignored. Moreover, using the terminal functional from (5) instead of corresponding integral functional [15–17,26] simplifies the subsequent analytical arguments.

The existence in the minimization problem (5) of the optimal controls $(u_*(t), v_*(t))$ and the corresponding optimal solutions $l_*(t)$, $k_*(t)$, $m_*(t)$ for System (1) follows from Lemma 1 and Theorem 4 ([27], Chapter 4).

Finally, based on the results from [17,23,26], we assume that the following assumption is true.

Assumption 1. *Let the inequalities:*

$$\gamma_1 \neq \gamma_2, \quad (\beta + \delta)\gamma_1 > \delta\gamma_2, \quad \lambda > \mu, \quad \lambda > \nu \tag{6}$$

be valid.

Let us introduce the constants:

$$\alpha = \gamma_1 \gamma_2^{-1}(\beta + \delta) - \delta, \quad \epsilon = \alpha(\lambda - \nu) + \delta(\lambda - \mu).$$

By this assumption, it is easy to see that these constants are positive.

3. Pontryagin Maximum Principle

We apply the Pontryagin maximum principle [28] to analyze the optimal controls $u_*(t)$, $v_*(t)$ and the corresponding optimal solutions $l_*(t)$, $k_*(t)$, $m_*(t)$. Firstly, let us define the Hamiltonian:

$$H(l, m, k, u, v, \psi_1, \psi_2, \psi_3) = (\sigma - \delta v l m - \gamma_1 u l k - \mu l)\psi_1$$
$$+ ((\beta + \delta) v l m + \gamma_2 u l k - \lambda k)\psi_2 + (\rho - \beta v l m - \nu m)\psi_3,$$

where ψ_1, ψ_2, ψ_3 are the adjoint variables.

Secondly, we calculate the required partial derivatives:

$$H'_l(l, m, k, u, v, \psi_1, \psi_2, \psi_3) = uk(\gamma_2\psi_2 - \gamma_1\psi_1)$$
$$+ vm(-\delta\psi_1 + (\beta + \delta)\psi_2 - \beta\psi_3) - \mu\psi_1,$$
$$H'_k(l, m, k, u, v, \psi_1, \psi_2, \psi_3) = ul(\gamma_2\psi_2 - \gamma_1\psi_1) - \lambda\psi_2,$$
$$H'_m(l, m, k, u, v, \psi_1, \psi_2, \psi_3) = vl(-\delta\psi_1 + (\beta + \delta)\psi_2 - \beta\psi_3) - \nu\psi_3,$$
$$H'_u(l, m, k, u, v, \psi_1, \psi_2, \psi_3) = lk(\gamma_2\psi_2 - \gamma_1\psi_1),$$
$$H'_v(l, m, k, u, v, \psi_1, \psi_2, \psi_3) = lm(-\delta\psi_1 + (\beta + \delta)\psi_2 - \beta\psi_3).$$

Then, in accordance with the Pontryagin maximum principle, for the optimal controls $u_*(t)$, $v_*(t)$ and the optimal solutions $l_*(t)$, $k_*(t)$, $m_*(t)$ there exists a vector-function $\psi_*(t) = (\psi_1^*(t), \psi_2^*(t), \psi_3^*(t))^\top$ such that:

- $\psi_*(t)$ is a nontrivial solution of the adjoint system:

$$\begin{cases} \psi_1^{*\prime}(t) = -u_*(t)k_*(t)(\gamma_2\psi_2^*(t) - \gamma_1\psi_1^*(t)) \\ \quad -v_*(t)m_*(t)(-\delta\psi_1^*(t) + (\beta + \delta)\psi_2^*(t) - \beta\psi_3^*(t)) + \mu\psi_1^*(t), \\ \psi_2^{*\prime}(t) = -u_*(t)l_*(t)(\gamma_2\psi_2^*(t) - \gamma_1\psi_1^*(t)) + \lambda\psi_2^*(t), \\ \psi_3^{*\prime}(t) = -v_*(t)l_*(t)(-\delta\psi_1^*(t) + (\beta + \delta)\psi_2^*(t) - \beta\psi_3^*(t)) + \nu\psi_3^*(t), \\ \psi_1^*(T) = 0, \ \psi_2^*(T) = -1, \ \psi_3^*(T) = 0; \end{cases} \quad (7)$$

- the controls $u_*(t)$ and $v_*(t)$ maximize the Hamiltonian

$$H(l_*(t), k_*(t), m_*(t), u, v, \psi_1^*(t), \psi_2^*(t), \psi_3^*(t))$$

with respect to variables $u \in [u_{\min}, 1]$ and $v \in [v_{\min}, 1]$ for almost all $t \in [0, T]$, and therefore they satisfy the relationships:

$$u_*(t) = \begin{cases} 1 & \text{, if } L_u(t) > 0, \\ \text{any } u \in [u_{\min}, 1] & \text{, if } L_u(t) = 0, \\ u_{\min} & \text{, if } L_u(t) < 0; \end{cases} \quad (8)$$

$$v_*(t) = \begin{cases} 1 & \text{, if } L_v(t) > 0, \\ \text{any } v \in [v_{\min}, 1] & \text{, if } L_v(t) = 0, \\ v_{\min} & \text{, if } L_v(t) < 0; \end{cases} \quad (9)$$

where the functions:

$$\begin{aligned} L_u(t) &= l_*(t)k_*(t)(\gamma_2\psi_2^*(t) - \gamma_1\psi_1^*(t)), \\ L_v(t) &= l_*(t)m_*(t)(-\delta\psi_1^*(t) + (\beta + \delta)\psi_2^*(t) - \beta\psi_3^*(t)) \end{aligned} \quad (10)$$

are the switching functions describing the behavior of the controls $u_*(t)$ and $v_*(t)$ in accordance with Formulas (8) and (9), respectively.

Analysis of Formulas (8) and (9) shows possible types of the optimal controls $u_*(t)$ and $v_*(t)$. They can only have a bang-bang type and switch between the corresponding values $u_{\min}$ and 1, $v_{\min}$ and 1. This occurs, when passing through the points at which the switching functions $L_u(t)$ and $L_v(t)$ are zero, the sign of these functions changes. Or, in addition to intervals of a bang-bang type, the controls $u_*(t)$ and $v_*(t)$ can also contain singular arcs [22,29]. This occurs when the switching functions $L_u(t)$ and $L_v(t)$ individually or both simultaneously vanish identically on certain subintervals of the interval $[0, T]$. Furthermore, such subintervals we will call as singular intervals. The following sections are devoted to a detailed study of singular arcs for the optimal controls $u_*(t)$ and $v_*(t)$.

Now, we establish the important property of the controls $u_*(t)$ and $v_*(t)$. Namely, by Lemma 1, the initial values of System (7), Formula (10), and the continuity of the switching functions $L_u(t)$ and $L_v(t)$, the following lemma is valid.

Lemma 2. *There exist such values $t_\star^u, t_\star^v \in [0, T]$ that the inequalities $L_u(t) < 0$ and $L_v(t) < 0$ hold for all t from the corresponding intervals $(t_\star^u, T]$ and $(t_\star^v, T]$.*

Corollary 1. *Lemma 2 and Formulas (8) and (9) imply the following relationships for the optimal controls $u_*(t)$ and $v_*(t)$:*

$$u_*(t) = u_{\min}, \ t \in (t_\star^u, T]; \qquad v_*(t) = v_{\min}, \ t \in (t_\star^v, T].$$

4. Differential Equations of the Switching Functions

Let us obtain differential equations for the switching functions $L_u(t)$ and $L_v(t)$. To do this, we draw on the concepts and notations of geometric control theory from [21,22].

Again, we consider the Euclidean space $\mathbb{R}^3$ in which the value $\langle \tau, s \rangle$ is the scalar product of its elements. Let $z = (l, k, m)^\top \in \mathbb{R}^3$. Then, we rewrite System (1) as follows

$$z'(t) = f(z(t)) + u(t)g(z(t)) + v(t)h(z(t)), \tag{11}$$

where

$$f(z) = \begin{pmatrix} \sigma - \mu l \\ -\lambda k \\ \rho - \nu m \end{pmatrix}, \quad g(z) = \begin{pmatrix} -\gamma_1 lk \\ \gamma_2 lk \\ 0 \end{pmatrix}, \quad h(z) = \begin{pmatrix} -\delta lm \\ (\beta + \delta)lm \\ -\beta lm \end{pmatrix} \tag{12}$$

with $f(z)$ the drift and $g(z)$, $h(z)$ the control vector fields of this system. Here $z(t)$ is the column vector consisting of the solutions $l(t)$, $k(t)$, $m(t)$ that correspond to the admissible controls $u(t)$ and $v(t)$, that is $z(t) = (l(t), k(t), m(t))^\top \in \mathbb{R}^3$. Let $Df(z)$, $Dg(z)$ and $Dh(z)$ be the Jacobian matrices of the vector functions $f(z)$, $g(z)$ and $h(z)$, respectively. By Formula (12), we find the following relationships:

$$Df(z) = \begin{pmatrix} -\mu & 0 & 0 \\ 0 & -\lambda & 0 \\ 0 & 0 & -\nu \end{pmatrix}, \quad Dg(z) = \begin{pmatrix} -\gamma_1 k & -\gamma_1 l & 0 \\ \gamma_2 k & \gamma_2 l & 0 \\ 0 & 0 & 0 \end{pmatrix},$$

$$Dh(z) = \begin{pmatrix} -\delta m & 0 & -\delta l \\ (\beta + \delta)m & 0 & (\beta + \delta)l \\ -\beta m & 0 & -\beta l \end{pmatrix}. \tag{13}$$

Then, using the introduced concepts and notations, it is easy to see that $z_*(t) = (l_*(t), k_*(t), m_*(t))^\top$ is the optimal trajectory for System (11) corresponding to the optimal controls $u_*(t)$ and $v_*(t)$; $\psi_*(t) = (\psi_1^*(t), \psi_2^*(t), \psi_3^*(t))^\top$ is the appropriate nontrivial solution for the adjoint System (7), or in the new notations:

$$\begin{cases} \psi_*'(t) = -\Big(Df(z_*(t)) + u_*(t)Dg(z_*(t)) + v_*(t)Dh(z_*(t))\Big)^\top \psi_*(t), \\ \psi_*(T) = (0, -1, 0)^\top. \end{cases} \tag{14}$$

The Hamiltonian

$$H(t) = H(l_*(t), k_*(t), m_*(t), u_*(t), v_*(t), \psi_1^*(t), \psi_2^*(t), \psi_3^*(t))$$

takes the form:

$$H(t) = \langle \psi_*(t), f(z_*(t)) \rangle + u_*(t)\langle \psi_*(t), g(z_*(t)) \rangle + v_*(t)\langle \psi_*(t), h(z_*(t)) \rangle.$$

The switching functions $L_u(t)$ and $L_v(t)$, defined by Formula (10), become the scalar products of the adjoint function $\psi_*(t)$ with the corresponding control fields $g(z_*(t))$ and $h(z_*(t))$:

$$L_u(t) = \langle \psi_*(t), g(z_*(t)) \rangle, \quad L_v(t) = \langle \psi_*(t), h(z_*(t)) \rangle. \tag{15}$$

Now, according to [21,22], we introduce for the drift and control vector fields $f(z)$, $g(z)$ and $h(z)$ the corresponding Lie brackets:

$$[f,g](z) = Dg(z)f(z) - Df(z)g(z), \tag{16}$$

$$[f,h](z) = Dh(z)f(z) - Df(z)h(z), \tag{17}$$

$$[g,h](z) = Dh(z)g(z) - Dg(z)h(z). \tag{18}$$

Using (11) and (14)–(18), the derivatives of the switching functions $L_u(t)$ and $L_v(t)$ can be obtained as follows

$$L_u'(t) = \langle \psi_*(t), [f,g](z_*(t)) \rangle + v_*(t)\langle \psi_*(t), [h,g](z_*(t)) \rangle, \tag{19}$$

$$L_v'(t) = \langle \psi_*(t), [f,h](z_*(t)) \rangle + u_*(t)\langle \psi_*(t), [g,h](z_*(t)) \rangle. \tag{20}$$

Now, using (12) and (13) in (16)–(18), we compute the Lie brackets $[f,g](z)$, $[f,h](z)$ and $[g,h](z)$. As a result, we find the formulas:

$$[f,g](z) = \begin{pmatrix} -\gamma_1 k(\sigma - \lambda l) \\ \gamma_2 k(\sigma - \mu l) \\ 0 \end{pmatrix}, \quad [g,h](z) = \begin{pmatrix} \gamma_1(\beta + \delta)l^2 m \\ -\gamma_2(\beta + \delta)l^2 m - \alpha\gamma_2 lkm \\ \gamma_1 \beta lkm \end{pmatrix},$$

$$[f,h](z) = \begin{pmatrix} -\delta((\rho l + \sigma m) - vlm) \\ (\beta + \delta)((\rho l + \sigma m) + (\lambda - \mu - v)lm) \\ -\beta((\rho l + \sigma m) - \mu lm) \end{pmatrix}. \tag{21}$$

Let us write these Lie brackets in a more convenient form for the subsequent analysis. For this, we introduce the following linearly independent vectors:

$$p = \begin{pmatrix} -\gamma_1 \\ \gamma_2 \\ 0 \end{pmatrix}, \quad q = \begin{pmatrix} -\delta \\ \beta + \delta \\ -\beta \end{pmatrix}, \quad r = \begin{pmatrix} 1 \\ 0 \\ 0 \end{pmatrix}.$$

It is easy to see that the following representations for the control vector fields $g(z)$ and $h(z)$ are valid:

$$g(z) = lk \cdot p, \qquad h(z) = lm \cdot q.$$

Using these relationships, Formula (15) for the switching functions $L_u(t)$ and $L_v(t)$ can be rewritten as

$$L_u(t) = l_*(t)k_*(t)\langle \psi_*(t), p \rangle, \quad L_v(t) = l_*(t)m_*(t)\langle \psi_*(t), q \rangle. \tag{22}$$

Now, let us find the decompositions of the Lie brackets $[f,g](z)$, $[f,h](z)$ and $[g,h](z)$, defined by (21), by the vectors p, q, r. As a result, we have the representations:

$$[f,g](z) = k(\sigma - \mu l) \cdot p + \gamma_1(\lambda - \mu)lk \cdot r, \tag{23}$$

$$[f,h](z) = \gamma_2^{-1}(\beta + \delta)(\lambda - v)lm \cdot p + (\rho l + \sigma m - \mu lm) \cdot q + \epsilon lm \cdot r, \tag{24}$$

$$[g,h](z) = -lm((\beta + \delta)l - \delta k) \cdot p - \gamma_1 lkm \cdot q. \tag{25}$$

Next, let us introduce the auxiliary function $G(t) = \langle \psi_*(t), r \rangle = \psi_1^*(t)$, and define the following absolutely continuous functions:

$$
\begin{aligned}
&a_u(t) = l_*^{-1}(t)(\sigma - \mu l_*(t)), \quad a_v(t) = \gamma_2^{-1}(\beta + \delta)(\lambda - \nu)k_*^{-1}(t)m_*(t), \\
&b_v(t) = l_*^{-1}(t)m_*^{-1}(t)(\rho l_*(t) + \sigma m_*(t) - \mu l_*(t)m_*(t)), \\
&c_u(t) = \gamma_1(\lambda - \mu)l_*(t)k_*(t), \quad c_v(t) = \epsilon l_*(t)m_*(t), \\
&d(t) = k_*^{-1}(t)m_*(t)((\beta + \delta)l_*(t) - \delta k_*(t)), \quad e(t) = \gamma_1 k_*(t).
\end{aligned}
\tag{26}
$$

It is easy to see that the functions $c_u(t)$ and $c_v(t)$ are positive on the interval $[0, T]$.

Finally, substituting (23)–(25) into (19), (20) and using (22), (26), we find the required differential equations for the switching functions $L_u(t)$ and $L_v(t)$:

$$
\begin{aligned}
L_u'(t) &= a_u(t)L_u(t) + c_u(t)G(t) + v_*(t)(d(t)L_u(t) + e(t)L_v(t)), \\
L_v'(t) &= a_v(t)L_u(t) + b_v(t)L_v(t) + c_v(t)G(t) \\
&\qquad - u_*(t)(d(t)L_u(t) + e(t)L_v(t)).
\end{aligned}
$$

We add to these equations the first equation of System (7), written with use of the functions $L_u(t)$, $L_v(t)$, $G(t)$:

$$
G'(t) = -u_*(t)l_*^{-1}(t)L_u(t) - v_*(t)l_*^{-1}(t)L_v(t) + \mu G(t),
$$

as well as the corresponding initial values:

$$
L_u(T) = -\gamma_2 l_*(T)k_*(T), \quad L_v(T) = -(\beta + \delta)l_*(T)m_*(T), \quad G(T) = 0.
$$

As a result, we obtain the Cauchy problem for the switching functions $L_u(t)$, $L_v(t)$ and the function $G(t)$:

$$
\left\{
\begin{aligned}
&L_u'(t) = a_u(t)L_u(t) + c_u(t)G(t) + v_*(t)(d(t)L_u(t) + e(t)L_v(t)), \\
&L_v'(t) = a_v(t)L_u(t) + b_v(t)L_v(t) + c_v(t)G(t) \\
&\qquad\qquad - u_*(t)(d(t)L_u(t) + e(t)L_v(t)), \\
&G'(t) = -u_*(t)l_*^{-1}(t)L_u(t) - v_*(t)l_*^{-1}(t)L_v(t) + \mu G(t), \\
&L_u(T) = -\gamma_2 l_*(T)k_*(T), \quad L_v(T) = -(\beta + \delta)l_*(T)m_*(T), \quad G(T) = 0,
\end{aligned}
\right.
\tag{27}
$$

which we will use to justify the properties of the functions $L_u(t)$ and $L_v(t)$.

Now, let us establish the properties of the switching functions $L_u(t)$ and $L_v(t)$. Firstly, the following lemma is true.

Lemma 3. *There is no subinterval of the interval $[0, T]$ at which both switching functions $L_u(t)$ and $L_v(t)$ are identically zero.*

Proof of Lemma 3. We suppose the contrary. Let there be the interval $\Delta_{u,v} \subset [0, T]$ on which the functions $L_u(t)$ and $L_v(t)$ identically equal to zero. Then, their derivatives $L_u'(t)$ and $L_v'(t)$ almost everywhere on this subinterval also vanish. From the first two differential equations of the Cauchy problem (27) we find that $G(t) = 0$ everywhere on the subinterval $\Delta_{u,v}$. Hence, on this subinterval the derivative $G'(t)$ is almost everywhere zero. Therefore, the third differential equation of the Cauchy problem (27) is also satisfied. Using the definition of the function $G(t)$, Lemma 1 and Formula (10), we find that the adjoint function $\psi_*(t)$ vanishes identically on the subinterval $\Delta_{u,v}$. Since the system of linear differential equations (7) is homogeneous, then $\psi_*(t) = 0$ identically everywhere on the interval $[0, T]$, which is contradictory. Hence, our assumption was wrong, and the subinterval $\Delta_{u,v} \subset [0, T]$ on which both switching functions $L_u(t)$ and $L_v(t)$ are identically zero, does not exist. This completes the proof. $\square$

Secondly, the following lemma holds.

Lemma 4. *Let the subinterval* $\Delta_u \subset [0,T]$ *be a singular interval of the optimal control* $u_*(t)$. *Then, everywhere on this subinterval, the optimal control* $v_*(t)$ *is constant and takes one of the values* $\{v_{\min}; 1\}$.

Proof of Lemma 4. On the subinterval Δ_u the switching function $L_u(t)$ is identically zero, and its derivative $L_u'(t)$ vanishes almost everywhere on this interval. Then, the first differential equation of the Cauchy problem (27) yields the equality:

$$c_u(t)G(t) + v_*(t)e(t)L_v(t) = 0, \quad t \in \Delta_u. \tag{28}$$

Let us suppose that the switching function $L_v(t)$ is zero at some point $t_0^u \in \Delta_u$, that is

$$L_v(t_0^u) = 0. \tag{29}$$

Then, (28) implies the equality:

$$G(t_0^u) = 0. \tag{30}$$

The identical equality to zero of the function $L_u(t)$ on the subinterval Δ_u and (29), (30) lead us to the equality $\psi_*(t_0^u) = 0$. A further repetition of the corresponding arguments from Lemma 3 gives a contradiction. Hence, our assumption was wrong, and the switching function $L_v(t)$ does not vanish at any point of the subinterval Δ_u. Therefore, it is sign-definite on this subinterval, and, by Formula (9), the optimal control $v_*(t)$ corresponding to it, is constant and takes one of the values $\{v_{\min}; 1\}$. This completes the proof. $\square$

Furthermore, performing arguments similar to the arguments of Lemma 4, one can show that the following lemma is valid.

Lemma 5. *Let the subinterval* $\Delta_v \subset [0,T]$ *be a singular interval of the optimal control* $v_*(t)$. *Then, everywhere on this subinterval, the optimal control* $u_*(t)$ *is constant and takes one of the values* $\{u_{\min}; 1\}$.

Now, we strengthen the result obtained in Lemma 4. Namely, the following lemma holds.

Lemma 6. *Let the subinterval* $\Delta_u = (t_1^u, t_2^u) \subset [0,T]$, *where* $t_1^u > 0$, *be a singular interval of the optimal control* $u_*(t)$. *Then, there exists a number* $\varepsilon_u > 0$ *such that on the interval* $(t_1^u - \varepsilon_u, t_2^u + \varepsilon_u)$ *the optimal control* $v_*(t)$ *is constant and takes one of the values* $\{v_{\min}; 1\}$.

Proof of Lemma 6. It suffices to show that the ends of the subinterval $\Delta_u = (t_1^u, t_2^u)$ cannot be zeros of the switching function $L_v(t)$, which is an absolutely continuous function. Indeed, if such a fact holds, that is the inequalities:

$$L_v(t_1^u) \neq 0, \quad L_v(t_2^u) \neq 0$$

are true, then Lemma 6 immediately follows from the Theorem on the stability of the sign of a continuous function [30]. Therefore, for definiteness, we consider the right end t_2^u of the subinterval Δ_u. Let us suppose the contrary. This means that the equality:

$$L_v(t_2^u) = 0 \tag{31}$$

is valid. Then, for the function $G(t)$ the following two cases are possible.

    Case 1. Let $G(t_2^u) = 0$. The definition of the function $G(t)$, Lemma 1 and the second formula of (10) imply the relationship:

$$(\beta + \delta)\psi_2^*(t_2^u) = \beta\psi_3^*(t_2^u). \tag{32}$$

If $\psi_2^*(t_2^u) = 0$, then (32) yields $\psi_3^*(t_2^u) = 0$. This fact leads to contradictory equality $\psi_*(t_2^u) = 0$, as the proofs of Lemmas 3 and 4 show. If $\psi_2^*(t_2^u) \neq 0$, then again the definition of the function $G(t)$, Lemma 1 and the first formula of (10) imply $L_u(t_2^u) \neq 0$. The absolutely continuity of the function $L_u(t)$ and the Theorem on the stability of the sign of a continuous function [30] lead to the existence of a left neighborhood of the point t_2^u at which the switching function $L_u(t)$ is sign-definite. This contradicts the fact that such a left neighborhood of the point t_2^u belongs to the subinterval Δ_u, which is a singular portion of the optimal control $u_*(t)$. Thus, Case 1 is impossible.

Case 2. Let $G(t_2^u) \neq 0$. For definiteness, we consider that the inequality:

$$G(t_2^u) > 0 \tag{33}$$

is true. As it was already noted, $L_v(t)$ and $G(t)$ are the absolutely continuous functions. In Lemma 4 it was established that the control $v_*(t)$ is constant on the subinterval Δ_u, that is $v_*(t) = v_*$. Therefore, by (31), (33) and the Theorem on the stability of the sign of a continuous function [30], there exists the interval $(\tilde{t}_2^u, t_2^u) \subset \Delta_u$ on which the inequality:

$$c_u(t)G(t) + v_*e(t)L_v(t) > 0 \tag{34}$$

is valid. Let $\tilde{t}_0^u$ be the midpoint of this interval. We rewrite the first differential equation from the Cauchy problem (27) as

$$L_u'(t) = (a_u(t) + v_*d(t))L_u(t) + (c_u(t)G(t) + v_*e(t)L_v(t)).$$

Then, we integrate this equation on the interval $(\tilde{t}_2^u, t_2^u)$ with the initial value $L_u(\tilde{t}_0^u) = 0$. As a result, the following formula can be found:

$$L_u(t) = \int_{\tilde{t}_0^u}^{t} e^{\int_s^t (a_u(\xi) + v_*d(\xi))d\xi} \left(c_u(s)G(s) + v_*e(s)L_v(s) \right) ds. \tag{35}$$

By (34) and (35), we obtain the positivity of the function $L_u(t)$ for $t > \tilde{t}_0^u$ and the negativity of this function for $t < \tilde{t}_0^u$ on the interval $(\tilde{t}_2^u, t_2^u)$. This is again contradictory. Hence, Case 2 is also impossible.

Thus, our assumption was wrong, and the switching function $L_v(t)$ does not vanish at the point t_2^u. This completes the proof. $\square$

Furthermore, carrying out arguments similar to the arguments of Lemma 6, we can show that the following lemma holds, which strengthens the result of Lemma 5.

Lemma 7. *Let the subinterval $\Delta_v = (t_1^v, t_2^v) \subset [0, T]$, where $t_1^v > 0$, be a singular interval of the optimal control $v_*(t)$. Then, there exists a number $\varepsilon_v > 0$ such that on the interval $(t_1^v - \varepsilon_v, t_2^v + \varepsilon_v)$ the optimal control $u_*(t)$ is constant and takes one of the values $\{u_{\min}; 1\}$.*

Remark 2. *From Lemmas 4–7 we conclude that when a singular arc occurs for one of the optimal controls, $u_*(t)$ or $v_*(t)$, System (1) becomes on the corresponding singular interval a system with one control, because the other control is constant.*

Next, in Sections 5 and 6, we separately study the existence of singular arcs of the optimal controls $u_*(t)$ and $v_*(t)$, applying the approach from [21]. It consists in the sequential differentiation on a singular interval of the corresponding switching function. We carry out this differentiation if the derivative of even order does not have a nonzero term containing control. If such a term appears in a second order derivative, then we say that a singular arc is of the order of one. If it occurs in a derivative of the fourth order, then it is considered that the singular arc is of the order of two. Then, the corresponding optimality condition of a singular arc is checked. If it is satisfied, we find the formulas of the optimal control and the corresponding optimal solutions of System (1) on a singular interval corresponding to such a singular arc. Also, the type of concatenation of this singular arc with non-singular intervals is studied, where the considered optimal control is bang-bang. Finally, we demonstrate a singular arc of the optimal control through the results of a numerical solution of the minimization problem (5), presented in Section 7.

5. Investigation of a Singular Arc of the Optimal Control $u_*(t)$

Let us study the existence of a singular arc of the optimal control $u_*(t)$. According to [22,29], this means the existence of a subinterval $\Delta_u \subset [0, T]$ on which the corresponding switching function $L_u(t)$ identically vanishes. By Lemma 4, everywhere on this subinterval the optimal control $v_*(t)$ is a constant function taking one of the values $\{v_{\min}; 1\}$, that is,

$$v_*(t) = v_* \in \{v_{\min}; 1\}, \quad t \in \Delta_u.$$

Then, Formula (19) of the first derivative $L_u^{(1)}(t)$ of the switching function $L_u(t)$ is rewritten in the form:

$$L_u^{(1)}(t) = \langle \psi_*(t), [f + v_* h, g](z_*(t)) \rangle, \quad t \in \Delta_u. \tag{36}$$

Let us transform the right-hand side of this formula. To do this, we rewrite (23) and (25) as

$$[f, g](z) = \theta_{f,g}(z)p + \chi_{f,g}(z)r, \quad [h, g](z) = \theta_{h,g}(z)p + \eta_{h,g}(z)q, \tag{37}$$

where

$$\theta_{f,g}(z) = (\sigma - \mu l)k, \quad \chi_{f,g}(z) = \gamma_1(\lambda - \mu)lk,$$
$$\theta_{h,g}(z) = lm((\beta + \delta)l - \delta k), \quad \eta_{h,g}(z) = \gamma_1 lkm.$$

Then, using (37), we can write the Lie bracket $[f + v_* h, g](z)$ as follows

$$[f + v_* h, g](z) = \theta_u(z)p + \eta_u(z)q + \chi_u(z)r, \tag{38}$$

where

$$\theta_u(z) = \theta_{f,g}(z) + v_* \theta_{h,g}(z) = (\sigma - \mu l)k + v_* lm((\beta + \delta)l - \delta k),$$
$$\eta_u(z) = v_* \eta_{h,g}(z) = \gamma_1 v_* lkm,$$
$$\chi_u(z) = \chi_{f,g}(z) = \gamma_1(\lambda - \mu)lk.$$

Substituting (38) into the right-hand side of (36), we obtain the relationship:

$$L_u^{(1)}(t) = \theta_u(z_*(t))\langle \psi_*(t), p \rangle + \eta_u(z_*(t))\langle \psi_*(t), q \rangle + \chi_u(z_*(t))\langle \psi_*(t), r \rangle, \quad t \in \Delta_u. \tag{39}$$

Now, differentiating (36), we find the formula for the second derivative $L_u^{(2)}(t)$ of the switching function $L_u(t)$:

$$\begin{aligned} L_u^{(2)}(t) = &\langle \psi_*(t), [f + v_* h, [f + v_* h, g]](z_*(t)) \rangle \\ &+ u_*(t)\langle \psi_*(t), [g, [f + v_* h, g]](z_*(t)) \rangle, \quad t \in \Delta_u. \end{aligned} \tag{40}$$

Let us transform the terms of this formula. First, we consider the second term and its factor

$$\langle \psi_*(t), [g, [f + v_* h, g]](z_*(t)) \rangle. \tag{41}$$

By analogy with (16)–(18), we have equality:

$$[g, [f + v_* h, g]](z) = D[f + v_* h, g](z)g(z) - Dg(z)[f + v_* h, g](z), \tag{42}$$

where $D[f + v_*h, g](z)$ is the Jacobi matrix of the vector function $[f + v_*h, g](z)$. Using (38), we find the relationship:

$$
\begin{aligned}
D[f + v_*h, g](z) &= D(\theta_u(z)p) + D(\eta_u(z)q) + D(\chi_u(z)r) \\
&= p(\nabla\theta_u(z))^\top + q(\nabla\eta_u(z))^\top + r(\nabla\chi_u(z))^\top,
\end{aligned}
\tag{43}
$$

where $\nabla\theta_u(z)$, $\nabla\eta_u(z)$, $\nabla\chi_u(z)$ are the column gradients of the functions $\theta_u(z)$, $\eta_u(z)$, $\chi_u(z)$, respectively. In addition, the representations:

$$
Dg(z) = p(\tau_g(z))^\top, \quad Dh(z) = q(\tau_h(z))^\top
\tag{44}
$$

are valid, where the vector functions $\tau_g(z)$, $\tau_h(z)$ are defined as

$$
\tau_g(z) = \begin{pmatrix} k \\ l \\ 0 \end{pmatrix}, \qquad \tau_h(z) = \begin{pmatrix} m \\ 0 \\ l \end{pmatrix}.
$$

We substitute (43) and the first representation of (44) into (42). After the necessary transformations, the following relationship can be obtained:

$$
\begin{aligned}
[g, [f + v_*h, g]](z) &= \Big(\langle \nabla\theta_u(z), g(z) \rangle - \langle [f + v_*h, g](z), \tau_g(z) \rangle \Big) p \\
&\quad + \langle \nabla\eta_u(z), g(z) \rangle q + \langle \nabla\chi_u(z), g(z) \rangle r.
\end{aligned}
$$

Substituting this expression into (41), we find the formula:

$$
\begin{aligned}
&\langle \psi_*(t), [g, [f + v_*h, g]](z_*(t)) \rangle \\
&= \Big(\langle \nabla\theta_u(z_*(t)), g(z_*(t)) \rangle - \langle [f + v_*h, g](z_*(t)), \tau_g(z_*(t)) \rangle \Big) \langle \psi_*(t), p \rangle \\
&\quad + \langle \nabla\eta_u(z_*(t)), g(z_*(t)) \rangle \langle \psi_*(t), q \rangle + \langle \nabla\chi_u(z_*(t)), g(z_*(t)) \rangle \langle \psi_*(t), r \rangle.
\end{aligned}
\tag{45}
$$

In (39) the function $\chi_u(z_*(t))$ is positive on the interval $[0, T]$. Therefore, let us express the scalar product $\langle \psi_*(t), r \rangle$ through the remaining terms as follows

$$
\begin{aligned}
\langle \psi_*(t), r \rangle &= \chi_u^{-1}(z_*(t)) L_u^{(1)}(t) - \chi_u^{-1}(z_*(t)) \theta_u(z_*(t)) \langle \psi_*(t), p \rangle \\
&\quad - \chi_u^{-1}(z_*(t)) \eta_u(z_*(t)) \langle \psi_*(t), q \rangle,
\end{aligned}
\tag{46}
$$

and then, we substitute this expression into (45). After the necessary transformations, the following relationship finally can be obtained:

$$
\begin{aligned}
&\langle \psi_*(t), [g, [f + v_*h, g]](z_*(t)) \rangle = \chi_u^{-1}(z_*(t)) \langle \nabla\chi_u(z_*(t)), g(z_*(t)) \rangle L_u^{(1)}(t) \\
&\quad + \chi_u^{-1}(z_*(t)) \Big\{ \langle \chi_u(z_*(t)) \nabla\theta_u(z_*(t)) - \theta_u(z_*(t)) \nabla\chi_u(z_*(t)), g(z_*(t)) \rangle \\
&\quad - \chi_u(z_*(t)) \langle [f + v_*h, g](z_*(t)), \tau_g(z_*(t)) \rangle \Big\} \langle \psi_*(t), p \rangle \\
&\quad + \chi_u^{-1}(z_*(t)) \langle \chi_u(z_*(t)) \nabla\eta_u(z_*(t)) - \eta_u(z_*(t)) \nabla\chi_u(z_*(t)), g(z_*(t)) \rangle \langle \psi_*(t), q \rangle.
\end{aligned}
\tag{47}
$$

On the subinterval Δ_u the switching function $L_u(t)$ vanishes identically, that is

$$
L_u(t) = l_*(t) k_*(t) \langle \psi_*(t), p \rangle = 0,
\tag{48}
$$

and therefore its first derivative $L_u^{(1)'}(t)$ is also zero everywhere on this subinterval:

$$L_u^{(1)'}(t) = 0. \tag{49}$$

Then, Formula (47) is simplified and takes the following form:

$$\begin{aligned}
&\langle \psi_*(t), [g, [f + v_*h, g]](z_*(t)) \rangle \\
&= \chi_u^{-1}(z_*(t)) \langle \chi_u(z_*(t)) \nabla \eta_u(z_*(t)) - \eta_u(z_*(t)) \nabla \chi_u(z_*(t)), g(z_*(t)) \rangle \langle \psi_*(t), q \rangle.
\end{aligned} \tag{50}$$

Let us calculate the right-hand side of this formula. For this, we consider the expression $(\chi_u(z) \nabla \eta_u(z) - \eta_u(z) \nabla \chi_u(z))$. Calculating the column gradients $\nabla \eta_u(z), \nabla \chi_u(z)$ of the corresponding functions $\eta_u(z), \chi_u(z)$, the following equality can be found:

$$\chi_u(z) \nabla \eta_u(z) - \eta_u(z) \nabla \chi_u(z) = \gamma_1(\lambda - \mu)v_* \begin{pmatrix} 0 \\ 0 \\ l^2 k^2 \end{pmatrix}. \tag{51}$$

Multiplying this expression scalarly by the vector function $g(z)$ from (12), we conclude that the scalar product

$$\langle \chi_u(z) \nabla \eta_u(z) - \eta_u(z) \nabla \chi_u(z), g(z) \rangle$$

is zero. Consequently, (50) has the form:

$$\langle \psi_*(t), [g, [f + v_*h, g]](z_*(t)) \rangle = 0, \quad t \in \Delta_u. \tag{52}$$

Now, let us consider the first term on the right-hand side of (40):

$$\langle \psi_*(t), [f + v_*h, [f + v_*h, g]](z_*(t)) \rangle. \tag{53}$$

By analogy with (42), we have the equality:

$$[f + v_*h, [f + v_*h, g]](z) = D[f + v_*h, g](z)(f + v_*h)(z) - D(f + v_*h)(z)[f + v_*h, g](z). \tag{54}$$

Let us find the Jacobi matrix $D(f + v_*h)(z)$. The Jacobi matrix $Dh(z)$ of the vector function $h(z)$ is given by the second formula of (44). The use of direct calculations allows us to find the representation of the Jacobi matrix $Df(z)$ from (13) in the following form:

$$Df(z) = p \begin{pmatrix} 0 \\ -\gamma_2^{-1}\lambda \\ -\gamma_2^{-1}\beta^{-1}(\beta + \delta)v \end{pmatrix}^\top + q \begin{pmatrix} 0 \\ 0 \\ \beta^{-1}v \end{pmatrix}^\top + r \begin{pmatrix} -\mu \\ -\gamma_1\gamma_2^{-1}\lambda \\ -\beta^{-1}\alpha v \end{pmatrix}^\top. \tag{55}$$

Then, we obtain the required representation of the Jacobi matrix $D(f + v_*h)(z)$ as

$$D(f + v_*h)(z) = p(\tau_p^u)^\top + q(\tau_q^u(z))^\top + r(\tau_r^u)^\top, \tag{56}$$

where

$$\tau_p^u = \begin{pmatrix} 0 \\ -\gamma_2^{-1}\lambda \\ -\gamma_2^{-1}\beta^{-1}(\beta + \delta)v \end{pmatrix}, \quad \tau_q^u(z) = \begin{pmatrix} v_*m \\ 0 \\ \beta^{-1}v + v_*l \end{pmatrix}, \quad \tau_r^u = \begin{pmatrix} -\mu \\ -\gamma_1\gamma_2^{-1}\lambda \\ -\beta^{-1}\alpha v \end{pmatrix}.$$

We substitute (43) and (56) into (54). After the necessary transformations, the following formula can be found:

$$
\begin{aligned}
[f + v_* h, [f + v_* h, g]](z) = & \left(\langle \nabla \theta_u(z), (f + v_* h)(z) \rangle - \langle [f + v_* h, g](z), \tau_p^u \rangle \right) p \\
& + \left(\langle \nabla \eta_u(z), (f + v_* h)(z) \rangle - \langle [f + v_* h, g](z), \tau_q^u(z) \rangle \right) q \\
& + \left(\langle \nabla \chi_u(z), (f + v_* h)(z) \rangle - \langle [f + v_* h, g](z), \tau_r^u \rangle \right) r.
\end{aligned}
$$

Substituting this formula into (53), we obtain the expression:

$$
\begin{aligned}
& \langle \psi_*(t), [f + v_* h, [f + v_* h, g]](z_*(t)) \rangle \\
& = \left(\langle \nabla \theta_u(z_*(t)), (f + v_* h)(z_*(t)) \rangle - \langle [f + v_* h, g](z_*(t)), \tau_p^u \rangle \right) \langle \psi_*(t), p \rangle \\
& + \left(\langle \nabla \eta_u(z_*(t)), (f + v_* h)(z_*(t)) \rangle - \langle [f + v_* h, g](z_*(t)), \tau_q^u(z_*(t)) \rangle \right) \langle \psi_*(t), q \rangle \\
& + \left(\langle \nabla \chi_u(z_*(t)), (f + v_* h)(z_*(t)) \rangle - \langle [f + v_* h, g](z_*(t)), \tau_r^u \rangle \right) \langle \psi_*(t), r \rangle.
\end{aligned} \tag{57}
$$

Finally, let us substitute (46) into (57). After the necessary transformations, we find the relationship:

$$
\begin{aligned}
& \langle \psi_*(t), [f + v_* h, [f + v_* h, g]](z_*(t)) \rangle \\
& = \chi_u^{-1}(z_*(t)) \Big\{ \langle \nabla \chi_u(z_*(t)), (f + v_* h)(z_*(t)) \rangle - \langle [f + v_* h, g](z_*(t)), \tau_r^u \rangle \Big\} L_u^{(1)}(t) \\
& + \chi_u^{-1}(z_*(t)) \Big\{ \langle \chi_u(z_*(t)) \nabla \theta_u(z_*(t)) - \theta_u(z_*(t)) \nabla \chi_u(z_*(t)), (f + v_* h)(z_*(t)) \rangle \\
& - \langle [f + v_* h, g](z_*(t)), \chi_u(z_*(t)) \tau_p^u - \theta_u(z_*(t)) \tau_r^u \rangle \Big\} \langle \psi_*(t), p \rangle \\
& + \chi_u^{-1}(z_*(t)) \Big\{ \langle \chi_u(z_*(t)) \nabla \eta_u(z_*(t)) - \eta_u(z_*(t)) \nabla \chi_u(z_*(t)), (f + v_* h)(z_*(t)) \rangle \\
& - \langle [f + v_* h, g](z_*(t)), \chi_u(z_*(t)) \tau_q^u(z_*(t)) - \eta_u(z_*(t)) \tau_r^u \rangle \Big\} \langle \psi_*(t), q \rangle.
\end{aligned} \tag{58}
$$

On the subinterval Δ_u Equalities (48) and (49) are valid. Then, Formula (58) is simplified and takes the following form:

$$
\begin{aligned}
& \langle \psi_*(t), [f + v_* h, [f + v_* h, g]](z_*(t)) \rangle \\
& = \chi_u^{-1}(z_*(t)) \Big\{ \langle \chi_u(z_*(t)) \nabla \eta_u(z_*(t)) - \eta_u(z_*(t)) \nabla \chi_u(z_*(t)), (f + v_* h)(z_*(t)) \rangle \\
& - \langle [f + v_* h, g](z_*(t)), \chi_u(z_*(t)) \tau_q^u(z_*(t)) - \eta_u(z_*(t)) \tau_r^u \rangle \Big\} \langle \psi_*(t), q \rangle.
\end{aligned} \tag{59}
$$

Let us calculate the right-hand side of this formula. The expression $(\chi_u(z) \nabla \eta_u(z) - \eta_u(z) \nabla \chi_u(z))$ is given by (51). Multiplying this expression scalarly by the vector function $(f + v_* h)(z)$, defined by the vector functions $f(z)$ and $h(z)$ from (12), we obtain the following formula for the first term in braces:

$$
\begin{aligned}
& \langle \chi_u(z) \nabla \eta_u(z) - \eta_u(z) \nabla \chi_u(z), (f + v_* h)(z) \rangle \\
& = \gamma_1^2 (\lambda - \mu) v_* l^2 k^2 (\rho - \nu m - \beta v_* l m).
\end{aligned} \tag{60}
$$

For the second term in braces, we first find the relationship:

$$
\chi_u(z) \tau_q^u(z) - \eta_u(z) \tau_r^u = \gamma_1 l k \begin{pmatrix} \lambda v_* m \\ \gamma_1 \gamma_2^{-1} \lambda v_* m \\ \beta^{-1}(\lambda - \mu)\nu + (\lambda - \mu)v_* l + \beta^{-1} \alpha \nu v_* m \end{pmatrix},
$$

and then, by (38), obtain the representation:

$$[f + v_*h, g](z) = \begin{pmatrix} \gamma_1\Big((\sigma - \mu l)k + v_*((\beta + \delta)l - \delta k)lm\Big) + \gamma_1(\lambda - \mu)lk - \gamma_1\delta v_*lkm \\ \gamma_2\Big((\sigma - \mu l)k + v_*((\beta + \delta)l - \delta k)lm\Big) + \gamma_1(\beta + \delta)v_*lkm \\ -\gamma_1\beta v_*lkm \end{pmatrix}.$$

Multiplying the last two expressions scalarly and taking into account (60), we have the following expression for the relationship in braces of (59):

$$-\gamma_1^2 v_* l^2 k^2 \Big(\alpha(\lambda - \nu)v_* m^2 + \lambda(\lambda - \mu)m - \rho(\lambda - \mu)\Big). \tag{61}$$

Let us define the quadratic function:

$$w_2(m) = \alpha(\lambda - \nu)v_* m^2 + \lambda(\lambda - \mu)m - \rho(\lambda - \mu).$$

We use this function, when substitute (61) together with the formula of the function $\chi_u(z)$ into (59). After the necessary transformations, this formula is written as follows

$$\langle \psi_*(t), [f + v_*h, [f + v_*h, g]](z_*(t)) \rangle$$
$$= -\gamma_1(\lambda - \mu)^{-1} v_* l_*(t) k_*(t) w_2(m_*(t)) \langle \psi_*(t), q \rangle, \quad t \in \Delta_u. \tag{62}$$

Finally, let us substitute (52) and (62) into (40). As a result, we have the formula of the second derivative $L_u^{(2)}(t)$ of the switching function $L_u(t)$:

$$L_u^{(2)}(t) = -\gamma_1(\lambda - \mu)^{-1} v_* l_*(t) k_*(t) w_2(m_*(t)) \langle \psi_*(t), q \rangle, \quad t \in \Delta_u. \tag{63}$$

On the subinterval Δ_u not only the switching function $L_u(t)$ itself and its first derivative $L_u^{(1)}(t)$ vanish, but also the second derivative $L_u^{(2)}(t)$. By Lemma 1, the linear independence of the vectors p, q, r and the non-triviality of the adjoint function $\psi_*(t)$, the relationship:

$$L_u^{(2)}(t) = 0, \quad t \in \Delta_u$$

implies equality:

$$w_2(m_*(t)) = 0, \quad t \in \Delta_u. \tag{64}$$

By Assumption 1, the discriminant of the quadratic function $w_2(m)$ is positive, and $w_2(0) < 0$, and therefore, it has a unique positive root m_{sing}, defined by the formula:

$$m_{\text{sing}} = \frac{-(\lambda - \mu)\lambda + \sqrt{(\lambda - \mu)^2\lambda^2 + 4\alpha\rho v_*(\lambda - \mu)(\lambda - \nu)}}{2\alpha(\lambda - \nu)v_*}.$$

This root is the value of the solution $m_*(t)$ on the subinterval Δ_u. We note the important properties of the value m_{sing}:

$$m_{\text{sing}} \in \left(0, \rho\nu^{-1}\right), \quad w_2'(m_{\text{sing}}) > 0. \tag{65}$$

Analyzing (63), we see that on the subinterval Δ_u the second derivative $L_u^{(2)}(t)$ of the switching function $L_u(t)$ does not contain the control $u_*(t)$. It means that the order of the singular arc is greater than one [22,29]. Therefore, we continue to differentiate the switching function $L_u(t)$ on this subinterval, and using (63) find its third derivative $L_u^{(3)}(t)$:

$$
\begin{aligned}
L_u^{(3)}(t) = & -\gamma_1(\lambda - \mu)^{-1}v_*\Big\{l_*(t)k_*(t)w_2'(m_*(t))m_*'(t)\langle\psi_*(t), q\rangle \\
& + w_2(m_*(t))\Big(l_*(t)k_*(t)\langle\psi_*(t), q\rangle\Big)'\Big\}.
\end{aligned}
\tag{66}
$$

On the subinterval Δ_u Equality (64) is valid, and therefore the second term in braces of (66) is zero. Substituting the formula for $m_*'(t)$ from System (1) into the first term of (66), we find the required formula:

$$
\begin{aligned}
L_u^{(3)}(t) = & -\gamma_1(\lambda - \mu)^{-1}v_*l_*(t)k_*(t)w_2'(m_{\text{sing}}) \\
& \times \Big(\rho - \nu m_{\text{sing}} - \beta v_*l_*(t)m_{\text{sing}}\Big)\langle\psi_*(t), q\rangle, \quad t \in \Delta_u.
\end{aligned}
\tag{67}
$$

On the subinterval Δ_u the third derivative $L_u^{(3)}(t)$ is also zero that leads to the equality:

$$
m_*'(t) = 0.
\tag{68}
$$

It allows us to find a value l_{sing} that is the value of the solution $l_*(t)$ on this subinterval:

$$
l_{\text{sing}} = \frac{\rho - \nu m_{\text{sing}}}{\beta v_* m_{\text{sing}}}.
\tag{69}
$$

We note that due to the inclusion of (65), the value l_{sing} is positive.

Finally, on the subinterval Δ_u let us calculate the fourth derivative $L_u^{(4)}(t)$ of the switching function $L_u(t)$ using (66). Some of the terms that are obtained with such a differentiation vanish by virtue of (64) and (68). As a result, the following relationship can be obtained:

$$
\begin{aligned}
L_u^{(4)}(t) = & \gamma_1(\lambda - \mu)^{-1}\beta v_*^2 l_{\text{sing}}k_*(t)m_{\text{sing}}w_2'(m_{\text{sing}}) \\
& \times \Big(\sigma - \mu l_{\text{sing}} - \delta v_*l_{\text{sing}}m_{\text{sing}} - \gamma_1 u_*(t)l_{\text{sing}}k_*(t)\Big)\langle\psi_*(t), q\rangle, \quad t \in \Delta_u,
\end{aligned}
\tag{70}
$$

which implies the expression:

$$
\begin{aligned}
\frac{\partial}{\partial u}L_u^{(4)}(t) = & -\gamma_1^2(\lambda - \mu)^{-1}\beta v_*^2 l_{\text{sing}}^2 k_*^2(t)m_{\text{sing}}w_2'(m_{\text{sing}})\langle\psi_*(t), q\rangle \\
= & -\gamma_1^2(\lambda - \mu)^{-1}\beta v_*^2 l_{\text{sing}}k_*^2(t)w_2'(m_{\text{sing}})L_v(t), \quad t \in \Delta_u.
\end{aligned}
\tag{71}
$$

Here we applied the second formula of (22). It is easy to see that this expression is sign-definite everywhere on the subinterval Δ_u. Therefore, firstly, the order of the singular arc equals two. Secondly, the necessary optimality condition of the singular arc, the Kelly-Cope-Moyer condition [29], is either carried out in a strengthened form, and then the singular arc exists, or it is not satisfied, and then the singular arc does not exist. By [29] and (71), the strengthened Kelly-Cope-Moyer condition leads to the inequality:

$$
\gamma_1^2(\lambda - \mu)^{-1}\beta v_*^2 l_{\text{sing}}k_*^2(t)w_2'(m_{\text{sing}})L_v(t) > 0, \quad t \in \Delta_u.
\tag{72}
$$

By Assumption 1, Lemma 1, and the inequality of (65), Relationship (72) implies the inequality $L_v(t) > 0$, which in turn, by Formula (9), implies $v_* = 1$ everywhere on the subinterval Δ_u.

Remark 3. *On the subinterval $\Delta_u \subset [0, T]$, which is the singular interval of the optimal control $u_*(t)$, the optimal control $v_*(t)$ is constant and takes the value 1. This leads to the corresponding correction of the formulations of Lemmas 4 and 6.*

Next, Formula (70) and the vanishing of the fourth derivative $L_u^{(4)}(t)$ of the switching function $L_u(t)$ yield the following relationship for the control $u_{\text{sing}}(t)$ and function $k_{\text{sing}}(t)$:

$$u_{\text{sing}}(t)k_{\text{sing}}(t) = \frac{\sigma - \mu l_{\text{sing}} - \delta l_{\text{sing}} m_{\text{sing}}}{\gamma_1 l_{\text{sing}}}, \tag{73}$$

which are the control $u_*(t)$ and solution $k_*(t)$ on the subinterval Δ_u. Formula (69) allows us to rewrite (73) as follows

$$u_{\text{sing}}(t)k_{\text{sing}}(t) = \frac{w_4(m_{\text{sing}})}{\gamma_1 l_{\text{sing}}}, \quad t \in \Delta_u. \tag{74}$$

Here $w_4(m)$ is the quadratic function given by the formula:

$$w_4(m) = \delta v m^2 + (\sigma \beta + v \mu - \delta \rho)m - \mu \rho.$$

By the inclusion of (65), we consider this function on the interval $\left[0, \rho v^{-1}\right]$. The relationships:

$$w_4(0) = -\mu \rho < 0, \quad w_4\left(\rho v^{-1}\right) = \rho v^{-1} \sigma \beta > 0$$

lead to the conclusion that the function $w_4(m)$ has exactly one zero $m_\star \in (0, \rho v^{-1})$. In turn, this fact implies the validity of the formula:

$$w_4(m) \begin{cases} < 0 & \text{, if } 0 \leq m < m_\star, \\ = 0 & \text{, if } m = m_\star, \\ > 0 & \text{, if } m_\star < m \leq \rho v^{-1}. \end{cases}$$

We apply this formula in analysis of (74). Positivity of the product on its left-hand side and the inclusion of (65) imply the validity of the inclusion:

$$m_{\text{sing}} \in \left(m_\star, \rho v^{-1}\right),$$

which is a necessary condition for the existence of the singular arc.

Finally, let us discuss the behavior of the optimal control $u_*(t)$ over the entire interval $[0, T]$. When the inclusion $u_{\text{sing}}(t) \in [u_{\text{min}}, 1]$ holds for all $t \in \Delta_u$, the control $u_{\text{sing}}(t)$ is admissible. Corollary 1 shows that the singular arc of the optimal control $u_*(t)$ is concatenated with the nonsingular interval, where this control is bang-bang. Let $\xi \in (0, T)$ be the time moment, where such a concatenation occurs. Then, as it follows from [22,29], when $u_{\text{sing}}(t) \in (u_{\text{min}}, 1)$ for all $t \in \Delta_u$, the nonsingular interval contains at least the countable number of switchings of the control $u_*(t)$, accumulating to the point ξ. This behavior of the optimal control $u_*(t)$ on nonsingular intervals is called a chattering [22,29], and will be observed on both sides of the subinterval Δ_u.

Thus, the above arguments of this section lead us to the validity of the following proposition.

Proposition 1. *The optimal control $u_*(t)$ on a singular interval can contain a singular arc of order two, which concatenates with bang-bang intervals of this control using chattering. On such an interval the optimal control $v_*(t)$ is constant and takes the value 1.*

6. Investigation of a Singular Arc of the Optimal Control $v_*(t)$

Now, let us carry out arguments similar to those presented in the previous section to study the existence of a singular arc for the optimal control $v_*(t)$. According to [22,29], this means the existence of a subinterval $\Delta_v \subset [0, T]$ on which the corresponding switching function $L_v(t)$ identically vanishes. By Lemma 5, everywhere on this subinterval the optimal control $u_*(t)$ is a constant function that takes one of the values $\{u_{min}; 1\}$, that is,

$$u_*(t) = u_* \in \{u_{min}; 1\}, \quad t \in \Delta_v.$$

Then, Formula (20) for the first derivative $L_v^{(1)}(t)$ of the switching function $L_v(t)$ is rewritten in the form:

$$L_v^{(1)}(t) = \langle \psi_*(t), [f + u_*g, h](z_*(t)) \rangle, \quad t \in \Delta_v. \tag{75}$$

Let us transform the right-hand side of this formula. To do this, we rewrite (24) and (25) as

$$[f, h](z) = \theta_{f,h}(z)p + \eta_{f,h}(z)q + \chi_{f,h}(z)r, \quad [g, h](z) = \theta_{g,h}(z)p + \eta_{g,h}(z)q, \tag{76}$$

where

$$\theta_{f,h}(z) = \gamma_2^{-1}(\beta + \delta)(\lambda - \nu)lm, \quad \eta_{f,h}(z) = \rho l + \sigma m - \mu lm, \quad \chi_{f,h}(z) = \epsilon lm,$$
$$\theta_{g,h}(z) = -\theta_{h,g}(z) = -lm((\beta + \delta)l - \delta k), \quad \eta_{g,h}(z) = -\eta_{h,g}(z) = -\gamma_1 lkm.$$

Then, using (76), we can write the Lie bracket $[f + u_*g, h](z)$ as follows

$$[f + u_*g, h](z) = \theta_v(z)p + \eta_v(z)q + \chi_v(z)r, \tag{77}$$

where

$$\theta_v(z) = \theta_{f,h}(z) + u_*\theta_{g,h}(z) = lm(\gamma_2^{-1}(\beta + \delta)(\lambda - \nu) - u_*((\beta + \delta)l - \delta k)),$$
$$\eta_v(z) = \eta_{f,h}(z) + u_*\eta_{g,h}(z) = \rho l + \sigma m - \mu lm - \gamma_1 u_* lkm,$$
$$\chi_v(z) = \chi_{f,h}(z) = \epsilon lm.$$

Substituting (77) into the right-hand side of (75), we obtain the relationship:

$$L_v^{(1)}(t) = \theta_v(z_*(t))\langle \psi_*(t), p \rangle + \eta_v(z_*(t))\langle \psi_*(t), q \rangle + \chi_v(z_*(t))\langle \psi_*(t), r \rangle, \quad t \in \Delta_v. \tag{78}$$

Now, differentiating (75), we find the formula for the second derivative $L_v^{(2)}(t)$ of the switching function $L_v(t)$:

$$L_v^{(2)}(t) = \langle \psi_*(t), [f + u_*g, [f + u_*g, h]](z_*(t)) \rangle$$
$$+ v_*(t)\langle \psi_*(t), [h, [f + u_*g, h]](z_*(t)) \rangle, \quad t \in \Delta_v. \tag{79}$$

Let us transform the terms of this formula. First, we consider the second term and its factor

$$\langle \psi_*(t), [h, [f + u_*g, h]](z_*(t)) \rangle. \tag{80}$$

By analogy with (16)–(18), we have equality:

$$[h, [f + u_*g, h]](z) = D[f + u_*g, h](z)h(z) - Dh(z)[f + u_*g, h](z), \tag{81}$$

where $D[f + u_*g, h](z)$ is the Jacobi matrix of the vector function $[f + u_*g, h](z)$. Using (77), we find the relationship:

$$D[f + u_*g, h](z) = D(\theta_v(z)p) + D(\eta_v(z)q) + D(\chi_v(z)r)$$
$$= p(\nabla\theta_v(z))^\top + q(\nabla\eta_v(z))^\top + r(\nabla\chi_v(z))^\top, \tag{82}$$

where $\nabla\theta_v(z)$, $\nabla\eta_v(z)$, $\nabla\chi_v(z)$ are the column gradients of the functions $\theta_v(z)$, $\eta_v(z)$, $\chi_v(z)$, respectively. We substitute expression (82) and the second representation of (44) into (81). After the necessary transformations, the following relationship can be obtained:

$$[h,[f+u_*g,h]](z) = \langle\nabla\theta_v(z),h(z)\rangle p$$
$$+\Big(\langle\nabla\eta_v(z),h(z)\rangle - \langle[f+u_*g,h](z),\tau_h(z)\rangle\Big)q + \langle\nabla\chi_v(z),h(z)\rangle r.$$

Substituting this expression into (80), we find the formula:

$$\langle\psi_*(t),[h,[f+u_*g,h]](z_*(t))\rangle = \langle\nabla\theta_v(z_*(t)),h(z_*(t))\rangle\langle\psi_*(t),p\rangle$$
$$+\Big(\langle\nabla\eta_v(z_*(t)),h(z_*(t))\rangle - \langle[f+u_*g,h](z_*(t)),\tau_h(z_*(t))\rangle\Big)\langle\psi_*(t),q\rangle \tag{83}$$
$$+\langle\nabla\chi_v(z_*(t)),h(z_*(t))\rangle\langle\psi_*(t),r\rangle.$$

In (78), the function $\chi_v(z_*(t))$ is positive on the interval $[0,T]$. Therefore, let us express the scalar product $\langle\psi_*(t),r\rangle$ through the remaining terms as follows

$$\langle\psi_*(t),r\rangle = \chi_v^{-1}(z_*(t))L_v^{(1)}(t) - \chi_v^{-1}(z_*(t))\theta_v(z_*(t))\langle\psi_*(t),p\rangle$$
$$- \chi_v^{-1}(z_*(t))\eta_v(z_*(t))\langle\psi_*(t),q\rangle, \tag{84}$$

and then, we substitute this expression into (83). After the necessary transformations, the following relationship finally can be obtained:

$$\langle\psi_*(t),[h,[f+u_*g,h]](z_*(t))\rangle = \chi_v^{-1}(z_*(t))\langle\nabla\chi_v(z_*(t)),h(z_*(t))\rangle L_v^{(1)}(t)$$
$$+\chi_v^{-1}(z_*(t))\langle\chi_v(z_*(t))\nabla\theta_v(z_*(t)) - \theta_v(z_*(t))\nabla\chi_v(z_*(t)),h(z_*(t))\rangle\langle\psi_*(t),p\rangle$$
$$+\chi_v^{-1}(z_*(t))\Big\{\langle\chi_v(z_*(t))\nabla\eta_v(z_*(t)) - \eta_v(z_*(t))\nabla\chi_v(z_*(t)),h(z_*(t))\rangle \tag{85}$$
$$- \chi_v(z_*(t))\langle[f+u_*g,h](z_*(t)),\tau_h(z_*(t))\rangle\Big\}\langle\psi_*(t),q\rangle.$$

On the subinterval Δ_v the switching function $L_v(t)$ vanishes identically, that is

$$L_v(t) = l_*(t)m_*(t)\langle\psi_*(t),q\rangle = 0, \tag{86}$$

and therefore its first derivative $L_v^{(1)}(t)$ is also zero everywhere on this subinterval:

$$L_v^{(1)}(t) = 0. \tag{87}$$

Then, Formula (85) is simplified and takes the following form:

$$\langle\psi_*(t),[h,[f+u_*g,h]](z_*(t))\rangle$$
$$= \chi_v^{-1}(z_*(t))\langle\chi_v(z_*(t))\nabla\theta_v(z_*(t)) - \theta_v(z_*(t))\nabla\chi_v(z_*(t)),h(z_*(t))\rangle\langle\psi_*(t),p\rangle. \tag{88}$$

Let us calculate the right-hand side of this formula. For this, we consider the expression $(\chi_v(z)\nabla\theta_v(z) - \theta_v(z)\nabla\chi_v(z))$. Calculating the column gradients $\nabla\theta_v(z)$, $\nabla\chi_v(z)$ of the corresponding functions $\theta_v(z)$, $\chi_v(z)$, the following equality can be found:

$$\chi_v(z)\nabla\theta_v(z) - \theta_v(z)\nabla\chi_v(z) = \epsilon u_* \begin{pmatrix} -(\beta+\delta)l^2m^2 \\ \delta l^2m^2 \\ 0 \end{pmatrix}. \tag{89}$$

Multiplying this expression scalarly by the vector function $h(z)$ from (12), we conclude that the following relationship holds:

$$\langle \chi_v(z) \nabla \theta_v(z) - \theta_v(z) \nabla \chi_v(z), h(z) \rangle = 2\epsilon \delta(\beta + \delta) u_* l^3 m^3.$$

Substituting this relationship together with the formula of the function $\chi_v(z)$ into (88), we see that it has the form:

$$\langle \psi_*(t), [h, [f + u_* g, h]](z_*(t)) \rangle = 2\delta(\beta + \delta) u_* l_*^2(t) m_*(t)^2 \langle \psi_*(t), p \rangle, \quad t \in \Delta_v. \tag{90}$$

Now, let us consider the first term on the right-hand side of (79):

$$\langle \psi_*(t), [f + u_* g, [f + u_* g, h]](z_*(t)) \rangle. \tag{91}$$

By analogy with (81), we have the equality:

$$[f + u_* g, [f + u_* g, h]](z) = D[f + u_* g, h](z)(f + u_* g)(z) - D(f + u_* g)(z)[f + u_* g, h](z). \tag{92}$$

Let us find the Jacobi matrix $D(f + u_* g)(z)$. The Jacobi matrices $Df(z)$ and $Dg(z)$ of the corresponding vector functions $f(z)$ and $g(z)$ from (12) are given by the representation (55) and the first formula of (44). Then, the required representation can be written as

$$D(f + u_* g)(z) = p(\tau_p^v(z))^\top + q(\tau_q^v)^\top + r(\tau_r^v)^\top, \tag{93}$$

where

$$\tau_p^v(z) = \begin{pmatrix} u_* k \\ -\gamma_2^{-1} \lambda + u_* l \\ -\gamma_2^{-1} \beta^{-1}(\beta + \delta) v \end{pmatrix}, \quad \tau_q^v = \begin{pmatrix} 0 \\ 0 \\ \beta^{-1} v \end{pmatrix}, \quad \tau_r^v = \tau_r^u = \begin{pmatrix} -\mu \\ -\gamma_1 \gamma_2^{-1} \lambda \\ -\beta^{-1} \alpha v \end{pmatrix}.$$

We substitute (82) and (93) into (92). After the necessary transformations, the following formula can be found:

$$[f + u_* g, [f + u_* g, h]](z) = \left(\langle \nabla \theta_v(z), (f + u_* g)(z) \rangle - \langle [f + u_* g, h](z), \tau_p^v(z) \rangle \right) p$$
$$+ \left(\langle \nabla \eta_v(z), (f + u_* g)(z) \rangle - \langle [f + u_* g, h](z), \tau_q^v \rangle \right) q$$
$$+ \left(\langle \nabla \chi_v(z), (f + u_* g)(z) \rangle - \langle [f + u_* g, h](z), \tau_r^v \rangle \right) r.$$

Substituting this formula into (91), we obtain the expression:

$$\langle \psi_*(t), [f + u_* g, [f + u_* g, h]](z_*(t)) \rangle$$
$$= \left(\langle \nabla \theta_v(z_*(t)), (f + u_* g)(z_*(t)) \rangle - \langle [f + u_* g, h](z_*(t)), \tau_p^v(z_*(t)) \rangle \right) \langle \psi_*(t), p \rangle$$
$$+ \left(\langle \nabla \eta_v(z_*(t)), (f + u_* g)(z_*(t)) \rangle - \langle [f + u_* g, h](z_*(t)), \tau_q^v \rangle \right) \langle \psi_*(t), q \rangle \tag{94}$$
$$+ \left(\langle \nabla \chi_v(z_*(t)), (f + u_* g)(z_*(t)) \rangle - \langle [f + u_* g, h](z_*(t)), \tau_r^v \rangle \right) \langle \psi_*(t), r \rangle.$$

Now, let us substitute (84) into (94). After the necessary transformations, we have the relationship:

$$
\begin{aligned}
&\langle \psi_*(t), [f + u_* g, [f + u_* g, h]](z_*(t)) \rangle \\
&= \chi_v^{-1}(z_*(t)) \Big\{ \langle \nabla \chi_v(z_*(t)), (f + u_* g)(z_*(t)) \rangle - \langle [f + u_* g, h](z_*(t)), \tau_r^v \rangle \Big\} L_v^{(1)}(t) \\
&+ \chi_v^{-1}(z_*(t)) \Big\{ \langle \chi_v(z_*(t)) \nabla \theta_v(z_*(t)) - \theta_v(z_*(t)) \nabla \chi_v(z_*(t)), (f + u_* g)(z_*(t)) \rangle \\
&\quad - \langle [f + u_* g, h](z_*(t)), \chi_v(z_*(t)) \tau_p^v(z_*(t)) - \theta_v(z_*(t)) \tau_r^v \rangle \Big\} \langle \psi_*(t), p \rangle \\
&+ \chi_v^{-1}(z_*(t)) \Big\{ \langle \chi_v(z_*(t)) \nabla \eta_v(z_*(t)) - \eta_v(z_*(t)) \nabla \chi_v(z_*(t)), (f + u_* g)(z_*(t)) \rangle \\
&\quad - \langle [f + u_* g, h](z_*(t)), \chi_v(z_*(t)) \tau_q^v - \eta_v(z_*(t)) \tau_r^v \rangle \Big\} \langle \psi_*(t), q \rangle.
\end{aligned}
\tag{95}
$$

On the subinterval Δ_v equalities (86) and (87) are valid. Then, formula (95) is simplified and takes the following form:

$$
\begin{aligned}
&\langle \psi_*(t), [f + u_* g, [f + u_* g, h]](z_*(t)) \rangle \\
&= \chi_v^{-1}(z_*(t)) \Big\{ \langle \chi_v(z_*(t)) \nabla \theta_v(z_*(t)) - \theta_v(z_*(t)) \nabla \chi_v(z_*(t)), (f + u_* g)(z_*(t)) \rangle \\
&\quad - \langle [f + u_* g, h](z_*(t)), \chi_v(z_*(t)) \tau_p^v(z_*(t)) - \theta_v(z_*(t)) \tau_r^v \rangle \Big\} \langle \psi_*(t), p \rangle.
\end{aligned}
\tag{96}
$$

Let us calculate the right-hand side of this formula. The expression $(\chi_v(z) \nabla \theta_v(z) - \theta_v(z) \nabla \chi_v(z))$ is given by (89). Multiplying this expression scalarly by the vector function $(f + u_* g)(z)$, defined by the vector functions $f(z)$ and $g(z)$ from (12), we obtain the following formula for the first term in braces:

$$
\begin{aligned}
&\langle \chi_v(z) \nabla \theta_v(z) - \theta_v(z) \nabla \chi_v(z), (f + u_* g)(z) \rangle \\
&= -\epsilon u_* l^2 m^2 ((\beta + \delta)(\sigma - \mu l - \gamma_1 u_* l k) + \delta k (\lambda - \gamma_2 u_* l)).
\end{aligned}
\tag{97}
$$

Now, substituting (77) into the second term in braces, we find the expression:

$$
\begin{aligned}
&\langle [f + u_* g, h](z), \chi_v(z) \tau_p^v(z) - \theta_v(z) \tau_r^v \rangle \\
&= \theta_v(z) \chi_v(z) \langle \tau_p^v(z), p \rangle + \eta_v(z) \chi_v(z) \langle \tau_p^v(z), q \rangle + \chi_v^2(z) \langle \tau_p^v(z), r \rangle \\
&\quad - \theta_v^2(z) \langle \tau_r^v, p \rangle - \theta_v(z) \eta_v(z) \langle \tau_r^v, q \rangle - \theta_v(z) \chi_v(z) \langle \tau_r^v, r \rangle.
\end{aligned}
\tag{98}
$$

Using the formulas of the vectors p, q, r, τ_r^v, the vector function $\tau_p^v(z)$ and the functions $\theta_v(z)$, $\eta_v(z)$, $\chi_v(z)$, we calculate all the terms of (98). As a result, the following expression can be obtained:

$$
\begin{aligned}
&\langle [f + u_* g, h](z), \chi_v(z) \tau_p^v(z) - \theta_v(z) \tau_r^v \rangle \\
&= l^2 m^2 \Big\{ \gamma_1 \gamma_2^{-2} (\lambda - \mu) \Big((\beta + \delta)(\lambda - \nu) - \gamma_2 u_* ((\beta + \delta) l - \delta k) \Big)^2 \\
&\quad - \gamma_2^{-1} \epsilon ((\lambda - \mu) - u_* (\gamma_2 l - \gamma_1 k)) \\
&\quad \times \Big((\beta + \delta)(\lambda - \nu) - \gamma_2 u_* ((\beta + \delta) l - \delta k) \Big) + \epsilon^2 u_* k \Big\}.
\end{aligned}
\tag{99}
$$

We substitute (97) and (99) into (96). After the necessary transformations in this formula and the substitution of the quadratic function $\Phi(l, k)$ defined by the relationship:

$$
\begin{aligned}
\Phi(l, k) &= \gamma_1 \gamma_2^{-2} (\lambda - \mu) \Big((\beta + \delta)(\lambda - \nu) - \gamma_2 u_* ((\beta + \delta) l - \delta k) \Big)^2 \\
&\quad - \gamma_2^{-1} \epsilon ((\lambda - \mu) - u_* (\gamma_2 l - \gamma_1 k)) \Big((\beta + \delta)(\lambda - \nu) - \gamma_2 u_* ((\beta + \delta) l - \delta k) \Big) \\
&\quad + \epsilon u_* \Big((\beta + \delta) \sigma - ((\beta + \delta) \mu l - (\epsilon + \delta \lambda) k) - u_* (\gamma_1 (\beta + \delta) + \gamma_2 \delta) l k \Big),
\end{aligned}
$$

as well as the formula of the function $\chi_v(z)$, we have for (96) the following expression:

$$\langle \psi_*(t), [f + u_* g, [f + u_* g, h]](z_*(t)) \rangle = -\epsilon^{-1} l_*(t) m_*(t) \Phi(l_*(t), k_*(t)) \langle \psi_*(t), p \rangle. \tag{100}$$

Now, substituting (90) and (100) into (79), we finally find the formula for the second derivative $L_v^{(2)}(t)$ of the switching function $L_v(t)$ as

$$
\begin{aligned}
L_v^{(2)}(t) = l_*(t) m_*(t) \Big(-\epsilon^{-1} \Phi(l_*(t), k_*(t)) \\
+ 2\delta(\beta + \delta) u_* v_*(t) l_*(t) m_*(t) \Big) \langle \psi_*(t), p \rangle, \quad t \in \Delta_v.
\end{aligned}
\tag{101}
$$

This formula implies the relationship:

$$
\begin{aligned}
\frac{\partial}{\partial v} L_v^{(2)}(t) &= 2\delta(\beta + \delta) u_* l_*^2(t) m_*^2(t) \langle \psi_*(t), p \rangle \\
&= 2\delta(\beta + \delta) u_* l_*(t) k_*^{-1}(t) m_*^2(t) L_u(t), \quad t \in \Delta_v.
\end{aligned}
\tag{102}
$$

Here we applied the first formula of (22). It is easy to see that this expression is sign-definite everywhere on the subinterval Δ_v. Therefore, firstly, the order of the singular arc equals one. Secondly, the necessary optimality condition of the singular arc, the Kelly condition [29], is either carried out in a strengthened form, and then the singular arc exists, or it is not satisfied, and then the singular arc does not exist. By [29] and (102), the strengthened Kelly condition leads to the inequality:

$$2\delta(\beta + \delta) u_* l_*(t) k_*^{-1}(t) m_*^2(t) L_u(t) > 0, \quad t \in \Delta_v. \tag{103}$$

By Lemma 1, Relationship (103) implies the inequality $L_u(t) > 0$, which in turn, by Formula (8), implies $u_* = 1$ everywhere on the subinterval Δ_v.

Remark 4. *On the subinterval $\Delta_v \subset [0, T]$, which is the singular interval of the optimal control $v_*(t)$, the optimal control $u_*(t)$ is constant and takes the value 1. This leads to the corresponding correction of the formulations of Lemmas 5 and 7.*

Next, Formula (101) and the vanishing of the second derivative $L_v^{(2)}(t)$ of the switching function $L_v(t)$ yield the following relationship for the control $v_{\text{sing}}(t)$:

$$v_{\text{sing}}(t) = \frac{\Phi(l_{\text{sing}}(t), k_{\text{sing}}(t))}{2\epsilon\delta(\beta + \delta) l_{\text{sing}}(t) m_{\text{sing}}(t)},$$

which is the control $v_*(t)$ on the subinterval Δ_v. Here the functions $l_{\text{sing}}(t)$, $k_{\text{sing}}(t)$, $m_{\text{sing}}(t)$ are the corresponding solutions $l_*(t)$, $k_*(t)$, $m_*(t)$ on this subinterval. When the inclusion $v_{\text{sing}}(t) \in [v_{\min}, 1]$ holds for all $t \in \Delta_v$, the control $v_{\text{sing}}(t)$ is admissible. Corollary 1 shows that the singular arc of the optimal control $v_*(t)$ is concatenated with the nonsingular interval, where this control is bang-bang. Let $\xi \in (0, T)$ be the time moment, where such a concatenation occurs. Then, as it follows from [22,29], when $v_{\text{sing}}(t) \in (v_{\min}, 1)$ for all $t \in \Delta_v$, such concatenations are allowed and will be observed on both sides of the subinterval Δ_v.

Thus, the above arguments of this section lead us to the validity of the following proposition.

Proposition 2. *The optimal control $v_*(t)$ on a singular interval can contain a singular arc of order one, which concatenates with bang-bang intervals of this control. On such an interval the optimal control $u_*(t)$ is constant and takes the value 1.*

7. Numerical Results

Here we demonstrate the results of a numerical solution of the minimization problem (5). For numerical calculations the following values of the parameters of System (1), the initial values (2) and the control constraints (3) taken from [17,26] were used:

$$
\begin{array}{llll}
\sigma = 15.0 & \rho = 3.6 & \beta = 0.4 & \delta = 0.005 \\
\lambda \in \{0.3; 1.2\} & \mu = 0.01 & \nu = 0.02 & T = 100.0 \\
\gamma_1 = 0.8 & \gamma_2 = 0.05 & u_{min} = 0.3 & v_{min} = 0.3 \\
l_0 = 100.0 & k_0 = 40.0 & m_0 = 50.0
\end{array}
\tag{104}
$$

These numerical calculations were conducted using "BOCOP–2.0.5" [31]. It is an optimal control interface, implemented in MATLAB, for solving optimal control problems with general path and boundary constraints, and free or fixed final time. By a time discretization, such problems are approximated by finite-dimensional optimization problems, which are then solved by well-known software IPOPT, using sparse exact derivatives computed by ADOL-C. IPOPT is an open-source software package for large-scale nonlinear optimization.

Considering the time interval of 100 days ($T = 100.0$), a time grid with 8000 nodes was created, i.e., for $t \in [0, 100.0]$ we get $\triangle t = 0.0125$. Since our problem is solved by a direct method and, consequently, using an iterative approach, we impose at each step the acceptable convergence tolerance of $\varepsilon_{rel} = 10^{-15}$. Moreover, we use the sixth-order Lobatto III C discretization rule. In this respect, for more details we refer to [31].

The corresponding results of the numerical calculations are presented in Figures 1 and 2. In each figure for a specific value of λ we give the graphs of the optimal controls $u_*(t)$, $v_*(t)$, the corresponding optimal solutions $l_*(t)$, $k_*(t)$, and $m_*(t)$; J_* is the minimum value of the functional $J(u, v)$ of (5).

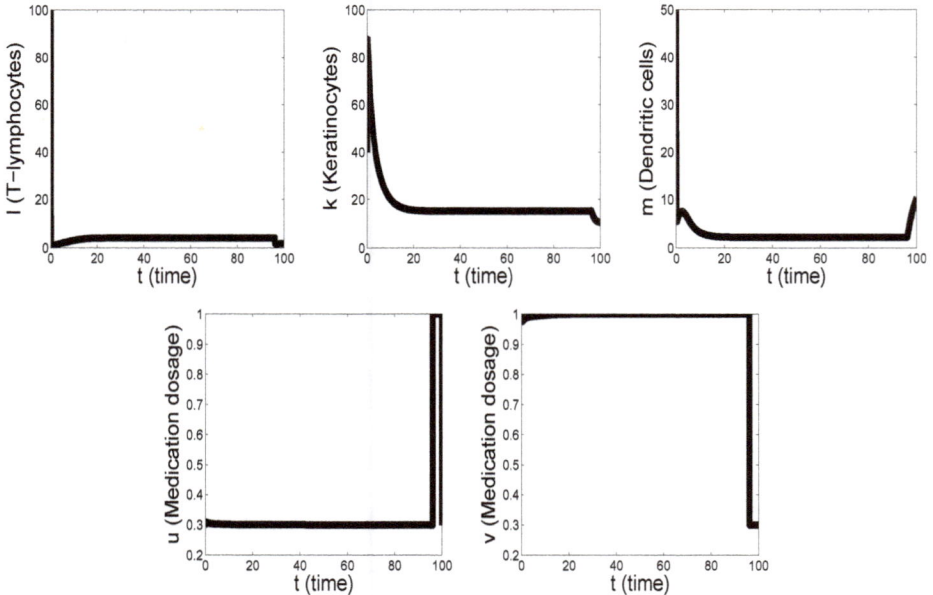

Figure 1. Optimal solutions and optimal controls for $\lambda = 0.3$: (**upper row**) $l_*(t)$, $k_*(t)$, $m_*(t)$; (**lower row**) $u_*(t)$, $v_*(t)$; $J_* = 10.48323$.

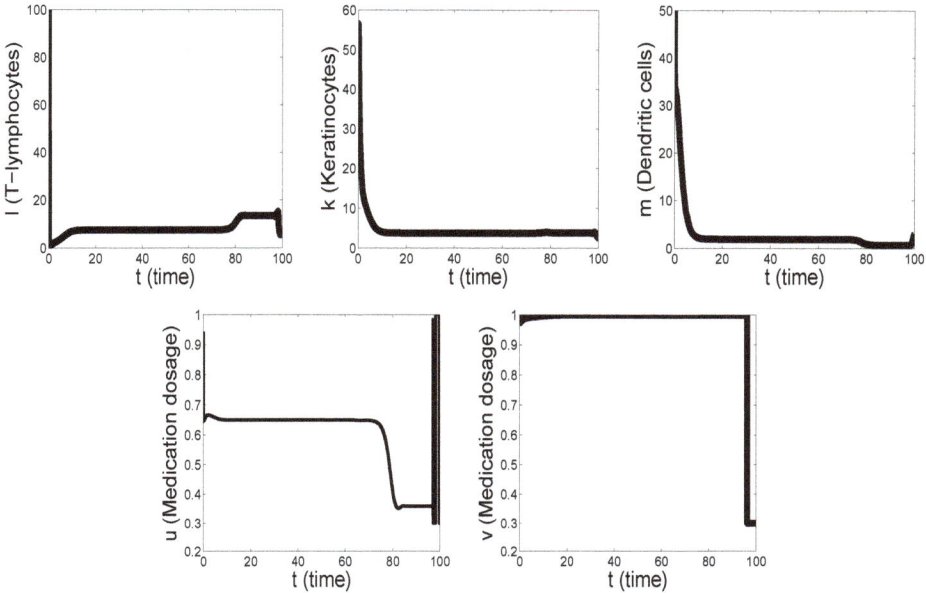

Figure 2. Optimal solutions and optimal controls for $\lambda = 1.2$: (**upper row**) $l_*(t)$, $k_*(t)$, $m_*(t)$; (**lower row**) $u_*(t)$, $v_*(t)$; $J_* = 2.643956$.

It is important to note that the controls $u(t)$ and $v(t)$ are auxiliary. They are introduced into System (1) to simplify analytical analysis. The corresponding actual controls $\tilde{u}(t)$ and $\tilde{v}(t)$ in the same system are related to the controls $u(t)$ and $v(t)$ by the formulas:

$$\tilde{u}(t) = 1 - u(t), \quad \tilde{v}(t) = 1 - v(t). \tag{105}$$

Therefore, where the auxiliary optimal control $u_*(t)$ has a maximum value of 1, the appropriate actual optimal control $\tilde{u}_*(t)$ takes a minimum value of 0, and vice versa. Similar remark is also valid for the auxiliary optimal control $v_*(t)$ and the appropriate actual optimal control $\tilde{v}_*(t)$. Moreover, the controls $\tilde{u}_*(t)$ and $\tilde{v}_*(t)$ are the optimal strategies of psoriasis treatment.

The conducted numerical calculations show the performance of software "BOCOP–2.0.5" in the study of such a complex (from computational point of view) phenomenon as chattering of the optimal control $u_*(t)$. We note that the corresponding actual optimal control $\tilde{u}_*(t)$ also has such behavior. This type of psoriasis treatment does not make sense. However, there is no reason for concern because there are approaches for chattering approximation presented, for example, in [21,32–34].

Taking into account Formula (105) related to the actual optimal controls $\tilde{u}_*(t)$ and $\tilde{v}_*(t)$ and the optimal controls $u_*(t)$ and $v_*(t)$ corresponding to them, and after analyzing the graphs of these controls (see Figures 1 and 2), we conclude that the medication intake schedule during 100 days of psoriasis treatment is as follows.

At small values of λ (for example, $\lambda = 0.3$), for most of the entire treatment period (95 days), a drug, which suppresses the interaction between T-lymphocytes and keratinocytes must be taken at the maximum dosage. In this case, a drug suppressing the interaction between T-lymphocytes and dendritic cells, must be taken at the minimum dosage. Then, on the 96th day, the medication schedule changes to the opposite one and for the next three days it looks like this: a medication suppressing the interaction between T-lymphocytes and keratinocytes is taken at the minimum dosage and a drug that suppresses the interaction between T-lymphocytes and dendritic cells is, on the contrary, taken at the maximum dosage. Finally, on the 99th day, the schedule of taking a drug, which suppresses the interaction between T-lymphocytes and keratinocytes, is again reversed. It is taken at the maximum dosage until the end of the treatment period. The schedule of taking a drug, which suppresses the interaction between T-lymphocytes and dendritic cells, does not change. It is still taken at the maximum dosage.

With the increase of the value of λ (for example, $\lambda = 1.2$), the actual optimal control $\tilde{u}_*(t)$ responsible for taking a drug that suppresses the interaction between T-lymphocytes and keratinocytes during most of the entire treatment period (95 days) has an interval corresponding to a smooth increase in the dosage of the used medication (singular arc). At the beginning and at the end of such an interval of psoriasis treatment there are the periods with increasing number of switchings from the lower intensity to the greatest intensity and vice versa (chattering). In addition, the whole process of this treatment ends with the interval of the greatest intensity (maximum dosage of the medication intake). Moreover, the schedule of taking a drug that suppresses the interaction between T-lymphocytes and dendritic cells does not qualitatively change. Almost during the entire period of psoriasis treatment (98 days), it is taken at the minimum dosage. Then, on the 99th day of the treatment period, the schedule for taking such a drug changes to the opposite. During the remaining two days it is taken at the maximum dosage.

Next, it was found that with the decrease or increase of the length of psoriasis treatment (for example, 10 days or 190 days), the above schedule of the drugs intake does not qualitatively change. It should be noted only that with the reduction in the duration of this period, the first change in the schedule of taking medications begins to occur earlier (see Figures 3 and 4 for 10 days), and with an increase in the duration of such a period, on the contrary, later (for example, 190 days).

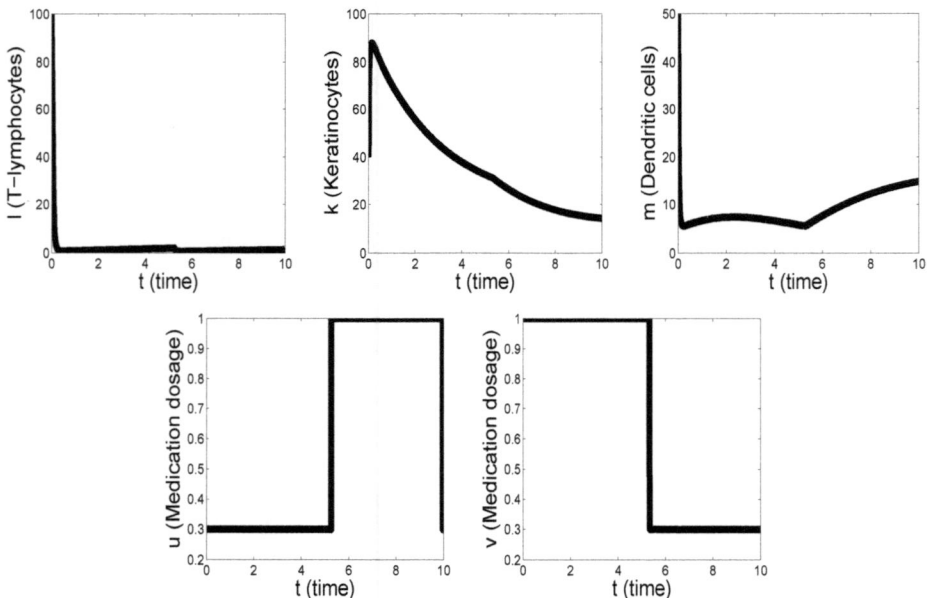

Figure 3. Optimal solutions and optimal controls for $\lambda = 0.3$: (**upper row**) $l_*(t)$, $k_*(t)$, $m_*(t)$; (**lower row**) $u_*(t)$, $v_*(t)$; $J_* = 14.09275$.

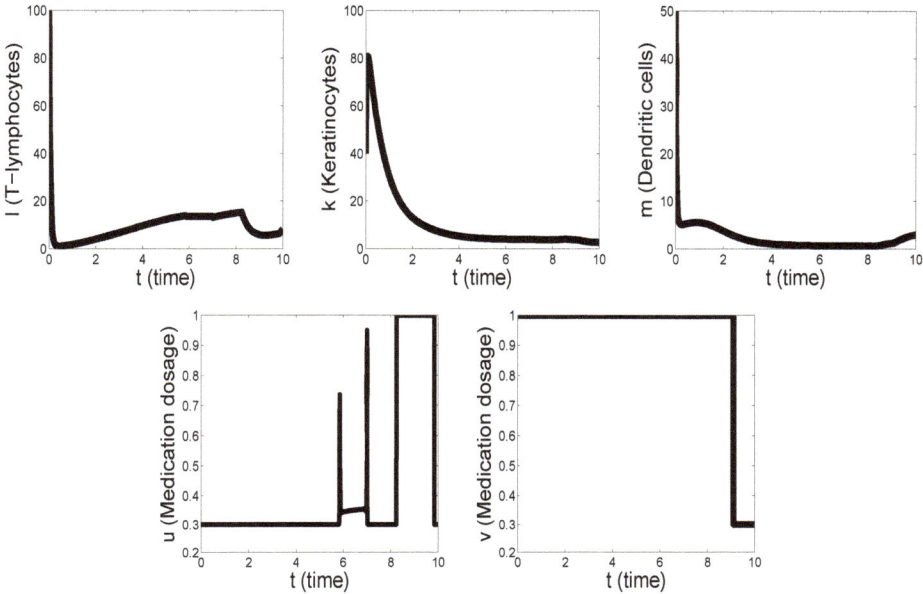

Figure 4. Optimal solutions and optimal controls for $\lambda = 1.2$: (**upper row**) $l_*(t)$, $k_*(t)$, $m_*(t)$; (**lower row**) $u_*(t)$, $v_*(t)$; $J_* = 2.644754$.

From the comparison of the schedule of the medication intake for 100 days psoriasis treatment, one can conclude that a drug that suppresses the interaction between T-lymphocytes and keratinocytes predominates over a drug that suppresses the interaction between T-lymphocytes and dendritic cells. To show that this predominance is not absolute in psoriasis treatment, we (along with the values of the parameters of System (1), the initial values (2) and the control constraints (3) presented in (104), use the following values taken from [13]:

$$
\begin{array}{llll}
\sigma = 9.0 & \rho = 14.0 & \beta = 0.065 & \delta = 0.01 \\
\lambda = 0.4 & \mu = 0.07 & \nu = 0.02 & T = 100.0 \\
\gamma_1 = 0.0032 & \gamma_2 = 0.0002 & u_{\min} = 0.3 & v_{\min} = 0.3 \\
l_0 = 100.0 & k_0 = 40.0 & m_0 = 50.0 &
\end{array}
$$

The corresponding results of the numerical calculations are given in Figure 5, which include the graphs of the optimal controls $u_*(t)$, $v_*(t)$, the corresponding optimal solutions $l_*(t)$, $k_*(t)$, and $m_*(t)$, and the minimum value of the functional $J(u,v)$ of (5). From the analysis of the graphs of controls $u_*(t)$, $v_*(t)$ and Formula (105), we conclude that for the majority of 100 days psoriasis treatment we use the maximum dosage of a drug suppressing the interaction between T-lymphocytes and dendritic cells. A drug that suppresses the interaction between T-lymphocytes and keratinocytes, should be used at the minimum dosage for 99 days and only on the last day the schedule of taking this medication should be changed to the maximum dosage. This is the first conclusion we draw from Figure 5. The second conclusion is that unlike the results shown in Figures 1 and 3, the first change in the schedule of taking drugs does not occur simultaneously in both, but only in one, which suppresses the interaction between T-lymphocytes and dendritic cells.

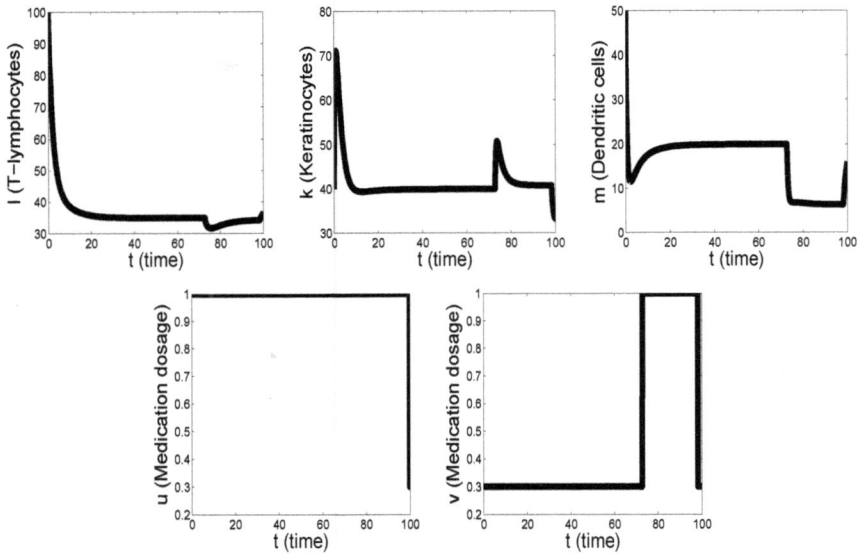

Figure 5. Optimal solutions and optimal controls for $T = 100.0$: (**upper row**) $l_*(t)$, $k_*(t)$, $m_*(t)$; (**lower row**) $u_*(t)$, $v_*(t)$; $J_* = 33.185501$.

Figure 6 shows that with increasing duration of the period of psoriasis treatment (for example, to 150 days), the actual optimal control $\tilde{v}_*(t)$ responsible for taking a drug that suppresses the interaction between T-lymphocytes and dendritic cells has a period with a smooth increase in the dosage of the used medication (singular arc).

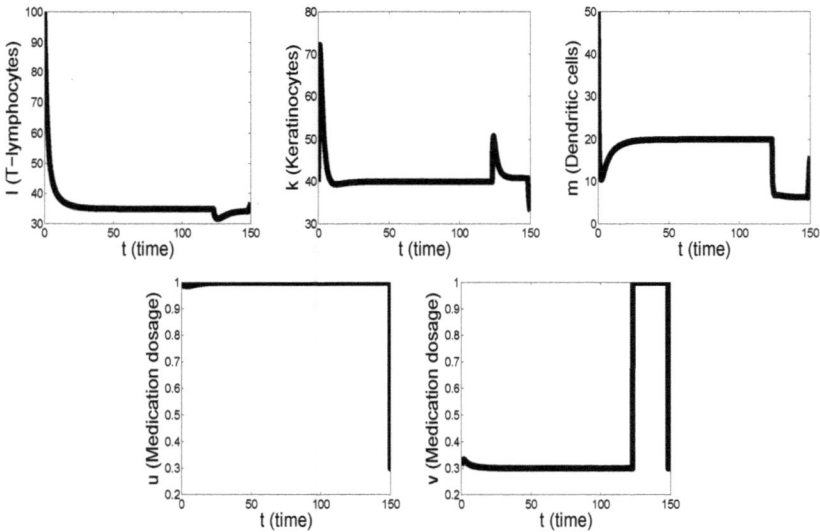

Figure 6. Optimal solutions and optimal controls for $T = 150.0$: (**upper row**) $l_*(t)$, $k_*(t)$, $m_*(t)$; (**lower row**) $u_*(t)$, $v_*(t)$; $J_* = 33.185511$.

Finally, the graphs of the optimal solution $k_*(t)$ from Figures 1–6 show that the optimal concentration of keratinocytes $k_*(t)$ reaches at the end T of the interval $[0, T]$ the level, which is the minimal for the entire period $[0, T]$ of psoriasis treatment. This fact is very important for the treatment of the disease.

8. Conclusions

On a given time interval, a nonlinear system of three differential equations describing psoriasis treatment was considered. It established the relationships between the concentrations of T-lymphocytes, keratinocytes, and dendritic cells, which were its phase variables. Also, two scalar bounded controls were introduced into the system that reflected the doses of drugs aimed at suppressing interactions between T-lymphocytes and keratinocytes, as well as between T-lymphocytes and dendritic cells. The problem of minimizing the concentration of keratinocytes at the final moment of the time interval was stated for such a system. To analyze the optimal solution of the minimization problem consisting of the optimal controls and the corresponding optimal solutions of the original system, the Pontryagin maximum principle was applied. It allowed to obtain the Cauchy problem for the switching functions describing the behavior of the optimal controls. An analysis of this Cauchy problem showed that such controls could be either bang-bang functions, or, in addition to bang-bang intervals, they could contain singular arcs. Next, the possibility of simultaneous existence of singular arcs of the optimal controls was studied. The orders of singular arcs of such controls were found, the corresponding necessary optimality conditions for them were checked. Also, the forms of concatenations of singular arcs with nonsingular intervals (where the optimal controls are bang-bang functions) were investigated. After that, the finding of specific optimal solution to the considered minimization problem was carried out numerically using "BOCOP–2.0.5". The corresponding results of calculations and their discussion were provided.

Author Contributions: All authors contributed equally to this work. All authors read and approve the final manuscript.

Acknowledgments: Evgenii Khailov was funded by Russian Foundation for Basic Research according to the research project 18-51-45003 IND_a.

Conflicts of Interest: The authors declare no conflict of interest.

References

1. Parisi, R.; Symmons, D.P.; Griffiths, C.E.; Ashcroft, D.M. Global epidemiology of psoriasis: A systematic review of incidence and prevalence. *J. Investig. Dermatol.* **2013**, *133*, 377–385. [CrossRef] [PubMed]
2. Garshick, M.K.; Kimball, A.B. Psoriasis and the life cycle of persistent life effects. *Dermatol. Clin.* **2015**, *33*, 25–39. [CrossRef] [PubMed]
3. Ryan, C.; Kirby, B. Psoriasis is a systemic disease with multiple cardiovascular and metabolic comorbidities. *Dermatol. Clin.* **2015**, *33*, 41–55. [CrossRef] [PubMed]
4. Eder, L.; Chandran, V.; Gladman, D.D. What have we learned about genetic susceptibility in psoriasis and psoriatic arthritis? *Curr. Opin. Rheumatol.* **2015**, *27*, 91–98. [CrossRef] [PubMed]
5. Menter, M.A.; Griffiths, C.E. Psoriasis: The future. *Drematol. Clin.* **2015**, *33*, 161–166. [CrossRef] [PubMed]
6. Oza, H.B.; Pandey, R.; Roper, D.; Al-Nuaimi, Y.; Spurgeon, S.K.; Goodfellow, M. Modelling and finite-time stability analysis of psoriasis pathogenesis. *Int. J. Control* **2017**, *90*, 1664–1677. [CrossRef]
7. Savill, N.J. Mathematical models of hierarchically structured cell populations under equilibrium with application to the epidermis. *Cell Prolif.* **2003**, *36*, 1–26. [CrossRef] [PubMed]
8. Niels, G.; Karsten, N. Simulating psoriasis by altering transit amplifying cells. *Bioinformatics* **2007**, *23*, 1309–1312.
9. Laptev, M.V.; Nikulin, N.K. Numerical modeling of mutial synchronization of auto-oscillations of epidermal proliferative activity in lesions of psoriasis skin. *Biophysics* **2009**, *54*, 519–524. [CrossRef]

10. Valeyev, N.V.; Hundhausen, C.; Umezawa, Y.; Kotov, N.V.; Williams, G.; Clop, A.; Ainali, C.; Ouzounis, G.; Tsoka, S.; Nestle, F.O. A systems model for immune cell interactions unravels the mechanism of inflammation in human skin. *PLoS Comput. Biol.* **2010**, *6*, e1001024. [CrossRef] [PubMed]
11. Gandolfi, A.; Iannelli, M.; Marinoschi, G. An agestructured model of epidermis growth. *J. Math. Biol.* **2011**, *62*, 111–141. [CrossRef] [PubMed]
12. Chattopadhyay, B.; Hui, N. Immunopathogenesis in psoriasis throuth a density-type mathematical model. *WSEAS Trans. Math.* **2012**, *11*, 440–450.
13. Roy, P.K.; Datta, A. Negative feedback control may regulate cytokines effect during growth of keratinocytes in the chronic plaque of psoriasis: A mathematical study. *Int. J. Appl. Math.* **2012**, *25*, 233–254.
14. Zhang, H.; Hou, W.; Henrot, L.; Schnebert, S.; Dumas, M.; Heusèle, C.; Yang, J. Modelling epidermis homoeostasis and psoriasis pathogenesis. *J. R. Soc. Interface* **2015**, *12*, 1–22. [CrossRef] [PubMed]
15. Cao, X.; Datta, A.; Al Basir, F.; Roy, P.K. Fractional-order model of the disease psoriasis: A control based mathematical approach. *J. Syst. Sci. Complex.* **2016**, *29*, 1565–1584. [CrossRef]
16. Datta, A.; Roy, P.K. T-cell proliferation on immunopathogenic mechanism of psoriasis: A control based theoretical approach. *Control Cybern.* **2013**, *42*, 365–386.
17. Roy, P.K.; Datta, A. Impact of cytokine release in psoriasis: a control based mathematical approach. *J. Nonlinear Evol. Equ. Appl.* **2013**, *2013*, 23–42.
18. Ledzewicz, U.; Schättler, H. On optimal singular controls for a general SIR-model with vaccination and treatment. *Discret. Contin. Dyn. Syst.* **2011**, *2*, 981–990.
19. Joshi, H.R.; Lenhart, S.; Hota, S.; Agusto, F. Optimal control of an SIR model with changing behavior through an educational campaign. *Electr. J. Differ. Equ.* **2015**, *2015*, 1–14.
20. De Pinho, M.R.; Kornienko, I.; Maurer, H. Optimal control of a SEIR model with mixed constraints and L_1 cost. In *Controllo'2014—Proceedings of the 11th Portuguese Conference on Automatic Control*; Lecture Notes in Electrical Engineering; Springer: Cham, Switzerland, 2015; Volume 321, pp. 135–145.
21. Schättler, H.; Ledzewicz, U. *Optimal Control for Mathematical Models of Cancer Therapies. An Application of Geometric Methods*; Springer: New York, NY, USA; Heidelberg, Germany; Dordrecht, The Netherlands; London, UK, 2015.
22. Schättler, H.; Ledzewicz, U. *Geometric Optimal Control: Theory, Methods and Examples*; Springer: New York, NY, USA; Heidelberg, Germany; Dordrecht, The Netherlands; London, UK, 2012.
23. Roy, P.K.; Bhadra, J.; Chattopadhyay, B. Mathematical modeling on immunopathogenesis in chronic plaque of psoriasis: A theoretical study. In Proceedings of the World Congress on Engineering, London, UK, 30 June–2 July 2010; Volume 1, pp. 550–555.
24. Datta, A.; Kesh, D.K.; Roy, P.K. Effect of $CD4^+$ T-cells and $CD8^+$ T-cells on psoriasis: A mathematical study. *Imhotep Math. Proc.* **2016**, *3*, 1–11.
25. Grigorieva, E.; Khailov, E.; Deignan, P. Optimal treatment strategies for control model of psoriasis. In Proceedings of the SIAM Conference on Control and its Applications (CT17), Pittsburgh, PA, USA, 10–12 July 2017; pp. 86–93.
26. Datta, A.; Li, X.-Z.; Roy, P.K. Drug therapy between T-cells and DCs reduces the excess production of keratinicytes: ausal effect of psoriasis. *Math. Sci. Int. Res. J.* **2014**, *3*, 452–456.
27. Lee, E.B.; Marcus, L. *Foundations of Optimal Control Theory*; John Wiley & Sons: New York, NY, USA, 1967.
28. Pontryagin, L.S.; Boltyanskii, V.G.; Gamkrelidze, R.V.; Mishchenko, E.F. *Mathematical Theory of Optimal Processes*; John Wiley & Sons: New York, NY, USA, 1962.
29. Zelikin, M.I.; Borisov, V.F. *Theory of Chattering Control: with Applications to Astronautics, Robotics, Economics and Engineering*; Birkhäuser: Boston, MA, USA, 1994.
30. Zorich, V.A. *Mathematical Analysis I*; Springer: Berlin/Heidelberg, Germany, 2004.
31. Bonnans, F.; Martinon, P.; Giorgi, D.; Grélard, V.; Maindrault, S.; Tissot, O.; Liu, J. BOCOP 2.0.5—User Guide. 8 February 2017. Available online: http://bocop.org (accessed on 3 September 018).
32. Zelikin, M.I.; Zelikina, L.F. The deviation of a functional from its optimal value under chattering decreases exponentially as the number of switchings grows. *Differ. Equ.* **1999**, *35*, 1489–1493.

33. Zelikin. M.I.; Zelikina, L.F. The asymptotics of the deviation of a functional from its optimal value when chattering is replaced by a suboptimal regime. *Russ. Math. Surv.* **1999**, *54*, 662–664. [CrossRef]
34. Zhu, J.; Trélat, E.; Cerf, M. Planar titling maneuver of a spacecraft: Singular arcs in the minimum time problem and chattering. *Discrete Contin. Dyn. Syst. B* **2016**, *21*, 1347–1388. [CrossRef]

*Mathematical
and Computational
Applications*

MDPI

Article

Optimal Control Analysis of a Mathematical Model for Breast Cancer

Segun Isaac Oke * , Maba Boniface Matadi and Sibusiso Southwell Xulu

Department of Mathematical Sciences, University of Zululand, Private Bag X1001, KwaDlangezwa 3886,
South Africa; matadim@unizulu.ac.za (M.B.M.); xuluss@unizulu.ac.za (S.S.X.)
* Correspondence: segunoke2016@gmail.com

Received: 26 March 2018; Accepted: 20 April 2018; Published: 24 April 2018

Abstract: In this paper, a mathematical model of breast cancer governed by a system of ordinary differential equations in the presence of chemotherapy treatment and ketogenic diet is discussed. Several comprehensive mathematical analyses were carried out using a variety of analytical methods to study the stability of the breast cancer model. Also, sufficient conditions on parameter values to ensure cancer persistence in the absence of anti-cancer drugs, ketogenic diet, and cancer emission when anti-cancer drugs, immune-booster, and ketogenic diet are included were established. Furthermore, optimal control theory is applied to discover the optimal drug adjustment as an input control of the system therapies in order to minimize the number of cancerous cells by considering different controlled combinations of administering the chemotherapy agent and ketogenic diet using the popular Pontryagin's maximum principle. Numerical simulations are presented to validate our theoretical results.

Keywords: breast cancer; optimal control; ketogenic diet; chemotherapy

1. Introduction

Cancer is a generic name that refers to a group of diseases in which normal cells divide uncontrollably, that is, grow more rapidly than normal cells, and may eventually spread to other parts of the body by a process called metastasis [1]. According to the National Cancer Registry [2], cancer kills more people than tuberculosis (TB), AIDs and malaria combined. Statistics show that cancer related deaths amounted to about 8.2 million in 2010. The mortality rate from cancer is projected to continue to rise, with an estimated 13 million deaths by 2030 [3]. The most common types of cancer include: breast cancer, prostate cancer, brain cancer, lung cancer and skin cancer among others.

According to the [3] report, breast cancer is the most common invasive cancer in females worldwide. The formation of breast cancer can occur in the inner lining of the milk ducts, known as ductal carcinoma, or in the lobules of the breast, known as lobular carcinoma [4]. Breast cancer is one of the most widely recognized obstructive diseases in females around the world. The disease has presently been named as the most dangerous cancer in women [3]. However, little is known on the causes of the ailment. There are three major breast cancer risk factors namely hormonal imbalance (estrogen), genetic (family history), and environmental (poor diet, alcohol consumption, smoking, exposure to toxin, etc.) [5]. Surgery, chemotherapy, radiation therapy, hormonal therapy, hyperthermia, targeted therapy and ketogenic diet [5,6] amongst other therapeutics are used to inhibit tumor growth or kill the tumor cells in the body. However, each treatment has side effects attributed to it, for example, hair loss, vomiting, nausea and fatigue. Adverse effects occur as a result of chemotherapy, which is not able to differentiate between normal cells and tumor cells, consequently killing both of them [3].

Several dietary components and supplements have been examined as possible cancer prevention agents. Until recently, a few studies, such as [6–8], investigated diet as a possible adjuvant to cancer treatment, which includes a ketogenic diet. A ketogenic diet consists of high edible fat with moderate or low protein content and very low carbohydrates, which forces the body to burn fat instead of glucose for adenosine triphosphate (ATP) synthesis [6,9].

It is well-known that a mathematical model is a capable device used to investigate the spread of non-infectious diseases and to provide important insights into disease behaviors and control [10,11]. Over the years, it has become an important tool in comprehending the dynamics of diseases and in decision making processes regarding a medical intervention program for controlling breast cancer in many nations [12]. For instance, [13] explored the role of mathematical modeling on the optimal delivery of a combination therapy for tumors and to improve on the delivery of anti-tumor drugs.

Old and recent studies such as [4,12,14–16] amongst others have shown that mathematical modeling is a widely used tool for resolving questions on public health. For instance, it was used during the time of Bernoulli (on modeling the dynamics of Smallpox) in 1760 [17]. Kermack and McKendrick [14,15] and some other recent studies by [4,12,13,18–21] show that mathematical modeling is useful in solving the problem of epidemiology. However, these studies reveal that much has not been done in terms of the mathematical modeling of a nutritional diet (ketogenic diet) as a control or therapy on tumor cells. Hence, we improved the model in [19] for this paper by incorporating time dependent control parameters (use of ketogenic diet, immune booster, and anti-cancer drugs) based on the assumption that there is an interaction between normal cells and tumor cells that is due to a mutation in DNA as a result of excess estrogen in the body system [4,19,22].

Furthermore, we analyzed and applied an optimal control to the improved model to determine the possible impacts of ketogenic-diet use and anti-cancer drugs as a treatment on tumor cells. We carried out a rigorous qualitative optimal control analysis of the resulting model and found the necessary conditions for optimal control of the disease using Pontryagin's maximum principle [11,23–25] in order to determine the optimal strategies for controlling the metastatic of the tumor cells.

This paper is organized as follows: In Section 2, four compartment models of ODEs to study the dynamics of breast cancer are developed. In Section 3, the existence of equilibria, their stabilities and basic reproductive numbers are discussed. In Section 4, an uncertainty and sensitivity analysis to check the most sensitive parameters in the model are discussed. In Section 5, an optimal control problem according to the model is proposed and an optimal solution is proffered. Numerical simulations are illustrated by implementing the forward and backward finite difference scheme in Section 6, while concluding remarks are provided in Section 7.

2. Model Formulation

Based on the existing model in [19], we developed a model by assuming the logistic (Verhulst) growth of a cell population and basic competition between normal cells and tumor cells. We considered the immune cells compartment to comprise Natural Killer cells (NK) and CD8+ T-cells as in [19] and we used a similar equation to model the immune response dynamic by introducing immune booster (ketone bodies) and anti-cancer drug efficacy.

We adapted an estrogen equation as presented in a model by [26]. Pinho and his co-workers in [26] considered that when a chemotherapy agent is continuously infused into the body and engulfed by different cell populations, natural death can occur. Excess estrogen was used in a similar way and assumed to be saturated daily through birth control (constant source rate) $(1 - k)$. This was introduced to serve as anti-cancer drug efficacy (e.g., Tamoxifen) in order to bind estrogen receptors positive and to reduce excess estrogen from promoting cancer growth [27].

In this study, a model that splits the entire population of cells of the human breast tissues at any given period of time $P(t)$ was reflected upon. Hence, normal cells compartment, represented by $N(t)$ in the form of epithelial cells that constitute the breast tissue is described. The cells are assumed to develop and die normally as they have unaltered DNA that control all cell activities. It was suggested that the normal cells and tumor cells compete for nutrients and other resources in a small volume, which is the competition model used by [28]. Normal cells are represented by

$$\frac{dN}{dt} = N\alpha_1 - \mu_1 N^2 - \phi_1 NT - (1-k)\lambda_1 NE. \tag{1}$$

The first term represents the logistic growth rate α_1 of the normal cells, which are breast tissues that are made-up of epithelial cells. The second term represents the natural death rate of normal cells. ϕ_1 represents the rate at which normal cells inhibit due to an alteration in DNA that is responsible for cancer cells having an uncontrolled cycle that normal cells do not have [19]. The final term describes the gene transactivation that can be a contributing growth factor responsible for the estrogen stimulation of breast cancer, which can result in damage of DNA. Thus, there will be a reduction in the population of normal cells $N(t) =$ being transformed into tumor cells by $\lambda_1 NE$ where λ_1 represents the tumor formation rate resulting from DNA mutation caused by the presence of excess estrogen [4]. However, $(1-k)$ represents the effectiveness of anti-cancer drugs (Tamoxifen).

The tumor cells compartment can be denoted by $T(t)$ in the form of an abnormal mass of tissue. Tumors are classic signs of inflammation, and can be benign or malignant (cancerous). Their names usually reflect the kind of tissue from where they arise, for example in breast or brain cancer, among others. There are about 51 breast cancer cell lines that mirror the 145 primary breast tumors [29]. These can be classified into two major branches: the Luminal, which has estrogen receptors ($ESR1 + ve$), and the Basal-like, which has no estrogen receptors ($ESR1 - ve$). A homogeneous luminal type of cancer cells in the form of MDAMB361, MCF-7, BT474, T47D and ZR75 of the cell lines [19] are then assumed to be

$$\frac{dT}{dt} = T\alpha_2 d - \mu_2 T^2 - \gamma_2 MT - \mu_5 T + (1-k)\lambda_1 NE. \tag{2}$$

The first term of the equation is a limited growth term for tumor cells that depends on the rate of parameter d (ketogenic diets). Although, if $d = 0$, tumor cells are automatically eradicated, but any DNA mutation that is caused by excess estrogen will repopulate the tumor cells again $\lambda_1 NE$. The induced death rate μ_5 is as a result of tumor starvation of nutrients, glucose and so on from the body system during the ketogenic diet, which alters nutrition. We assumed that γ_2 is the rate at which tumor cells are being removed due to the effectiveness of immune response.

The immune response compartment is represented by $M(t)$ in the form of natural killer (NK) cells and CD8+ T cells. Their growth may be stimulated by the presence of the tumor and they can destroy tumor cells through the kinetics process. We also assumed that the presence of a detectable tumor in a body system does not necessarily imply that the tumor has completely escaped active immunosurveillance. However, a tumor is immunogenic. It is possible that the immune response may not be sufficient on its own to completely combat the rapid growth of the tumor cells population and their eventual development into a tumor.

$$\frac{dM}{dt} = s\beta + \frac{\rho MT}{\omega + T} - \gamma_3 MT - \mu_3 M - \left((1-k)\frac{\lambda_3 ME}{g+E}\right) \tag{3}$$

The constant source parameter s denotes the source rate of immune response fully infused in the body daily. We introduced immune booster β (a supplement such as ketone bodies) to assist immune response whenever tumor cells overpower immune cells in order to activate the immune response and fight the cancer cells. The next term is a nonlinear growth term for immune response where ρ the rate of immune response is and ω is the immune cell threshold [12]. We denoted γ_3 as the rate at which immune response is inactivated upon interacting with tumor cells while μ_3 represents the immune cells natural death rate as a result of necrosis. The final term explains a limited rate at which estrogen suppresses immune cells activation where λ_3 is the rate of immune suppression and g is the estrogen threshold [19].

Finally, we considered estrogen compartment denoted by $E(t)$. Estrogen is a female steroid hormone that is produced by the ovaries in lesser amounts, and by the adrenal cortex, placenta and male testes. Estrogen helps to control and guide sexual development, including the physical changes associated with puberty [11,30]. However, an increase in estrogen levels can lead to the growth of the tumor cells. It also serves as a mitogen by triggering cell division in breast tissue [30]. Estrogen acts as a carcinogen by directly damaging DNA, forcing healthy epithelial cells to have a higher likelihood of malignant conversion [5,30].

$$\frac{dE}{dt} = (1-k)\epsilon - \mu_4 E \tag{4}$$

The process of constantly replenishing excess estrogen is denoted by ϵ. We assumed that the majority of cancer cells are estrogen-receptor positive and only a small proportion of epithelial cells are estrogen-receptor positive, which can only be blocked by the anti-cancer drug $(1-k)$ Tamoxifen. μ_4 is the rate at which estrogen is being washed out from the body system. Thus, system (5) is our modified model.

$$
\begin{aligned}
\frac{dN}{dt} &= N(\alpha_1 - \mu_1 N - \phi_1 T) - (1-k)(\lambda_1 NE) \\
\frac{dT}{dt} &= T(\alpha_2 d - \mu_2 T) - \gamma_2 MT - \mu_5 T + (1-k)(\lambda_1 NE) \\
\frac{dM}{dt} &= s\beta + \frac{\rho MT}{\omega + T} - \gamma_3 MT - \mu_3 M - \left((1-k)\frac{\lambda_3 ME}{g+E}\right) \\
\frac{dE}{dt} &= (1-k)\epsilon - \mu_4 E
\end{aligned}
\tag{5}
$$

3. Model Analysis

3.1. Boundedness and Positivity of Solutions

The system of Equation (5) has an initial condition by

$$N(0) = N_0 \geq 0, \, T(0) = T_0 \geq 0, \, M(0) = M_0 \geq 0, \text{ and } E(0) = E_0 \geq 0$$

since our model is to investigate cellular populations, therefore all the variables and parameters of the model are non-negative. Based on the biological finding, the system of Equation (5) will be studied in the following region such as:

$$\Delta = \left\{ (N, T, M, E) \in \Re_+^4 \right\}$$

The following theorem assures that the system of Equation (5) is well-posed such that solutions with non-negative initial conditions remain non-negative for all $0 < t < \infty$, and therefore makes the variable biologically meaningful. Hence, we have the following result:

Theorem 1. *The region $\Delta \subset \Re_+^4$ is positively invariant with respect to the system of Equation (5) and non-negative solution exists for all time $0 < t < \infty$.*

Proof: Let $\Delta = \Delta_c \subset \Re_+^4$ with $\Delta = \{(N, T, M, E) \in \Re_+^4 : N \leq \frac{\alpha_1}{\mu_1}\}$, then the solutions (N (t), T(t), M(t), E(t)) of system (5) are positive $\forall t \geq 0$. It is obvious from the first compartment of system (5) that

$$\frac{dN}{dt} \leq N(t)\alpha_1 - \mu_1 N^2(t).$$

Solving with Bernoulli method and taking $N(0) = N_0$, we have

$$N(t) \leq \frac{\alpha_1}{\mu_1 + k\alpha_1 e^{-\alpha_1 t}}$$

with

$$k = \frac{\alpha_1 - N_0 \mu_1}{N_0 \alpha_1}$$

$$N_0 = \frac{\alpha_1}{\mu_1 + k\alpha_1}.$$

Then,

$$N(t) \leq \frac{\alpha_1}{\mu_1 + \left(\frac{\alpha_1 - N_0 \mu_1}{N_0}\right)e^{-\alpha_1 t}}$$

$$N(t) \leq \frac{\alpha_1}{\mu_1} \text{ as } t \to \infty$$

hence, $N(t) > 0, \forall t > 0$ and if and only if $(1 - k) \geq 0$ [31].

Consequently, it can be shown that $T(t) > 0$, $M(t) > 0$, and $E(t) > 0 \, \forall t > 0$. This completes the proof. □

3.2. The Equilibrium Points of System (5)

The steady states occur by setting the left hand side (LHS) of system (5) to zero, i.e.,

$$\frac{dN}{dt} = \frac{dT}{dt} = \frac{dM}{dt} = \frac{dE}{dt} = 0$$

The model system admits six steady states in which there are four dead equilibria, one tumor-free equilibrium point and one co-existing equilibrium point $P = (N^*, T^*, M^*, E^*)$ where N^*, T^*, M^*, E^* represent the tumor-free equilibrium values for the normal cells, tumor cells, immune cells and estrogen hormone respectively. We have $N^* > 0$, $M^* > 0$, $E^* > 0$ since cell populations are non-negative and real. Therefore, all parameters $s, \beta, g, \mu_1, \mu_3, \mu_4, \epsilon, \lambda_3, k, \alpha_1$, and λ_1 are positive.

Tumor-Free equilibrium point

$$P_0 = \left(\frac{\alpha_1 \mu_4 - (1-k)^2 \lambda_1 \epsilon}{\mu_1 \mu_4}, 0, \frac{s\beta(g\mu_4 + (1-k)\epsilon)}{\mu_3(g\mu_4 + (1-k)\epsilon) + (1-k)^2 \lambda_3 \epsilon}, \frac{(1-k)\epsilon}{\mu_4}\right)$$

Type 1 Dead equilibrium point

$$P_{d1} = \left(0, 0, \frac{s\beta(g\mu_4 + (1-k)\epsilon)}{\mu_3(g\mu_4 + (1-k)\epsilon) + (1-k)^2 \lambda_3 \epsilon}, \frac{(1-k)\epsilon}{\mu_4}\right)$$

Type 2 Dead equilibrium point

$$P_{d2} = \left(0, \frac{d\alpha_2 - \gamma_2 m_1^* - \mu_5}{\mu_2}, m_1^*, \frac{(1-k)\epsilon}{\mu_4}\right)$$

Type 3 Dead equilibrium point

$$P_{d3} = \left(0, \frac{d\alpha_2 - \gamma_2 m_2^* - \mu_5}{\mu_2}, m_2^*, \frac{(1-k)\epsilon}{\mu_4}\right)$$

Type 4 Dead equilibrium point

$$P_{d4} = \left(0, \frac{d\alpha_2 - \gamma_2 m_3^* - \mu_5}{\mu_2}, m_3^*, \frac{(1-k)\epsilon}{\mu_4}\right)$$

Co-existing equilibrium point

$$P_e = (N_4^*, T_4^*, M_4^*, E_4^*)$$

3.3. The Reproductive Number and Tumor-Free Equilibrium Point

In this section, we mainly analyzed the stability behaviors of system (5) by means of eigenvalues. We apply Hartman–Grobman Theorem which states that in the neighborhood of a hyperbolic equilibrium point, a nonlinear dynamical system is topologically equivalent to its linearization [32].

Theorem 2. *The tumor-free equilibrium (TFE) point P_0 of system (5) is locally asymptotically stable if $R_0 < 1$, otherwise unstable.*

Proof. Linearizing system (5) around TFE P_0, we obtained the following Jacobian matrix $J(P_0)$

$$
J = \begin{pmatrix}
\frac{2\mu_1\lambda_1(1-k)^2\epsilon - \alpha_1\mu_1\mu_4 - (1-k)^2\lambda_1\mu_1\epsilon}{\mu_1\mu_4} & \frac{(1-k)^2\lambda_1\Phi_1\epsilon - \Phi_1\alpha_1\mu_4}{\mu_1\mu_4} & 0 & -B_6 \\
\frac{(1-k)^2\lambda_1\epsilon}{\mu_4} & B_3 & 0 & B_6 \\
0 & B_4 & -B_5 & -B_7 \\
0 & 0 & 0 & -\mu_4
\end{pmatrix}
$$

$$
J(P_0) = \begin{pmatrix}
B_0 & B_2 & 0 & -B_6 \\
B_1 & B_3 & 0 & B_6 \\
0 & B_4 & -B_5 & -B_7 \\
0 & 0 & 0 & -\mu_4
\end{pmatrix}
$$

$$
|J(P_0)| = \begin{vmatrix}
B_0 - \delta & B_2 & 0 & -B_6 \\
B_1 & B_3 - \delta & 0 & B_6 \\
0 & B_4 & -B_5 - \delta & -B_7 \\
0 & 0 & 0 & -\mu_4 - \delta
\end{vmatrix} = 0
$$

Then the characteristic equation at P_0 of the linearized system of the model (5) is given below. Obviously, there exists two negative characteristic roots

$$
\delta_1 = -\mu_4, \ \delta_2 = -B_5
$$

However, we only need to consider

$$
\delta^2 - (B_0 + B_3)\delta + B_0 B_3 - B_1 B_2 = 0
$$

$$
\delta^2 - (B_0 + B_3)\delta + B_0 B_3 \left(1 - \frac{B_1 B_2}{B_0 B_3}\right) = 0 \tag{6}
$$

from (6), we have basic reproduction number

$$
R_0 = \frac{B_1 B_2}{B_0 B_3} \tag{7}
$$

$$
\delta^2 - (B_0 + B_3)\delta + B_0 B_3 (1 - R_0) = 0,
$$

where

$$
B_0 = \frac{2\mu_1\lambda_1(1-k)^2\epsilon - \alpha_1\mu_1\mu_4 - (1-k)^2\lambda_1\mu_1\epsilon}{\mu_1\mu_4}, \ B_1 = \frac{(1-k)^2\lambda_1\epsilon}{\mu_4}, \ B_2 = \frac{(1-k)^2\lambda_1\Phi_1\epsilon - \Phi_1\alpha1\mu_4}{\mu_1\mu4}
$$

$$
B_3 = \frac{\alpha_2\mu_3 d\psi^* + \alpha_2\lambda_3 d\epsilon(1-k)^2 - \gamma_2 s\beta\psi^* - \mu_5\mu_3\psi^* - \mu_5\lambda_3\epsilon(1-k)^2}{\mu_3\psi^* + (1-k)^2\lambda_3\epsilon}
$$

where

$$\psi^* = (g\mu_4 + (1-k)\epsilon)$$

$$-B_4 = \frac{s\beta(g\mu_4 + (1-k)\epsilon(\rho - \gamma_3\omega)}{\omega\mu_3(g\mu_4 + (1-k)\epsilon)}$$

$$-B_5 = \frac{\mu_3\mu_4 g + \mu_3(1-k)\epsilon + \lambda_3(1-k)^2\epsilon}{g\mu_4 + (1-k)\epsilon}, -B_6 = -\left(\frac{(1-k)\lambda_1\alpha_1\mu_4 - (1-k)^3\lambda_1^2\epsilon}{\mu_1\mu_4}\right)$$

$$B_6 = \frac{(1-k)\lambda_1\alpha_1\mu_4 - (1-k)^3\lambda_1^2\epsilon}{\mu_1\mu_4}, -B_7 = -\left(\frac{\mu_4^2(1-k)\lambda_3 g s\beta(g\mu_4 + (1-k)\epsilon)}{(g\mu_4 + (1-k)\epsilon)^2\left[\mu_3(g\mu_4 + (1-k)\epsilon) + (1-k)^2\lambda_3\epsilon\right]}\right)$$

Here, we can apply the Routh-Hurwitz criterion namely,

$$\text{(i) } Tr(A) < 0 \quad \text{(ii) } Det(A) > 0$$

provided

$$a_0 = 1 > 0, a_1 = (B_0 + B_3) < 0, \ B_0 B_3(1 - R_0) > 0 \text{ if } R_0 < 1,$$
$$B_0 B_3 > B_1 B_2, \ B_1 > 0, \ B_2 > 0, \ B_3 < 0, \ B_0 < 0$$

Since the Routh–Hurwitz criterion holds, all the eigenvalues are negative, i.e., $\delta_3 < 0$ and $\delta_4 < 0$. Therefore, the TFE point of system (5) is locally asymptotically stable if (7) $R_0 < 1$ otherwise unstable.
□

The epidemiological implication of the above result is that the tumor cells that are governed by system (5) can be eliminated from the population (normal cells or breast tissues) whenever an influx by tumor cells into the normal cells is small, such that $R_0 < 1$. Therefore, the existence of a tumor-free equilibrium in this case depends on the estrogen level.

Theorem 3. *The Type 1 Dead equilibrium point P_{d1} of system (5) is locally asymptotically stable if*

$$\left(\frac{(1-k)^2\lambda_1\epsilon}{\alpha_1\mu_4}\right) > 1,$$

otherwise unstable.

Proof. Linearizing system (5) around the Type 1 Dead free equilibrium point P_{d1}, we obtained the following Jacobian matrix $J(P_{d1})$

$$J = \begin{pmatrix} \alpha_1 - (1-k)\lambda_1 E_0^* & 0 & 0 & 0 \\ (1-k)\lambda_1 E_0^* & d\alpha_2 - \gamma_2 M_0^* - \mu_5 & 0 & 0 \\ 0 & \frac{\rho\omega M_0^* - \gamma_3 M_0^*\omega^2}{\omega^2} & -\left(\frac{\mu_3(g+E_0^*) + (1-k)\lambda_3 E_0^*}{(g+E_0^*)}\right) & \frac{\lambda_3 g M_0^*(1-k)}{(g+E_0^*)^2} \\ 0 & 0 & 0 & -\mu_4 \end{pmatrix}$$

$$J(P_{d1}) = \begin{pmatrix} C_0 & 0 & 0 & 0 \\ C_1 & C_2 & 0 & 0 \\ 0 & C_3 & -C_4 & -C_5 \\ 0 & 0 & 0 & -\mu_4 \end{pmatrix}$$

$$|J(P_1)| = \begin{vmatrix} C_0 - \delta & 0 & 0 & 0 \\ C_1 & C_2 - \delta & 0 & 0 \\ 0 & C_3 & -C_4 - \delta & -C_5 \\ 0 & 0 & 0 & -\mu_4 - \delta \end{vmatrix} = 0$$

Clearly, two eigenvalues of the system (5) at P_{d1} are negative and real

$$\delta_1 = -\mu_4$$

and

$$\delta_2 = -C_4 \Rightarrow -\left(\frac{\mu_3(g\mu_4 + (1-k)\epsilon) + (1-k)^2\lambda_3\epsilon}{g\mu_4 + (1-k)\epsilon} \right)$$

while the remaining two eigenvalues are obtained from 2×2 matrix.

$$A = \begin{pmatrix} C_0 & 0 \\ C_1 & C_2 \end{pmatrix}$$

Applying the Routh-Hurwitz criterion stated above; we have

(i) $Tr(A) = C_0 + C_2 \Rightarrow \left(\frac{\alpha_1\mu_4 - (1-k)^2\lambda_1\epsilon}{\mu_4} + \frac{(\mu_3\alpha_2 d - \gamma_2 s\beta - \mu_5\mu_3)A^* + d\alpha_2\lambda_3(1-k)^2\epsilon - (1-k)^2\lambda_3\mu_5\epsilon}{\mu_3 A^* + (1-k)^2\lambda_3\epsilon} \right) < 0$

$\quad$ *if* $\alpha_1\left(1 - \frac{(1-k)^2\lambda_1\epsilon}{\alpha_1\mu_4} \right) > 0, \Rightarrow \left(\frac{(1-k)^2\lambda_1\epsilon}{\alpha_1\mu_4} \right) > 1$

Therefore, $Tr(A) < 0$

(ii) $Det(A) = C_0 C_2$

$$\left(\left(\frac{\alpha_1\mu_4 - (1-k)^2\lambda_1\epsilon}{\mu_4} \right) \left(\frac{(\mu_3\alpha_2 d - \gamma_2 s\beta - \mu_5\mu_3)A^* + d\alpha_2\lambda_3(1-k)^2\epsilon - (1-k)^2\lambda_3\mu_5\epsilon}{\mu_3 A^* + (1-k)^2\lambda_3\epsilon} \right) \right) > 0$$

$\quad$ *if* $\alpha_1\left(1 - \frac{(1-k)^2\lambda_1\epsilon}{\alpha_1\mu_4} \right) > 0$ *provided* $\left(\frac{(1-k)^2\lambda_1\epsilon}{\alpha_1\mu_4} \right) > 1$

and

$$\left\{ \left(\alpha_1\left(1 - \frac{(1-k)^2\lambda_1\epsilon}{\alpha_1\mu_4} \right) \right) \left(\left(\frac{(\mu_3\alpha_2 d - \gamma_2 s\beta - \mu_5\mu_3)A^* + d\alpha_2\lambda_3(1-k)^2\epsilon - (1-k)^2\lambda_3\mu_5\epsilon}{\mu_3 A^* + (1-k)^2\lambda_3\epsilon} \right) \right) \right\} > 0$$

implies that $Det(A) > 0$. Thus, the remaining eigenvalues δ_3 *and* δ_4 are negative and real since R-H criterion has been satisfied. Hence, the type 1 Dead equilibrium point P_{d1} of the system (5) is locally asymptotically stable if $\left(\frac{(1-k)^2\lambda_1\epsilon}{\alpha_1\mu_4} \right) > 1$. $\square$

Epidemiologically it is implied that the net growth of the tumor cells must be more than the immune cells values in order to have the tumor cells overpower the normal cells as the reactivation of the immune cells is due to the estrogen effects that are greater than the reactivation of the immune cells due to the tumor effect. However, ketogenic diet is inactive at the type 1 Dead equilibrium point.

Theorem 4. *The Type 2 Dead equilibrium point P_{d2} of system (5) is locally asymptotically stable if*

$$\left(\frac{(1-k)^2\lambda_1\epsilon}{\alpha_1\mu_4} \right) > 1$$

$$\omega > \frac{A^*}{\mu_2}\left(\frac{\mu_2 C^* \rho}{\gamma_3 A^* C^* + \mu_2\mu_3 C^* + (1-k)^2 \lambda_3 \epsilon} - 1\right)$$

otherwise unstable.

Proof. We linearized system (5) around the Type 2 Dead free equilibrium point P_{d2}, we obtained the following Jacobian matrix $J(P_{d2})$ at $P_{d2} = \left(0, \frac{d\alpha_2 - \gamma_2 m_1^* - \mu_5}{\mu_2}, m_1^*, \frac{(1-k)\epsilon}{\mu_4}\right)$

$$J = \begin{pmatrix} \alpha_1 - (1-k)\lambda_1 E_1^* & 0 & 0 & 0 \\ (1-k)\lambda_1 E_1^* & Q_2 & -\gamma_2 T_1^* & 0 \\ 0 & Q_3 & Q_5 & \frac{\lambda_3 g M_1^*(1-k)}{(g+E_1^*)^2} \\ 0 & 0 & 0 & -\mu_4 \end{pmatrix}$$

where

$$Q_2 = (d\alpha_2 - 2\mu_2 T_1^* - \gamma_2 M_1^* - \mu_5), Q_3 = \left(\frac{\rho\omega\mu_2^2 M_1^* - \gamma_3 M_1^*(\omega\mu_2 + d\alpha_2 - \gamma_2 M_1^* - \mu_5)^2}{(\omega\mu_2 + d\alpha_2 - \gamma_2 M_1^* - \mu_5)^2}\right), Q_5 = \left(\frac{\rho T_1^*}{\omega + T_1^*} - \gamma_3 T_1^* - \mu_3 - \frac{(1-k)\lambda_3 E_1^*}{g+E_1^*}\right)$$

$$|J(P_2)| = \begin{pmatrix} Q_0 & 0 & 0 & 0 \\ Q_1 & Q_2 & -Q_4 & 0 \\ 0 & Q_3 & Q_5 & Q_6 \\ 0 & 0 & 0 & -\mu_4 \end{pmatrix}$$

$$|J(P_2)| = \begin{vmatrix} Q_0 - \delta & 0 & 0 & 0 \\ Q_1 & Q_2 - \delta & -Q_4 & 0 \\ 0 & Q_3 & Q_5 - \delta & Q_6 \\ 0 & 0 & 0 & -\mu_4 - \delta \end{vmatrix} = 0$$

Clearly, one of the eigenvalues of the system (5) at $|J(P_2)|$ is negative and real, i.e., $\delta_1 = -\mu_4$. However, the remaining can be analyzed by simple calculation.

$$(Q_0 - \delta)(Q_2 - \delta)(Q_5 - \delta) = 0$$
$$\Rightarrow Q_5 = \delta_2, \; Q_2 = \delta_3, \; Q_0 = \delta_4$$

where

$$Q_0 = \frac{\alpha_1\mu_4 - (1-k)^2\lambda_1\epsilon}{\mu_4}, Q_2 = \gamma_2 M_1^* - \alpha_2 d + \mu_5$$

$$Q_5 = \frac{(A^*\rho\mu_2^2 - A^* C^* \gamma_3)(d\alpha_2 - \gamma_2 M_1^* - \mu_5) - C^*\mu_2(\mu_3 A^* + (1-k)^2\lambda_3\epsilon)}{\mu_2 A^* C^*}$$

where $A^* = (g\mu_4 - (1-k)\epsilon)$ and $C^* = (\omega\mu_2 + d\alpha_2 - \gamma_2 M_1^* - \mu_5)$.

It follows the following conditions

(i) $Q_0 < 0$ if, $\alpha_1 \leq 1, 0 \leq k < 1$ and $\left(\frac{(1-k)^2\lambda_1\epsilon}{\alpha_1\mu_4}\right) > 1$;

(ii) $Q_5 < 0$ provided $A^* > 0, 0 \leq k \leq 1$ and $\omega > \frac{A^*}{\mu_2}\left(\frac{\mu_2 C^* \rho}{\gamma_3 A^* C^* + \mu_2\mu_3 C^* + (1-k)^2\lambda_3\epsilon} - 1\right)$. $\square$

3.4. Co-Existing Equilibrium Points

Theorem 5. *The co-existing equilibrium point P_e of system (5) is stable if the following Routh–Hurwitz criterion is satisfied,*

$$Trace(A) = (V_0 + V_3 + V_6 - \mu_4) < 0$$
$$Det(A) = (-\mu_4(V_0 V_6 V_3 + V_0 V_4 V_5 + V_1 V_2 V_6)) > 0,$$

otherwise unstable.

Proof. We analyzed and linearized system (5) around the co-existing equilibrium point P_e, we obtained the following Jacobian matrix $J(P_e)$ at $P_e = (N_4^*, T_4^*, M_4^*, E_4^*)$ where N_4^*, T_4^*, M_4^* & E_4^* represent the coexisting equilibrium values for normal cells, tumor cells, immune cells, and estrogen levels respectively.

A co-existing equilibrium state exists when all cells populations would have survived the competition.

$$N_4^* = \frac{2(1-k)^4\lambda_1^4\mu_1\mu_4\epsilon^2 + \phi_1\alpha_1^2\mu_4^2\mu_1 - 2(1-k)^2\mu_1\mu_4^2\alpha_1\lambda_1\phi_1\epsilon - 2\alpha_1\phi_1^2\mu_1\mu_4^3 - 2(1-k)^2\alpha_1\mu_1\mu_4^2\lambda_1\epsilon}{2\phi_1\alpha_1\mu_1^2\mu_4^3 - 2(1-k)^2\mu_1^2\mu_4^2\lambda_1\phi_1\epsilon}$$

$$T_4^* = \frac{\alpha_1^2\mu_1\mu_4^2 + 2\alpha_1\mu_1\mu_4^2\phi_1}{2\phi_1\alpha_1\mu_1\mu_4^2 - 2(1-k)^2\mu_1\mu_4\lambda_1\phi_1\epsilon}$$

$$M_4^* = \frac{G^{*2}Z^*(1-k)^2\lambda_1\epsilon + (\alpha_1^2\alpha_2\mu_1\mu_4^3 d + 2\alpha_1\alpha_2\mu_1\mu_4^3\phi_1 d - \mu_4^3\mu_5\alpha_1^2 - 2\mu_4^3\mu_1\mu_5\alpha_1\phi_1)G^* - \mu_3\alpha_1^4\mu_1^2\mu_4^5 - 4\alpha_1^2\mu_1^2\mu_2\mu_4^5\phi_1 - 4\phi_1^2\alpha_1^2\mu_1^2\mu_4^5\mu_2}{G^{*2}Q^*\mu_4}$$

$$E_4^* = \frac{(1-k)\epsilon}{\mu_4}$$

where

$$G^{*2} = 2\phi_1\alpha_1\mu_1\mu_4^2 - 2(1-k)^2\mu_1\mu_4\lambda_1\phi_1\epsilon$$

$$Z^* = \frac{2(1-k)^4\lambda_1^2\mu_1\mu_4\epsilon^2 + \phi_1\alpha_1^2\mu_4^2\mu_1 - 2(1-k)^2\mu_1\mu_4^2\alpha_1\lambda_1\phi_1\epsilon - 2\alpha_1\phi_1^2\mu_1\mu_4^3 - 2(1-k)^2\alpha_1\mu_1\mu_4^2\lambda_1\epsilon}{2\phi_1\alpha_1\mu_1^2\mu_4^3 - 2(1-k)^2\mu_1^2\mu_4^2\lambda_1\phi_1\epsilon}$$

$$Q^* = \frac{\alpha_1^2\mu_1\mu_4^2\gamma_2 - 2\alpha_1\mu_1\mu_4^2\phi_1\gamma_2}{2\phi_1\alpha 1\mu_1\mu_4^2 - 2(1-k)^2\mu_1\mu_4\lambda_1\phi_1\epsilon}$$

$$J = \begin{pmatrix} (\alpha_1 - 2\mu_1 N_4^* - (1-k)\lambda_1 E_4^*) & -N_4^*\phi & 0 & -V_7 \\ (1-k)\lambda_1 E_4^* & (d\alpha_2 - 2\mu_2 T_4^* - \gamma_2 M_4^* - \mu_5) & -\gamma_2 T_4^* & V_7 \\ 0 & V_4 & V_6 & \frac{\lambda_3 g M_4^*(1-k)}{(g+E_4^*)^2} \\ 0 & 0 & 0 & -\mu_4 \end{pmatrix}$$

$$A = \begin{pmatrix} V_0 & -V_2 & 0 & -V_7 \\ V_1 & V_3 & -V_5 & V_7 \\ 0 & V_4 & V_6 & V_8 \\ 0 & 0 & 0 & -\mu_4 \end{pmatrix}$$

$$|A| = \begin{vmatrix} V_0 & -V_2 & 0 & -V_7 \\ V_1 & V_3 & -V_5 & V_7 \\ 0 & V_4 & V_6 & V_8 \\ 0 & 0 & 0 & -\mu_4 \end{vmatrix} = 0$$

We need to show that $Trace(A) < 0$, that is

$$Tr(A) = (V_0 + V_3 + V_6 - \mu_4) < 0$$

$$= \alpha_1(1 - A_0) - 2\mu_1 N_4^* + d\alpha_2(1 - \mu_5) - \mu_4 + \frac{T_4^*(\rho - \gamma_3(\omega - T_4^*))}{\omega + T_4^*} - \mu_3 - \frac{(1-k)^4\lambda_3\epsilon}{(g\mu_4 + (1-k)\epsilon}$$

Thus,

$$Tr(A) < 0, \text{ if } A_0 > 1, \ \mu_5 > 1, \ \rho < \gamma_3(\omega + T_4^*) \text{ with } A_0 = \frac{(1-k)^2\lambda_3\epsilon}{\alpha_1\mu_4}$$

To show that,

$$|A| = (-\mu_4(V_0 V_3 V_6 + V_0 V_4 V_5 + V_1 V_2 V_6)) > 0$$

Let $\zeta_1 = -\mu_4 V_0 V_3 V_6$, $\zeta_2 = -\mu_4 V_0 V_4 V_5$, $\zeta_3 = -\mu_4 V_1 V_2 V_6$

$$\zeta_1 = \left(\alpha_1(1-A_0) - 2\mu_1 N_4^*\right)\left(d\alpha_2(1-\mu_5) - 2\mu_2 T_4^* - \gamma_2 \mu_4^*\right)\left(\frac{T_4^*(\rho - \gamma_3(\omega - T_4^*))}{\omega + T_4^*} - \mu_3 - \frac{(1-k)^4 \lambda_3 \epsilon}{(g\mu_4 + (1-k)\epsilon)}\right).$$

This implies that, $\zeta_1 > 0$ is a positive, if $A_0 > 1$, $\mu_5 > 1$, $\rho < \gamma_3(\omega + T_4^*)$ with $A_0 = \frac{(1-k)^2 \lambda_3 \epsilon}{\alpha_1 \mu_4}$

$$\zeta_2 = \left(\alpha_1(1-A_0) - 2\mu_1 N_4^*\right)\left(\frac{M_4^*}{(\omega + T_4^*)^2}(\rho\omega - \gamma_3(\omega + T_4^*))\right)(-\gamma_2 T_4^*)$$

This implies that, $\zeta_2 > 0$ is a positive, if $A_0 > 1$, $\mu_5 > 1$, $\rho\omega < \gamma_3(\omega + T_4^*)^2$, with $A_0 = \frac{(1-k)^2 \lambda_3 \epsilon}{\alpha_1 \mu_4}$

$$\zeta_3 = -\mu_4(A_0)(-\phi_1 N_4^*)\left(\frac{T_4^*(\rho - \gamma_3(\omega - T_4^*))}{\omega + T_4^*} - \mu_3 - \frac{(1-k)^4 \lambda_3 \epsilon}{(g\mu_4 + (1-k)\epsilon)}\right)$$

This implies that $\zeta_3 < 0$ is a negative and by Routh-Hurwitz criterion the system cannot be stable. Thus the co-existing equilibrium point is always unstable if the cells coexist where

$$V_0 = \frac{\alpha_1 \mu_4 - 2\mu_1 \mu_4 N_4^* - (1-k)^2 \lambda_1 \epsilon}{\mu_4}, V_1 = \frac{(1-k)^2 \lambda_1 \epsilon}{\mu_4}, V_2 = -\phi_1 N_4^*,$$

$$V_3 = (d\alpha_2 - 2\mu_2 T_4^* - \gamma_2 M_4^* - \mu_5), V_4 = \frac{\rho M_4^* \omega - \gamma_3 M_4^*(\omega + T_4^*)^2}{(\omega + T_4^*)^2}, V_5 = -\gamma_2 T_4^*$$

$$V_6 = \frac{\rho T_4^*(g\mu_4 + (1-k)\epsilon) - \gamma_3 T_4^*(\omega + T_4^*)(g\mu_4 + (1-k)\epsilon) - \mu_3(\omega + T_4^*)(g\mu_4 + (1-k)\epsilon) - (1-k)^2(\omega + T_4^*)\lambda_3 \epsilon}{(\omega + T_4^*)(g\mu_4 + (1-k)\epsilon)}$$

$$- V_7 = -(1-k)\lambda_1 N_4^*, V_7 = (1-k)\lambda_1 N_4^*, V_8 = \frac{\lambda_3 \mu_4^2 g M_4^*(1-k)}{(g\mu_4 + (1-k)\epsilon)^2}$$

□

4. Uncertainty and Sensitivity Analysis

In this section, we explore the dependence of the model solutions on the parameter values. We are able to figure-out a feasible range of parameter values and determine the most critical parameters in the model. We employed a similar method, which is discussed in detail by [20,33], using Latin Hypercube Sampling (LHS) for studying the uncertainty analysis and the Partial Rank Correlation Coefficient (PRCC) for analyzing the sensitivity analysis indexes of the parameters. LHS/PRCC was ran and analyzed with a sample size of 100. The choice of this sample size is due to the fact that PRCC produces accurate results for a lower sample size compared to other technique, such as eFAST [33].

Uncertainty and sensitivity analysis were performed on all non-dimensional system parameters in the system (5) with the aim of determining the most sensitive parameters to the model. The parameter baseline values in Table 1 were varied in the range of 25%. Figure 1 displays a bar graph of PRCCs plotted against the homogeneous parameter value with tumor compartment as the baseline dependent variable. The parameters that are significantly positively correlated with tumor cells, at $P < 0.05$ level of significance, are α_1, g while μ_1, γ_3, and ω are significantly negatively correlated. An increase in the production of normal cells α_1, leads to higher numbers of normal cells, thus the higher the α_1, the higher the normal cells. While Figure 2 displays a bar graph of PRCCs plotted against the homogeneous parameter value with tumor compartment as the baseline dependent variable. The most sensitive parameters are shown to be $P-values$ of s, γ_2, μ_3 and ρ are less than 0.01.

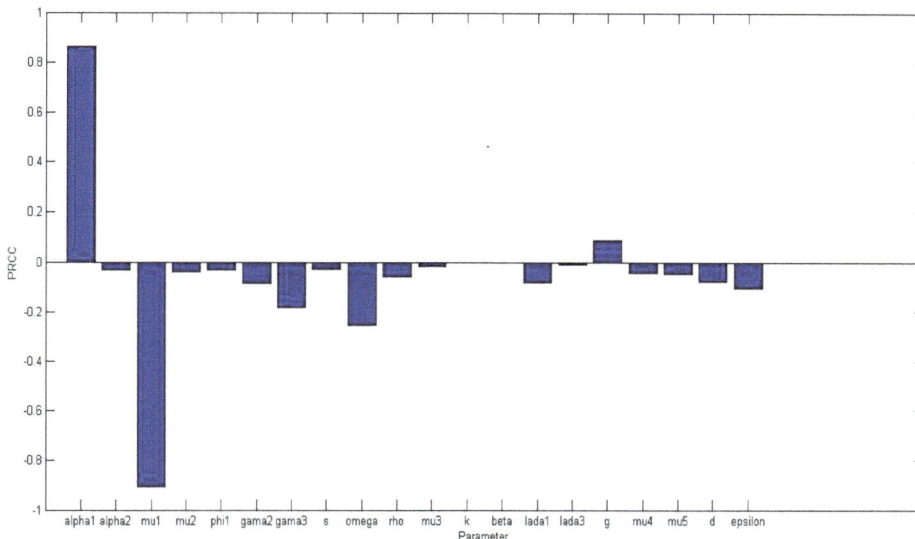

Figure 1. PRCCs of homogeneous model parameters with the tumor cells as the baseline variable. All parameter values were varied in 25% of their baseline values in Table 1. The most sensitive parameters are shown to be $P-values$ of $\alpha_1, g, \mu_1, \gamma_3$ and ω are less than 0.01.

Table 1. Description of parameters in the model.

Parameter	Symbol	Value	Unit	Refs
Per capita growth rate of normal cells	α_1	0.70	day^{-1}	[12]
Per capita growth rate of tumor cells	α_2	0.514	day^{-1}	[5]
Natural death rate of normal cells	μ_1	0.00003	day^{-1}	Assumed
Natural death rate of tumor cells	μ_2	0.01	day^{-1}	[7]
Rate of inhibition of normal cells	ϕ_1	6×10^{-8}	day^{-1}	[1]
Tumor cells death rate due to immune response	γ_2	3×10^{-6}	day^{-1}	[12]
Interaction coefficient rate with immune response	γ_3	1×10^{-7}	day^{-1}	[5]
Source rate of immune cells	s	1.3×10^4	day^{-1}	[12]
Source rate of estrogen	ϵ	1.3×10^4	day^{-1}	est
Immune threshold rate	ω	3×10^5	day^{-1}	[5]
Immune response rate	ρ	0.20	day^{-1}	[13]
Natural death rate of immune cells	μ_3	0.29	day^{-1}	[5]
Efficacy of anti-cancer drug	k	0–1	day^{-1}	Assumed
Supplement for immune booster	β	0.01	day^{-1}	est
Tumor formation rate as a result of DNA damage by excess estrogen	λ_1	0.20	$(Pg/mL)^{-1}day^{-1}$	est
Immune suppression rate due to excess estrogen	λ_3	0.002	day^{-1}	est
Assume constant of value of decay factor	g	0.1	day^{-1}	est
Natural death rate of estrogen	μ_4	0.97	day^{-1}	[19]
Death rate due to ketogenic diet	μ_5	2.0	day^{-1}	est
Constant rate of ketogenic diet	d	0.5	day^{-1}	est

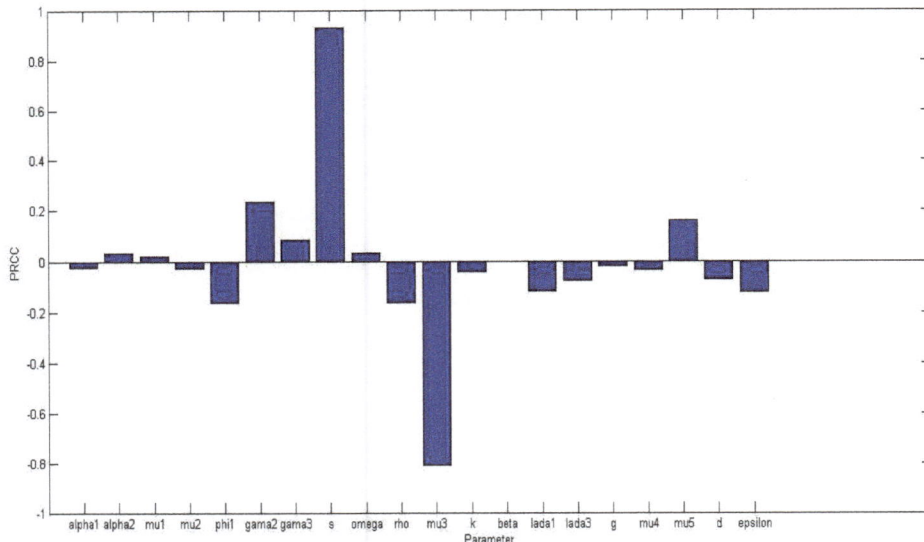

Figure 2. PRCCs of homogeneous model parameters with the tumor cells as the baseline variable. All parameter values were varied in 25% of their baseline values in Table 1. The most sensitive parameters are shown to be $P-values$ of s, γ_2, μ_3 and ρ are less than 0.01.

5. Analysis of Optimal Control

In this section, we formulated a corresponding optimal control problem for the model in the system (5) considering ketogenic-diet and anti-cancer drugs as control interventions to minimize the breast cancer and tumor burden at final time. The units of cells were normalized in order for the carrying capacity of normal cells to be kept above threshold of $0 \le t \le t_f$ [34–36]. On the other hand, the aim is to reduce the tumor-size which indicates the degree of the disease in the body and it requires the application of as much anti-cancer drugs as much as possible. However, it also minimized the systemic cost, which is based on the quantities of anti-cancer drugs, since large drug concentrations can be harmful and cause toxic side effects. In brief, the drug doses were minimized because the smaller the dose, the better. Then, we formulated the objective functional J_1

$$J_1(u_1, u_2) = \int_0^{T_f} \left(A_1 T(t) + A_2 E(t) + \frac{1}{2} A_3 u_1^2(t) + \frac{1}{2} A_4 u_2^2(t) \right) dt \tag{8}$$

System equations (5) is subject to:

$$\begin{aligned}
\frac{dN}{dt} &= N\alpha_1 - \mu_1 N^2 - \phi_1 TN - (1 - u_1(t))(\lambda_1 NE) \\
\frac{dT}{dt} &= (1 - u_2(t))T\alpha_2 - \mu_2 T^2 - \gamma_2 MT - \mu_5 T + (1 - u_1(t))(\lambda_2 NE) \\
\frac{dM}{dt} &= s\beta + \frac{\rho MT}{\omega + T} - \gamma_3 MT - \mu_3 M - \left((1 - u_1(t)) \frac{\lambda_3 ME}{g + E} \right) \\
\frac{dE}{dt} &= (1 - u_1(t))\epsilon - \mu_4 E
\end{aligned} \tag{9}$$

J_1 involves a quadratic control. In [37–41], it was established that quadratic control in the treatment terms has the added benefit of keeping the tumor in check both when it is small or large in size. The authors further explained that the quadratic control allows for a weaker treatment to minimize the toxic side-effects while permitting the system to maintain a low tumor size.

Furthermore, for us to address the tumor-to-therapy trade-off, we established the existence of an optimal control; by following the approach in [37,41,42], which required an analysis of the super-solutions (that is, the upper bounds on solutions) of the system (5). As soon as we were able to show that the system is bounded, we established the existence of an optimal control using a result from [43]. In addition, we proved that there exists an optimal control that minimizes the objective functional; using the established approach of [37–40,44]. We use the fact that super-solutions $\overline{N}, \overline{T}, \overline{M},$ $\overline{E}$ of

$$\frac{d\overline{N}}{dt} = N\alpha_1, \frac{d\overline{T}}{dt} = T\alpha_2(1 - u_2),$$
$$\frac{d\overline{M}}{dt} = s\beta + \frac{\rho MT}{\omega + T}, \frac{d\overline{E}}{dt} = 1 \tag{10}$$

are bounded on a finite time interval. Since the sub-solutions are zero, the result obtained shows that our system is bounded. Since we have a bounded system, our next task was to establish the existence of the optimal control using a result from [43].

Existence of an Optimal Control

Theorem 6. *Given the objective functional in (8), where* $U = \{u_i^*(t),$ *Lebesgue measure* $: 0 \leq u_i^*(t) \leq 1, \forall t \in [0, t]\}$ *subject to system (9) with* $N(0) = N_0,$ $T(0) = T_0,$ $M(0) = M_0,$ *and* $E(0) = E_0,$ *then there exists an optimal control* $\overline{u_i^*}$ *such that* $min_{\overline{u_i^*}(t) \in [0,1]} J_1\overline{(u_i^*)} = J_1(u_i^*(t))$ *if the following conditions holds:*

- f is not empty;
- The admissible control set U is closed and convex;
- Each right hand side of the state system is continuous, is bounded above by the sum of the bounded control and the state, and can be written as a linear function of $\overline{u_i^*(t)}$ with coefficients depending on time and the state;
- The integrand of $J_1\overline{(u_i^*)}$ is convex on U and is bounded below by $-c_2 + c_1\overline{u}^2$ with $c_1 > 0$.

Proof. Since the system (9) has bounded coefficients and the solutions are bounded on the finite time interval, we can apply the result of [45], to obtain the existence of the solution of the system (9). Furthermore, we note that U is closed and convex by definition. For the third conditions, the right hand side of the system (9) must be continuous. The right hand side is continuous since the denominators of all fractions from the right hand side of the system consists solely of positive entities. We let $\overleftarrow{\phi}(t, \overleftarrow{X})$ be right hand side of the system (9) except for the terms of $\overline{u_i^*}$ and define.

$$|\overleftarrow{f}(t, \overleftarrow{X}, u_i^*)| = \overleftarrow{\phi}(t, \overleftarrow{X}) + \begin{pmatrix} 0 \\ \lambda_1 NE \\ 0 \\ u_1 \end{pmatrix}, with \overleftarrow{X} = \begin{pmatrix} N \\ T \\ M \\ E \end{pmatrix}$$

using the boundedness of the solutions (10), we have

$$|\overleftarrow{f}(t, \overleftarrow{X}, u_i^*)| \leq \left| \begin{pmatrix} \alpha_1 & 0 & 0 & 0 \\ 0 & \alpha_2(1 - u_2) & 0 & 0 \\ 0 & 0 & \rho & 0 \\ 0 & 0 & 0 & 0 \end{pmatrix} \begin{pmatrix} N \\ T \\ M \\ E \end{pmatrix} \right| + \left| \begin{pmatrix} 0 \\ (1 - u_1)\lambda_1 NE \\ s\beta \\ -u_1 \epsilon \end{pmatrix} \right| \leq c_1 \left(|\overleftarrow{X}| + |\overleftarrow{u_i^*}| \right)$$

where c_1 depends on the coefficients of the system. For the fourth condition, we need to show

$$J(t, T, E, (1 - P_i)u_i + P_iV_i) \leq (1 - P_i)J(t, T, E, u_i) + P_iJ(t, T, E, V_i)$$

we analyze the difference of

$$J(t, T, E, (1 - P_i)u_i + P_iV_i) - [(1 - P_i)J(t, T, E, u_i) + P_i(t, T, E, V_i)]$$

$$= T(t) + E(t) + \frac{\epsilon}{2}(u_i^2 - 2P_iu_i^2 + P_i^2u_i^2 + P_i^2V_i^2 - 2P_i^2V_i^2u_i^2 + 2P_iV_iu_i)$$
$$- (T(t) + E(t) + \frac{\epsilon}{2}u_i^2 - \frac{\epsilon}{2}P_iu_i^2 + \frac{\epsilon}{2}P_iV_i^2)$$

$$= \frac{\epsilon}{2}(P_i^2 - P_i)(u_i - V_i)^2$$

since, $P_i \in (0,1)$ implies $(P_i^2 - P_i) < 0$ and $(u_i - V_i)^2 > 0$ but $(P_i^2 - P_i) < 0$, which implies $\frac{\epsilon}{2}(P_i^2 - P_i)(u_i - V_i)^2$ is negative. This implies that,

$$J(t, T, E, (1 - P_i)u_i + P_iV_i) \le (1 - P_i)J(t, T, E, u_i) + P_i(t, T, E, V_i)$$

Lastly,

$$T(t) + E(t) + \frac{\epsilon}{2}u_i^2(t) \ge \frac{\epsilon}{2}u_i^2(t) \ge -c + \frac{\epsilon}{2}u_i^2(t)$$

which gives $-c + \frac{\epsilon}{2}u_i^2(t)$ as the lower bound. With the existence of the optimal control established, we now characterized the optimal control using the Pontryagin's maximum principle [11]. The constants A_1, A_2, A_3 and A_4 are a measure of the relative cost of the interventions over $[0, T]$. The optimal control problem is that of finding optimal functions $(u_1^*(t), u_2^*(t))$ such that

$$J_1(u_1^*(t), u_2^*(t)) = \min_{\Omega} J_1(u_1(t), u_2(t)) \tag{11}$$

where

$$\Omega = \left\{ u_1(t) \& u_2(t) : 0 \le u_1(t) \le u_{1max}, 0 \le u_2(t) \le u_{2max}, t \in [0, T_f] \right\}$$

□

Three different control strategies are explored. This approach can be used to test various options. However, we only looked at the following three alternatives:

- Strategy 1: Anti-cancer drug treatment control on tumor cells (control $u_1(t)$ only);
- Strategy 2: Ketogenic diet control on excess estrogen and tumor cells (control $u_2(t)$ only);
- Strategy 3: Anti-cancer drug and ketogenic diet treatment combined control on tumor cells growth and excess estrogen (controls $u_1(t)$ and $u_2(t)$).

Thus, strategies (1–3) use the objective functional (8). We assumed that there are practical limitations on the maximum rate at which the anti-cancer treatment may be applied in a given time period. We defined the positive constant u_{max} accordingly. We also define the set Ω of admissible controls to be all Lebesgue measurable functions that take on values in the control set [13,46,47] $u = [0, u_{max}]$ almost everywhere on $[0, T]$. We sought an optimal control $u^* \in \Omega$ in (11) [13]. In order to find the optimal solutions, we first traced the Lagrangian and Hamiltonian for the optimal control problem (8) and (9). The Lagrangian of the optimal control problem is given by:

$$L(N, T, M, E, u_1, u_2) = A_1T(t) + A_2E(t) + \frac{1}{2}A_3u_1^2(t) + \frac{1}{2}A_4u_2^2(t) \tag{12}$$

For the purpose of the necessary conditions for optimal control functions with the help of Pontryagin's maximum principle [11]. We define the Hamiltonian, H for the control problem of the system (8) and (9)

$$H = L(N, T, M, E, u_1, u_2) + \theta_1N' + \theta_2T' + \theta_3M' + \theta_4E' \tag{13}$$

where L is the Lagrangian function (12),

$$H = \begin{pmatrix} A_1 T(t) + A_2 E(t) + \frac{1}{2} A_3 u_1^2(t) + \frac{1}{2} A_4 u_2^2(t) \\ + \theta_1 \left(N\alpha_1 - \mu_1 N^2 - \phi_1 TN - (1 - u_1(t))(\lambda_1 NE) \right) \\ + \theta_2 \left((1 - u_2(t)) T\alpha_2 - \mu_2 T^2 - \gamma_2 MT - \mu_5 T + (1 - u_1(t))(\lambda_1 NE) \right) \\ + \theta_3 \left(s\beta + \frac{\rho MT}{\omega + T} - \gamma_3 MT - \mu_3 M - \left((1 - u_1(t)) \frac{\lambda_3 ME}{g + E} \right) \right) \\ + \theta_4 ((1 - u_1(t))\epsilon - \mu_4 E) \end{pmatrix}$$

where $\theta_1, \theta_2, \theta_3, \theta_4$ are the adjoints variables for the states N, T, M, E. However, with the help of Pontryagin's Maximum Principle, we obtained a minimized Hamiltonian that minimizes the objective function or cost functional. We applied Pontryagin's Maximum Principle [11], to characterize the optimal control pair u_1^* & u_2^* in the following result.

Theorem 7. *Given optimal control variables u_1^* & u_2^* and N^*, T^*, M^* & E^* are corresponding optimal state variables of the control system (8) and (9). Then there exists the adjoint variable $\theta_i = (\theta_1, \theta_2, \theta_3, \theta_4) \in \Re_+^4$ that satisfies the following equations.*

$$\begin{aligned} \frac{d\theta_1}{dt} &= 2\theta_1 \mu_1 N + \phi_1 \theta_1 T + (\theta_1 + \theta_2)(1 - u_2(t))\lambda_1 E - \alpha_1 \theta_1 \\ \frac{d\theta_2}{dt} &= -A_1 + \theta_1 \phi_1 N + \theta_2 (2T\mu_2 + \gamma_2 M + \mu_5 - \alpha_2(1 - u_2)) + \theta_3 \left(\gamma_3 M - \frac{\rho \omega M}{(\omega + T)^2} \right) \\ \frac{d\theta_3}{dt} &= \theta_2 \gamma_2 T - \rho \theta_3 T + \gamma_3 \theta_3 T + \mu_3 \theta_3 + \theta_3 \left((1 - u_1) \frac{\lambda_1 E}{g + E} \right) \\ \frac{d\theta_4}{dt} &= -A_2 + (\theta_1 - \theta_2)(1 - u_1)\lambda_1 N - \theta_3 \left((1 - u_1) \frac{\lambda_3 Mg}{(g + E)^2} \right) - \theta_4 \mu_4 \end{aligned} \tag{14}$$

with transversality conditions

$$\theta_1(T_f) = \theta_2(T_f) = \theta_3(T_f) = \theta_4(T_f) = 0$$

The corresponding optimal controls u_1^* & u_2^* are given as,

$$u_1^* = min \left\{ max \left\{ 0, \frac{1}{A_3} \left(\theta_2 \lambda_1 N^* E^* + \theta_3 \epsilon - \theta_1 \lambda_1 N^* E^* - \frac{\theta_3 \lambda_3 M^* E^*}{g + E^*} \right) \right\}, 1 \right\} \tag{15}$$

and

$$u_2^* = min \left\{ max \left\{ 0, \frac{1}{A_4} (\theta_2 \alpha_2 T^*) \right\}, 1 \right\} \tag{16}$$

Proof. Let u_1^* & u_2^* be the given optimal control functions and N^*, T^*, M^* & E^* be the corresponding optimal state variables of the system (9) that minimize the cost functional or objective (8). Then by Pontryagin's maximum principle [11], there exists adjoint variables (14) $\theta_1, \theta_2, \theta_3$ & θ_4 which satisfy the following equations

$$\frac{d\theta_1}{dt} = -\frac{\partial H}{\partial N}, \frac{d\theta_2}{dt} = -\frac{\partial H}{\partial T}, \frac{d\theta_3}{dt} = -\frac{\partial H}{\partial M}, \frac{d\theta_4}{dt} = -\frac{\partial H}{\partial E}$$

with transversality conditions

$$\theta_1(T_f) = \theta_2(T_f) = \theta_3(T_f) = \theta_4(T_f) = 0$$

where H is the Hamiltonian and defined as

$$H(N, T, M, E, u_1, u_2, \theta) = L(N, T, M, E, u_1, u_2) + \theta_1 N' + \theta_2 T' + \theta_3 M' + \theta_4 E'$$

$$
H = \begin{pmatrix}
A_1 T(t) + A_2 E(t) + \frac{1}{2} A_3 u_1^2(t) + \frac{1}{2} A_4 u_2^2(t) \\
+\theta_1 \left(N\alpha_1 - \mu_1 N^2 - \phi_1 TN - (1 - u_1(t))(\lambda_1 NE) \right) \\
+\theta_2 \left((1 - u_2(t)) T\alpha_2 - \mu_2 T^2 - \gamma_2 MT - \mu_5 T + (1 - u_1(t))(\lambda_1 NE) \right) \\
+\theta_3 \left(s\beta + \frac{\rho MT}{\omega + T} - \gamma_3 MT - \mu_3 M - \left((1 - u_1(t)) \frac{\lambda_3 ME}{g + E} \right) \right) \\
+\theta_4 \left((1 - u_1(t))\epsilon - \mu_4 E \right)
\end{pmatrix}
$$

from the optimality condition, we have

$$
\frac{\partial H}{\partial u_1} = 0, \text{ at } u_1 = u_1^* \text{ and } \frac{\partial H}{\partial u_2} = 0, \text{ at } u_2 = u_2^*
$$

which implies that

$$
0 = \frac{\partial H}{\partial u_1} = A_3 u_1 + \theta_1 \lambda_1 NE - \theta_2 \lambda_1 NE + \theta_3 \frac{\lambda_3 ME}{g + E} - \theta_4 \epsilon
$$

$$
0 = \frac{\partial H}{\partial u_1} = A_4 u_2 - \theta_2 \alpha_2 T
$$

Hence, we obtain (see [10])

$$
u_1^* = \frac{1}{A_3} \left\{ \theta_1 \lambda_1 NE + \theta_4 \epsilon - \theta_1 \lambda_1 NE - \theta_3 \frac{\lambda_3 ME}{g + E} \right\} \tag{17}
$$

$$
u_2^* = \frac{1}{A_4} \left\{ \theta_2 \alpha_2 T \right\} \tag{18}
$$

Thus we have (17) and (18).

By standard control arguments involving the bounds on the controls, we conclude that (15) and (16) can be written in this form

$$
u_1^* = \begin{pmatrix}
0 & \text{if } \frac{1}{A_3} \left(\theta_1 \lambda_1 NE + \theta_4 \epsilon - \theta_1 \lambda_1 NE - \theta_3 \frac{\lambda_3 ME}{g+E} \right) < 0 \\
\frac{1}{A_3} \left(\theta_1 \lambda_1 NE + \theta_4 \epsilon - \theta_1 \lambda_1 NE - \theta_3 \frac{\lambda_3 ME}{g+E} \right) & \text{if } 0 \le \frac{1}{A_3} \left(\theta_1 \lambda_1 NE + \theta_4 \epsilon - \theta_1 \lambda_1 NE - \theta_3 \frac{\lambda_3 ME}{g+E} \right) \le 1 \\
1 & \text{if } \frac{1}{A_3} \left(\theta_1 \lambda_1 NE + \theta_4 \epsilon - \theta_1 \lambda_1 NE - \theta_3 \frac{\lambda_3 ME}{g+E} \right) > 1
\end{pmatrix}
$$

and

$$
u_2^* = \begin{pmatrix}
0 & \text{if } \frac{1}{A_4} (\theta_2 \alpha_2 T^*) < 0 \\
\frac{1}{A_4} (\theta_2 \alpha_2 T^*) & \text{if } 0 \le \frac{1}{A_4} (\theta_2 \alpha_2 T^*) \le 1 \\
1 & \text{if } \frac{1}{A_4} (\theta_2 \alpha_2 T^*) > 1
\end{pmatrix}
$$

□

However, we discuss the numerical solution of the optimality system and the corresponding results of varying the optimal controls u_1 & u_2 the parameter choices, and the interpretations from various cases.

6. Numerical Simulations and Discussion

A picture of the dynamical behavior of breast cancer cells in the presence of normal cells, tumor cells, immune cells, and estrogen is given by the numerical simulations of the model (5). The optimal control is acquired by solving the optimality system of four ordinary differential equations from the state variables and the adjoint system. An iterative scheme is used to solve the optimality system. All the numerical simulations were executed in MAPLE 18. We employed the forward-backward scheme method, beginning with an initial guess for optimal controls and solved the optimal state system forward in time and after that solved the adjoint state system backward in forward using the finite difference scheme in MAPLE. The two controls were then updated by using a convex combination of the previous controls as well as the characterization (17) and (18). The entire process was repeated until the values of the unknown at the previous iterations were closed to the one at the current iteration [39,41]. Key parameters are also noted in stabilizing the model in system (5), for example: ketogenic diet, anti-cancer, and immune booster. The initial values of variables are N(0) = 2000, T(0) = 800, M(0) = 500, E(0) = 20 and $s = 1.3 \times 10^4$ adopted from [12]. All parameter values used for the numerical simulation are stated in Table 1 above.

Figure 3, indicates that the introduction of a ketogenic diet results in a reduction of activities of cancer cells and we also note that too much of a ketogenic diet will result in ketoacidosis. Ketoacidosis is the combination of ketosis and acidosis. Ketosis is the accumulation of substances called ketone bodies and acidosis is the increased acidity of the blood which can cause frequent urination (Polyuria), poor appetite, and a loss of consciousness. Therefore, our ketogenic diet's parameter rate is best at $d = 0.6$ and it can complement the activity of the anti-cancer drug (Tamoxifen). Figure 4, shows the impact of anti-cancer drugs in reducing the production of excess estrogen in the system, but when there is less production of estrogen there will not be a rapid activation of the growth factor that expresses breast normal cells. However, the rapid production of estrogen results in abnormal breast cells expression, which will lead to breast cancer. Figure 5 shows the obvious effectiveness of anti-cancer drugs on tumor cells when there is no supply of nutrient or glucose to cancer cells.

Furthermore, Figure 6 illustrates that the red line $\beta = 0$ shows that during cancer formation the activities of both innate and adaptive reduces drastically, which is due to the expression of other proteins apart from those proteins that are responsible for the activation of the immune response, such as an immune booster introduced to the system, which reactivates the activities of the immune response towards the cancer cells.

The presence of abnormal estrogen level without anti-cancer drugs or a ketogenic diet will lead the system into critical condition and became unstable as shown in Figure 7. However, the system became stable as we introduced treatments, such as chemotherapy and the ketogenic diet as represented in Figure 8. In addition, Figure 9, indicates that there is DNA damage at $\lambda_1 = 0$, which occurs naturally as a result of metabolic or hydrolytic processes. It is as a result of the Tumor Suppressor Gene (TSG), which is able to control the activity of DNA gene repair successfully. On the other hand, at $\lambda_1 = 0.2, 0.4, 0.6$ showed that TSG (such as BRCA 1, BRCA 2, P53) compromised the pathway that leads cells to grow uncontrollably and later form a tumor or it leads to accelerated aging.

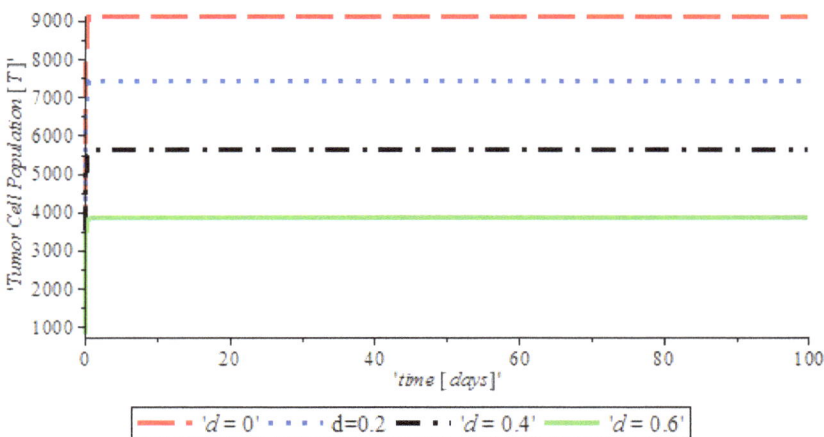

Figure 3. The variation of proportion of Tumor cell population for different values of *d* with other parameters fixed.

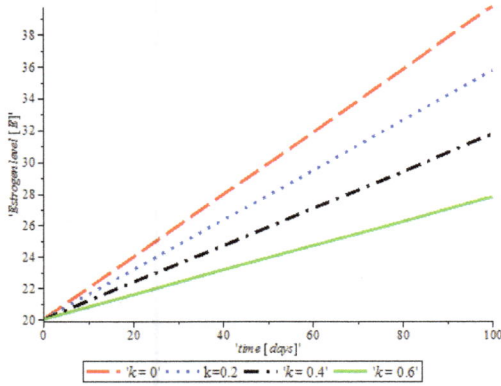

Figure 4. The variation of proportion of Estrogen level population for different values of k with other parameters fixed.

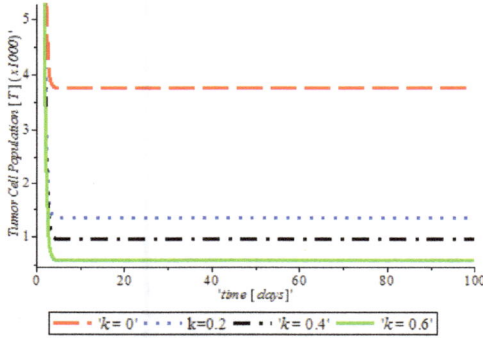

Figure 5. The variation of proportion of Tumor cell population for different values of k with other parameters fixed.

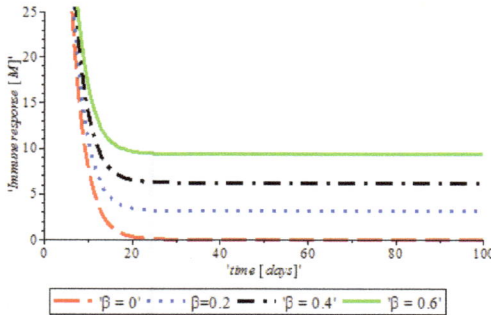

Figure 6. The variation of proportion of Immune booster population for different values of β with other parameters fixed.

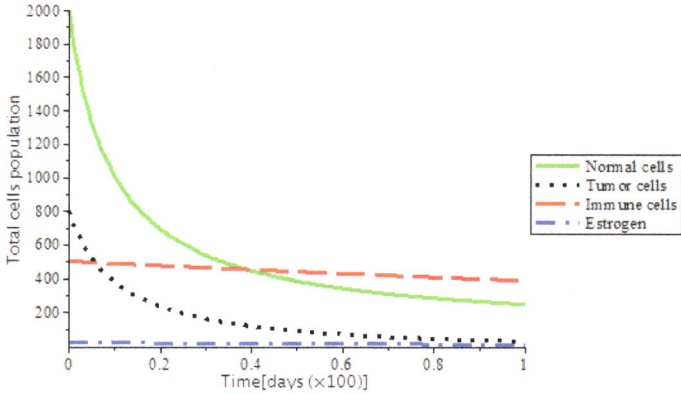

Figure 7. The variation of Total cells population depicted as locally asymptotically unstable.

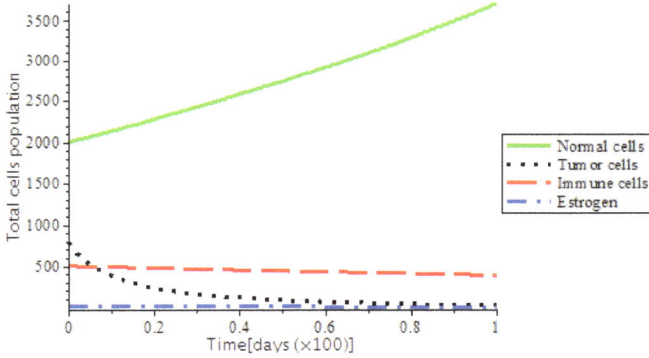

Figure 8. The variation of Total cells population depicted as locally asymptotically stable.

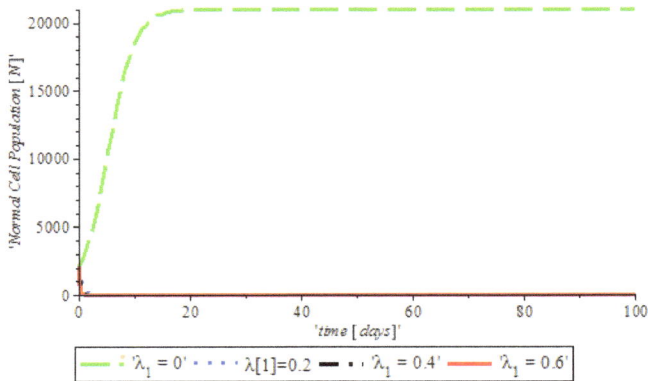

Figure 9. The variation of proportion of Normal cell population for different values of λ_1 with other parameters fixed.

However, the mathematical analysis of the model produced six equilibrium points. All the points have epidemiological implications in relation to explaining the dynamics of breast cancer growth. P_0 represents the situation where there is tumor-free equilibrium, that is when only tumor cell population has died off due to competition with other cells. P_{d1} represents Type 1 dead equilibrium point where both normal cells and tumor cells die-off as a result of breast tissue removal through mastectomy surgery or death. This is because overtime the cancer cells which are depending on estrogen to develop into independent cells that grow regardless of estrogen receptors. P_{d2} could be described by Type 2 dead equilibrium point where normal cells were only forced to extinction leaving the tumor cells surviving. P_{d3} represent Type 3 dead equilibrium point which means immune system is weak and it cannot fight the tumor cells which eventually overpower normal cells and forced it to extinction. P_{d4} show that Type 4 dead equilibrium point where ketogenic diet is not effective, immune booster is not active which lead to tumor cell over-compete normal cells as a result of infusion of excess estrogen to the body system.

We categorised this as "dead" because biologically there is no recovery of damaged normal cells since they have died off of the cell population. It could be as a result of anti-cancer drug that destroy red blood cells which affected normal cells.

Effects of Control on the System (9)

By numerical simulation, optimal single control of anti-cancer drugs measure u_1 and ketogenic-diet optimal control measure u_2 are shown in Figure 10a,b respectively; where (red dots line) represented tumor cells and (solid green line) represented normal cells. Figure 10c is the use of combination of two control therapies which have significant impact on the increase of normal cells population against time. However, all the strategies are effectively restrain the tumor growth, they cannot totally eliminate a large tumor in 100 days. In Figure 11, optimal control using anti-cancer drugs and ketogenic diet as we optimized the system (54) with the objective function J for breast cancer model. It was observed that the combination of the two controls resulted in appreciable decreases in the number of tumor cells population in the presence of control (solid green line) while (dots red line) in the case of uncontrolled. However, tumor growth is driven to a very low but non-zero level.

Furthermore, it was noticed from Figure 12, that the level of estrogen was reduced drastically in the presence of controls (solid green line) against the constant increase level of estrogen (dots red line) in uncontrolled cases. However, anti-cancer drugs (for example Tamoxifen) blocks estrogen receptors on breast cells, that is, it stops estrogen from connecting to the cancer cells while tamoxifen also acts like an anti-estrogen in breast cells; it acts like an estrogen in other tissues like the uterus and the bones [48]. In addition, ketosis also regulating hormonal imbalance [8,27]. On the other hand, Figure 13, shows the effect of immune response with and without controls. Immune response can help to fight cancer cells while immune system recognize cancer cells as abnormal and kill them. However, this may not be enough to eliminate cancer cells from the body.

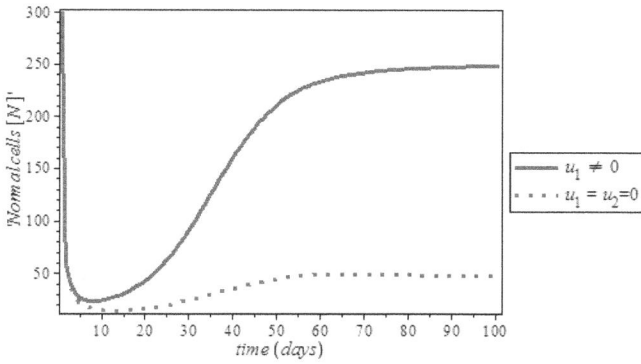

(a)

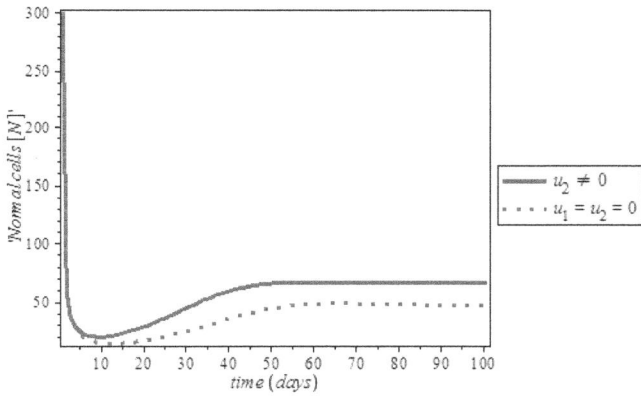

(b)

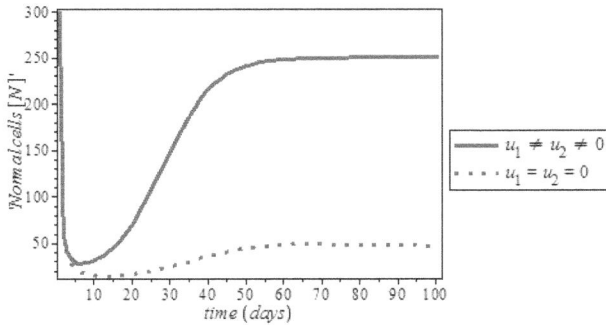

(c)

Figure 10. Simulation result of the model (9), showing normal cell population against time with and without control.

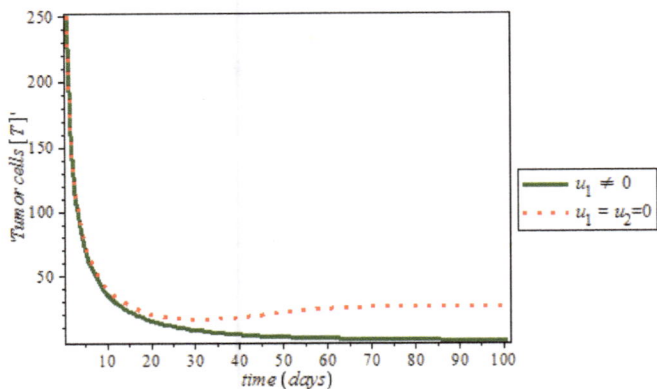

(a)

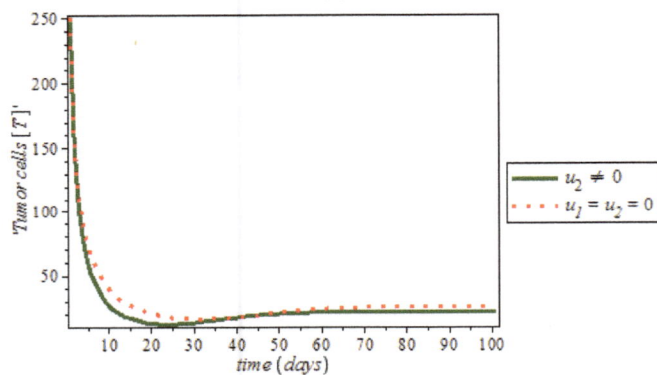

(b)

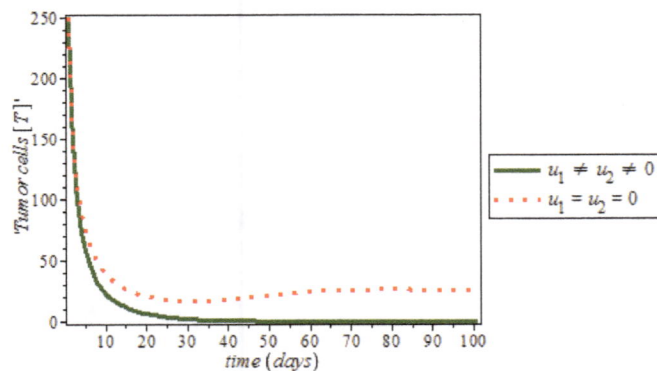

(c)

Figure 11. Simulation result of the model (9), showing tumor cell population against time with and without control.

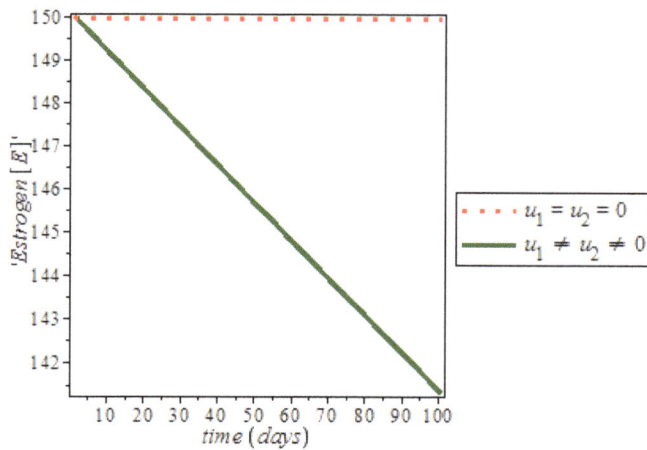

(a)

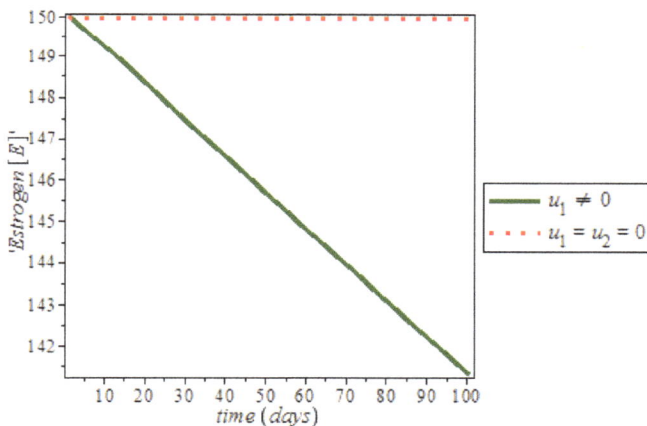

(b)

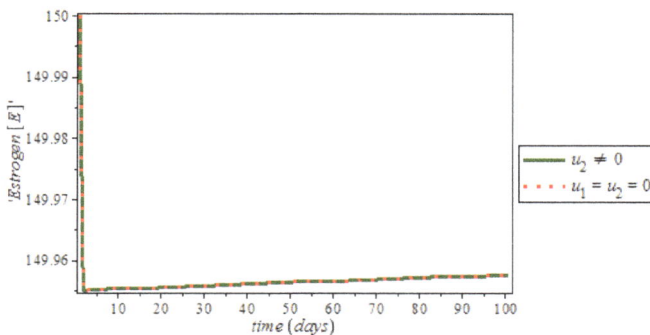

(c)

Figure 12. Simulation result of the model (9), showing estrogen level against time with and without control.

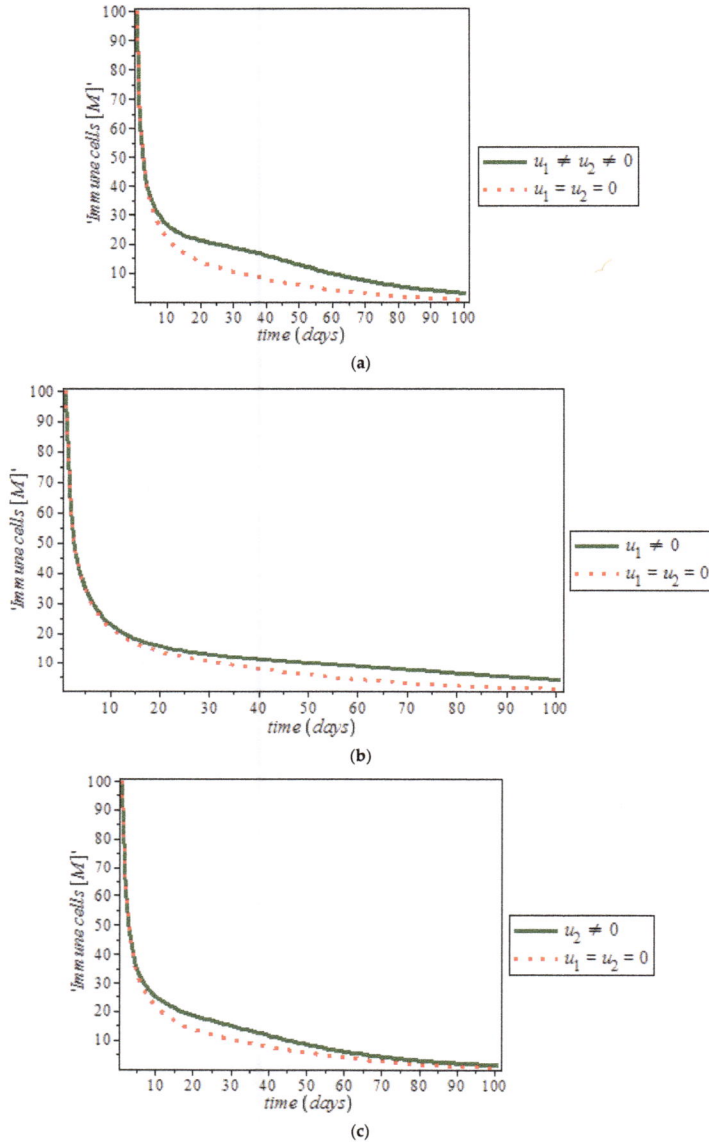

Figure 13. Simulation result of the model (9), showing immune response against time with and without control.

7. Conclusions

A four-dimensional compartmental deterministic model was designed and used to monitor the dynamics of breast cancer. The existing model in [19] was extended to incorporate treatments, ketogenic diet, and an immune booster. The system (5) was rigorously analyzed to gain insight into their dynamical behaviors. The study shows the following:

- The conditions of stability of the tumor-free equilibrium (TFE) was established and the system is only local asymptotically stable (LAS) if a certain threshold quantity, known as the reproductive number, is less than unity ($R_0 < 1$). It implies that the number of tumor cells in the body will be brought to zero if proper treatments and a ketogenic diet that can force make the threshold to a value less than unity are monitored.
- An individual has the chance of developing breast cancer depending on the level of the immune system (s), the efficacy of the anti-cancer drug (k) and the rate at which the ketogenic diet (d) is being taken to fight tumor cells. We also found out that the presence of excess estrogen in system makes it unstable, as depicted in Figure 7. This implies that any additional estrogen quantity introduced into the body through the birth control, and hormone replacement therapy (HRT) enhances the rate of tumor formation. Thus, the development of breast cancer is certain.
- The transition from normal cells class to tumor cells class plays a crucial role in breast cancer dynamics (λ_1). More tumor is formed if the DNA is damaged or altered as a result of excess estrogen, which reduces the number of normal cells being produced by red blood cells.

Furthermore, the results show that tumor cell formation depend on the level of excess estrogen introduced into the body system. It must be noted that the ability to resist changes in structure and amount of estrogen released during natural biological processes is dependent on an individual's DNA. Such biological processes include: premenopausal and menopause stages. Other risk factors may also be incorporated in the model for future work, which might generate different results.

However, the focus of this study has been identifying the advantages that come with the process of breast cancer relief policies that combined anti-cancer drugs and ketogenic diet procedures to knit the circumstances of unlimited and limited resources. The effort to moderate the effect of breast cancer on the body can be fruitful, especially if our basic reproductive number R_0 is properly analyzed. In addition, moderation is conceivable if the planning of intercessions is sufficiently quick and if the arrangement includes the utilization of more than one therapy procedure. No therapy (ketogenic diet and anti-cancer drug) is possible, unless minimal resources are accessible.

8. Further Research

Breast cancer is a health challenge disease, especially among women world-wide. This study explored the use of a quadratic control law to formulate the optimal control problem for the objective function. Hence, the authors hope to conduct further research into the application of a switching function and to investigate the side-effects of anti-cancer drugs by employing a linear control law to formulate the optimal control problem for further study.

Author Contributions: The authors contributed towards mathematical model formulation; Segun Isaac Oke analyzed the model both analytical and numerical simulation.

Acknowledgments: The corresponding author appreciate National Research Foundation (NRF), South Africa for the grant towards my PhD; Grant Number: 109824. The authors also acknowledges the support of Research Office of University of Zululand for providing the funds for the publication. The authors are grateful to Adeniyi Michael (LASPOTECH, Nigeria) and Chinaza Uleanya (Unizulu) for their useful comments in the preparation of the manuscript. The authors are grateful to the anonymous Reviewers and the Handling Editor for their constructive comments, which have enhanced the paper.

Conflicts of Interest: The authors declare no conflicts of interest.

References

1. Evans, C.W. The invasion and metastatic behaviour of malignant cells. In *The Metastatic Cell: Behavior and Biochemistry*; Chapman and Hall: London, UK, 1991; pp. 137–214.
2. National Cancer Registry of South Africa. Available online: http://www.cansa.org.za/files/2015/10/NCR_Final_2010_tables1.pdf (accessed on 24 April 2018).
3. World Health Organization. *Global Action Plan for the Prevention and Control on NCDs*; World Health Organization: Geneva, Switzerland, 2014.

4.   Abernathy, K.; Abernathy, Z.; Baxter, A.; Stevens, M. Global Dynamics of a Breast Cancer Competition Model. *Differ. Equ. Dyn. Syst.* **2017**, *3*, 1–15. [CrossRef]
5.   Patel, M.I.; Nagl, S. *The Role of Model Integration in Complex Systems Modelling: An Example from Cancer Biology*; Springer: Berlin, Germany, 2010.
6.   Allen, B.G.; Bhatia, S.K.; Anderson, C.M.; Eichenberger-Gilmore, J.M.; Sibenaller, Z.A.; Mapuskar, K.A.; Schoenfeld, J.D.; Buatti, J.M.; Spitz, D.R.; Fath, M.A. Ketogenic diets as an adjuvant cancer therapy: History and potential mechanism. *Redox Biol.* **2014**, *2*, 963–970. [CrossRef] [PubMed]
7.   Gilbert, D.L.; Pyzik, P.L.; Freeman, J.M. The ketogenic diet: Seizure control correlates better with serum β-hydroxybutyrate than with urine ketones. *J. Child Neurol.* **2000**, *15*, 787–790. [CrossRef] [PubMed]
8.   Westman, E.C.; Yancy, W.S.; Mavropoulos, J.C.; Marquart, M.; McDuffie, J.R. The effect of a low-carbohydrate, ketogenic diet versus a low-glycemic index diet on glycemic control in type 2 diabetes mellitus. *Nutr. Metab.* **2008**, *5*, 36. [CrossRef] [PubMed]
9.   Kareva, I.; Berezovskaya, F. Cancer immunoediting: A process driven by metabolic competition as a predator–prey–shared resource type model. *J. Theor. Biol.* **2015**, *380*, 463–472. [CrossRef] [PubMed]
10.   Lenhart, S.; Workman, J.T. *Optimal Control Applied to Biological Models*; CRC Press: Boca Raton, FL, USA, 2007.
11.   Pontryagin, L.S. *Mathematical Theory of Optimal Processes*; CRC Press: Boca Raton, FL, USA, 1987.
12.   De Pillis, L.G.; Radunskaya, A. A mathematical tumor model with immune resistance and drug therapy: An optimal control approach. *Comput. Math. Methods Med.* **2001**, *3*, 79–100. [CrossRef]
13.   D'Onofrio, A.; Ledzewicz, U.; Maurer, H.; Schättler, H. On optimal delivery of combination therapy for tumors. *Math. Biosci.* **2009**, *222*, 13–26. [CrossRef] [PubMed]
14.   Kermack, W.O.; McKendrick, A.G. A contribution to the mathematical theory of epidemics. In Proceedings of the Royal Society of London A: Mathematical, Physical and Engineering Sciences, London, UK, 1 August 1927; The Royal Society: London, UK, 1927; Volume 115, pp. 700–721.
15.   Kermack, W.O.; McKendrick, A.G. Contributions to the mathematical theory of epidemics: V. Analysis of experimental epidemics of mouse-typhoid; a bacterial disease conferring incomplete immunity. *Epidemiol. Infect.* **1939**, *39*, 271–288. [CrossRef]
16.   Cai, P.; Tang, J.S.; Li, Z.B. Analysis and controlling of Hopf Bifurcation for chaotic Van der Pol-Duffing system. *Math. Comput. Appl.* **2014**, *19*, 184–193. [CrossRef]
17.   Bernoulli, D. Essai d'une nouvelle analyse de la mortalité causée par la petite vérole et des avantages de l'inoculation pour la prévenir. *Hist. Acad. R. Sci. (Paris) Mém. Math. Phys. Mém.* **1760**, *1*, 1–45.
18.   De Pillis, L.G.; Radunskaya, A.E.; Wiseman, C.L. A validated mathematical model of cell-mediated immune response to tumor growth. *Cancer Res.* **2005**, *65*, 7950–7958. [CrossRef] [PubMed]
19.   Mufudza, C.; Walter, S.; Chiyaka, E.T. Assessing the effects of estrogen on the dynamics of breast cancer. *Comput. Math. Methods Med.* **2012**, *2012*, 473572. [CrossRef] [PubMed]
20.   Malinzi, J.; Eladdadi, A.; Sibanda, P. Modelling the spatiotemporal dynamics of chemovirotherapy cancer treatment. *J. Biol. Dyn.* **2017**, *11*, 244–274. [CrossRef] [PubMed]
21.   Swierniak, A.; Krzeslak, M.; Student, S.; Rzeszowska-Wolny, J. Development of a population of cancer cells: Observation and modeling by a mixed spatial evolutionary games approach. *J. Theor. Biol.* **2016**, *405*, 94–103. [CrossRef] [PubMed]
22.   Wu, C.H.; Motohashi, T.; Abdel-Rahman, H.A.; Flickinger, G.L.; Mikhail, G. Free and protein-bound plasma estradiol-17β during the menstrual cycle. *J. Clin. Endocrinol. Metab.* **1976**, *43*, 436–445. [CrossRef] [PubMed]
23.   Kumar, A.; Srivastava, P.K. Vaccination and treatment as control interventions in an infectious disease model with their cost optimization. *Commun. Nonlinear Sci. Numer. Simul.* **2017**, *44*, 334–343. [CrossRef]
24.   Kimmel, M.; Swierniak, A. Control theory approach to cancer chemotherapy: Benefiting from phase dependence and overcoming drug resistance. In *Tutorials in Mathematical Biosciences III*; Springer: Berlin/Heidelberg, Germany, 2006; pp. 185–221.
25.   Buonomo, B. Modeling itns usage: Optimal promotion programs versus pure voluntary adoptions. *Mathematics* **2015**, *3*, 1241–1254. [CrossRef]
26.   Pinho, S.T.R.D.; Freedman, H.I.; Nani, F. A chemotherapy model for the treatment of cancer with metastasis. *Math. Comput. Model.* **2002**, *36*, 773–803. [CrossRef]
27.   American Cancer Society. *Breast Cancer*; American Cancer Society: Atlanta, GA, USA, 2013.
28.   Gatenby, R.A. The potential role of transformation-induced metabolic changes in tumor-host interaction. *Cancer Res.* **1995**, *1766*, 4151–4156.

29. Neves, A.A.; Brindle, K.M. Assessing responses to cancer therapy using molecular imaging. *Biochim. Biophys. Acta* **2006**, *1766*, 242–261. [CrossRef] [PubMed]
30. Valayannopoulos, V.; Bajolle, F.; Arnoux, J.B.; Dubois, S.; Sannier, N.; Baussan, C.; Petit, F.; Labrune, P.; Rabier, D.; Ottolenghi, C.; et al. Successful treatment of severe cardiomyopathy in glycogen storage disease type III With D, L-3-hydroxybutyrate, ketogenic and high-protein diet. *Pediatr. Res.* **2011**, *70*, 638–641. [CrossRef] [PubMed]
31. De Leenheer, P.; Aeyels, D. Stability properties of equilibria of classes of cooperative systems. *IEEE Trans. Autom. Control* **2001**, *46*, 1996–2001. [CrossRef]
32. Perko, L. *Differential Equations and Dynamical Systems*; Springer: Berlin, Germany, 2013; Volume 7.
33. Marino, S.; Hogue, I.B.; Ray, C.J.; Kirschner, D.E. A methodology for performing global uncertainty and sensitivity analysis in systems biology. *J. Theor. Biol.* **2008**, *254*, 178–196. [CrossRef] [PubMed]
34. Acar, E.; Aplak, H.S. A Model Proposal for a Multi-Objective and Multi-Criteria Vehicle Assignment Problem: An Application for a Security Organization. *Math. Comput. Appl.* **2016**, *21*, 39. [CrossRef]
35. Madhi, M.; Mohamed, N. An Initial Condition Optimization Approach for Improving the Prediction Precision of a GM (1,1) Model. *Math. Comput. Appl.* **2017**, *22*, 21. [CrossRef]
36. Zhuang, K. Spatiotemporal Dynamics of a Delayed and Diffusive Viral Infection Model with Logistic Growth. *Math. Comput. Appl.* **2017**, *22*, 7. [CrossRef]
37. De Pillis, L.G.; Gu, W.; Fister, K.R.; Head, T.A.; Maples, K.; Murugan, A.; Neal, T.; Yoshida, K. Chemotherapy for tumors: An analysis of the dynamics and a study of quadratic and linear optimal controls. *Math. Biosci.* **2007**, *209*, 292–315. [CrossRef] [PubMed]
38. De Pillis, L.G.; Fister, K.R.; Gu, W.; Head, T.; Maples, K.; Neal, T.; Murugan, A.; Kozai, K. Optimal control of mixed immunotherapy and chemotherapy of tumors. *J. Biol. Syst.* **2008**, *16*, 51–80. [CrossRef]
39. Kirschner, D.; Lenhart, S.; Serbin, S. Optimal control of the chemotherapy of HIV. *J. Math. Biol.* **1997**, *35*, 775–792. [CrossRef] [PubMed]
40. Swan, G.W. Role of optimal control theory in cancer chemotherapy. *Math. Biosci.* **1990**, *101*, 237–284. [CrossRef]
41. Ratajczyk, E.; Ledzewicz, U.; Schättler, H. Optimal Control for a Mathematical Model of Glioma Treatment with Oncolytic Therapy and TNF-α Inhibitors. *J. Optim. Theory Appl.* **2018**, 456–477. [CrossRef]
42. Di Liddo, A. Optimal Control and Treatment of Infectious Diseases. The Case of Huge Treatment Costs. *Mathematics* **2016**, *4*, 21. [CrossRef]
43. Fleming, W.H.; Rishel, R.W. *Deterministic and Stochastic Optimal Control*; Springer: Berlin, Germany, 2012; Volume 1.
44. Ghaddar, C.K. Novel Spreadsheet Direct Method for Optimal Control Problems. *Math. Comput. Appl.* **2018**, *23*, 6. [CrossRef]
45. Lukes, D.L. *Differential Equations: Classical to Controlled*; Elsevier: New York, NY, USA, 1982.
46. Schattler, H.; Ledzewicz, U. *Optimal Control for Mathematical Models of Cancer Therapies*; Springer: New York, NY, USA, 2015.
47. Otieno, G.; Koske, J.K.; Mutiso, J.M. Cost effectiveness analysis of optimal malaria control strategies in kenya. *Mathematics* **2016**, *4*, 14. [CrossRef]
48. Davies, C.; Pan, H.; Godwin, J.; Gray, R.; Arriagada, R.; Raina, V.; Abraham, M.; Medeiros Alencar, V.H.; Badran, A.; Bonfill, X.; et al. Long-term effects of continuing adjuvant tamoxifen to 10 years versus stopping at 5 years after diagnosis of oestrogen receptor-positive breast cancer: ATLAS, a randomised trial. *Lancet* **2013**, *381*, 805–816, Erratum in **2013**, *381*, 804. [CrossRef]

Mathematical and Computational Applications

MDPI

Article

Cost-Effective Analysis of Control Strategies to Reduce the Prevalence of Cutaneous Leishmaniasis, Based on a Mathematical Model

Dibyendu Biswas [1], Suman Dolai [1], Jahangir Chowdhury [1], Priti K. Roy [1] and Ellina V. Grigorieva [2,*]

[1] Centre for Mathematical Biology and Ecology, Department of Mathematics, Jadavpur University, Kolkata 700032, India; dbiswasju@gmail.com (D.B.); suman.dolai18@gmail.com (S.D.); jahangirchowdhury.ju@gmail.com (J.C.); pritiju@gmail.com (P.K.R.)
[2] Department of Mathematics and Computer Sciences, Texas Womans University, Denton, TX 76204, USA
* Correspondence: egrigorieva@twu.edu

Received: 17 May 2018; Accepted: 16 July 2018; Published: 25 July 2018

Abstract: Leishmaniasis is a neglected tropical vector-borne epidemic disease, and its transmission is a complex process. Zoonotic transmission to humans or animals occurs through the bites of female *Phlebotominae* sand flies. Here, reservoir is considered as a major source of endemic pathogen pool for disease outbreak, and the role of more than one reservoir animal becomes indispensable. To study the role of the reservoir animals on disease dynamics, a mathematical model was constructed consisting of susceptible and infected populations of humans and two types of reservoir (animal) and vector populations, respectively. Our aim is to prevent the disease by applying a control theoretic approach, when more than one type of reservoir animal exists in the region. We use drugs like sodium stibogluconate and meglumine antimoniate to control the disease for humans and spray insecticide to control the sand fly population. Similarly, drugs are applied for infected reservoir animals of Types A and B. We calculated the cost-effectiveness of all possible combinations of the intervention and control policies. One of our findings is that the most cost-effective case for *Leishmania* control is the spray of insecticides for infected sand fly vector. Alternate strategic cases were compared to address the critical shortcomings of single strategic cases, and a range of control strategies were estimated for effective control and economical benefit of the overall control strategy. Our findings provide the most innovative techniques available for application to the successful eradication of cutaneous leishmaniasis in the future.

Keywords: vector borne disease; cutaneous leishmaniasis (CL); transmission probability; reservoir population; insecticide spraying; cost-effectiveness

---

1. Introduction

The disease leishmaniasis is caused by protozoan parasites from the genus *Leishmania* (Kinetoplastida: Trypanosomatidae) in their vertebrate hosts, including humans. Leishmaniasis is a neglected tropical disease [1] in the WHO list. *Leishmania* parasites are transmitted to other mammalian species through the vector bites of infected female phlebotomine sand flies [2,3]. Seventy animal species, including humans, have been found as natural reservoir hosts of *Leishmania* parasites [4]. Currently, the disease is endemic in eighty-eight countries [5,6]. These countries (e.g., Afghanistan, Algeria, Iran, Iraq, Pakistan, Brazil, Peru, etc.) account for more than 90% of the global cases of cutaneous leishmaniasis [7,8]. It is estimated that 12 million cases, comprising 1.5 to 2 million new cases, occur globally each year [9]. In India, Bihar and Rajasthan are the main affected states. The parasite's life-cycle occurs alternatively between a mammalian host and insect vectors. These vectors are phlebotomine sand flies (Diptera: Psychodidae, subfamily Phlebotominae). The *Leishmania* parasite thrives and spends a part of its life cycle within the female sand flies. The parasites are found alternatively as flagellated, motile promastigotes in the alimentary tract of phlebotomine sand flies, or as obligate intracellular aflagellate amastigotes in the phagolysosomes of mammalian host macrophages. Outside the vertebrate host, the *Leishmania* life cycle is confined to the digestive tract of sand flies, which become aggressively active during the warmer months in humid environments. It is established that mammals of several orders can be infected by the *Leishmania* sp. Natural *Leishmania* infections are found in a range of non-human mammal hosts (mainly marsupials, rodents, edentates, and carnivores). Reservoir implication is difficult because it is often specific to the nature of the local domain of animal context, and it depends on many variables (e.g., host abundance and distribution, infectiousness to the sand fly vector), which are rarely investigated. Domestic animals such as dogs can serve as reservoirs for the parasite. Transmission can occur from dog to sand fly, and from sand fly to human. Another important reservoir is the rodent population, which can serve as the cryptic reservoir for the persistence of the endemic state of infection, as recently suggested by many new research works. Recent evidence suggests that increasing species richness can lower or enhance the infection rate, which must be accounted for in this endemic state of disease despite the measures taken to control it [10]. Note that the prevalence of hosts in the affected region can influence disease transmission rates. We thus have considered disease transmission by the cryptic reservoirs, which are normally beyond control program coverage. The presence of more than two reservoir animals apart from the human host can exert a significant dilution effect due to selective pressure of host preference. It is widely believed that rodents can serve as the cryptic reservoir host in both urban and rural areas where they can outnumber the domestic animal population swiftly and help in the survival of the *Leishmania* pathogen. Recent evidence of this has come from Tunisia [11], where rodents are a potential reservoir of *Leishmania* pathogens. Presently, four clinical forms for leishmaniasis exist. These are cutaneous leishmaniasis (CL), mucocutaneous leishmaniasis (known as espundia), visceral leishmaniasis (known as kala-azar), and post-kala-azar dermal leishmaniasis (PKDL) [2,12]. Transmission modes are of zoonotic and/or anthroponotic type for cutaneous leishmaniasis (CL). In anthroponotic-type transmission, the sand flies are infected by a human during a blood meal, while in zoonotic transmission cycles animals serve as potential reservoir hosts [13].

The clinical symptom of CL is painless skin ulcers. Dermal changes may appear in only one to two weeks after being bitten by the sand fly. However, sometimes symptoms will not appear for months or years. The disease starts as an erythematous papule which gradually increases in size and turns into a nodule. It ultimately ulcerates and crusts over. The edge is frequently raised and distinct [9]. These are primarily ulcers in the mouth and nose, or on the lips. Other symptoms are stuffy or runny nose, nose bleeds, asphyxia. The fundamental prevention therapy for cutaneous leishmaniasis is pentavalent antimonial compounds. Presently, oral and topical prevention management are in practice.

Leishmania amazonensis is an intracellular protozoan parasite responsible for chronic cutaneous leishmaniasis. Cutaneous leishmaniasis is often self-healing, particularly in infection with *L. major* and *L. mexicana*. Therefore, prevention is not always recommended. However, if lesions do not spontaneously heal within six months or if the lesions are especially disfiguring and in a cosmetically sensitive area, prevention is indicated. Even though lesions may heal eventually in the absence of prevention, the process is often long and produces significant scarring, thereby justifying the use of chemotherapy. The goal of preventing cutaneous leishmaniasis is twofold: the eradication of amastigotes as well as reducing the size of the lesions so that healing will take place with minimal scarring.

Mathematical models can serve as a new tool to investigate the fate of infection dynamics with a multi-host environment. However, theoretical works on this topic are very rare, motivating us to study such dynamics with two different reservoir animals, as well as human and vector populations. Bacaer et al. [14] proposed a mathematical model taking the seasonal fluctuations into account in order to formulate an age-structured model and find the basic reproduction number based on periodic backgrounds. They suggested that the epidemic could be prevented if the vector population was reduced. Chaves et al. [3] studied a mathematical model for cutaneous leishmaniasis in the Americas and found conditions for the commencement of the infection. They presented a simple model to represent the dynamics of transmission densities of infected incidental hosts, infected reservoir hosts, and infected vectors. Miller and Huppert [10] studied multiple hosts of vector-borne infectious diseases from a significant fraction of the global infectious disease burden. They explored the relationship between host diversity, vector behavior, and disease risk. They developed a new dynamic model which included two distinct host species and one vector species with variable host preferences. They discussed the role of more than one reservoir population and how it could affect the disease transmission depending on host preferences and biting intensity. Biswas et al. [15] developed different models of cutaneous leishmaniasis consisting of different aspects of the disease transmission. We have previously considered susceptible and infected human and vector populations with the target of reducing the vector population so that the disease can be controlled. Then, considering the role of the macrophage for developing the disease intracellularly, we considered another stage of infection, as there is a transformation from the promastigote stage to a mastigote stage. Finally, we modelled the application of optimal drug dose to the infected macrophage cell and parasite populations in order to control the parasite population in the macrophage cells [16]. We have also studied a model through an impulsive strategy in a fixed time interval to observe perfect drug adherence behavior. The model has been analyzed to determine the threshold time interval and minimum effectiveness of drugs and also to observe the effect of an impulsive strategy in a non-fixed time interval on the system [17]. Recently, Biswas et al. [18] developed a model for evaluating the utility of awareness in controlling cutaneous leishmaniasis in affected regions where social mass media is present.

A recent mathematical model has focused on the transmission dynamics for anthroponotic cutaneous leishmaniasis in human populations and its control [19]. However, cutaneous leishmaniasis studies involving two reservoir animals has not yet been explored in the epidemiological literature. In this article, our study was motivated by the work of Huppert [10] and Chaves [3] using a set of ordinary differential equations as the foundation of the mathematical study of cutaneous leishmaniasis with two reservoir populations. We modelled the use of drugs for the human host, therapeutics for the animal reservoir, and insecticide application for the vector population, and studied the system using an optimal control technique. The aim of the optimal control problem was to minimize infection with cumulative control strategies to exert maximum benefit to the affected hosts. We also considered different control strategies and examined the impact of different combinations of these measures in controlling the disease. We used Pontryagin's minimum principle to derive the necessary conditions for the optimal control of the disease. By calculating the cost of drugs in each of the different strategies along with investigating the cost-effectiveness of the four control strategies under consideration, we determined the most effective strategy for eliminating leishmaniasis with minimum costs. The next section describes the formulation of the general model.

2. Model Formulation through Schematic Diagram and Its Validation

To formulate the mathematical model of cutaneous leishmaniasis, two types of animal reservoirs were considered: domestic and wild. For example, dog is the domestic animal reservoir for leishmaniasis (e.g., domestic dogs play the role of a reservoir host of *Leishmania donovani* in eastern Sudan), and some species of rodent (e.g., great gerbil, the crab-eating fox (*Cerdocyon thous*), opossums (*Didelphis* species), etc.) are wild animals worth consideration. Humans are in closer contact with domestic animals than they are with wild animals. We considered rodents to be cryptic reservoirs in nature, so they are away from human contact with respect to the domestic animals (e.g., dog). Furthermore, the recruitment rates and natural death rates of domestic animals and wild animals are different. For this reason, we considered two different classes of reservoirs: Type A and Type B.

We considered the transmission of the disease between four distinct populations: the human host population, Type A animals, Type B animals, and the vector population. The total human population was considered as:

$$S_H(t) + I_H(t),$$

where $S_H(t)$ denotes the susceptible individuals and $I_H(t)$ denotes the individuals infected with cutaneous leishmaniasis.

Reservoir classes are of two types. They are susceptible animals of Type A (i.e., $S_A(t)$) and susceptible animals of Type B (i.e., $S_B(t)$), and the corresponding infected classes are denoted by $I_A(t)$ and $I_B(t)$, respectively. Herein, the total reservoir population was considered as:

$$S_A(t) + I_A(t)$$

and

$$S_B(t) + I_B(t).$$

We considered the vector (sand fly) population to be of two categories: the susceptible vector population $S_F(t)$ and the infected vector population $I_F(t)$. The total vector (sand fly) population was considered as:

$$S_F(t) + I_F(t).$$

Here λ_H, λ_A, λ_B, and λ_F are the constant recruitment rates of humans, Type A animals, Type B animals, and sand fly. μ_H, μ_A, μ_B, and μ_F are their respective natural death rates.

For disease transmission, a susceptible human becomes infected through mass action after interaction with the infected vector, where β is the per capita biting rate of vector on human and π is the transmission probability per bite per human [14]. Thus, the infection term is frequency-dependent [20], and is described as:

$$\beta \pi I_F \frac{S_H}{S_H + I_H}.$$

A susceptible Type A animal (S_A) becomes infected by the bite (bite rate is α) of an infected sand fly (I_F) with the transmission probability ω. Here, the transmission of the disease is frequency-dependent [20] and is defined by:

$$\alpha \omega I_F \frac{S_A}{S_A + I_A}.$$

A susceptible Type B animal (S_B) becomes infected by the bite (bite rate is α) of an infected sand fly (I_F) with the transmission probability τ. The infection spreads as a frequency-dependent transmission [20], and is given as:

$$\alpha \tau I_F \frac{S_B}{S_B + I_B}.$$

A susceptible vector becomes infected after an interaction with an infected human at the rate β with transmission probability γ per bite from human to sand fly. Further, a susceptible sand fly (S_F) becomes infected by biting (per capita biting rate is α) infected animals of Type A (I_A) and Type B (I_B) with the transmission probability κ and ζ, respectively. Thus, a susceptible sand fly can be infected with the accumulation term:

$$\beta \gamma S_F \frac{I_H}{S_H + I_H} + \alpha \kappa S_F \frac{I_A}{S_A + I_A} + \alpha \zeta S_F \frac{I_B}{S_B + I_B}.$$

Abubakar et al. [21] found that leishmaniasis occurrence is a seasonal phenomenon in the regions of Africa. In the region, low transmission happens in the middle of the year and high transmission occurs in September. We can take biting rate as of the form: $\beta(t) = \beta_0(1 + \delta_r sin\frac{2\pi t}{365})$. The biting rate $\beta(t)$ is based on a period of 365 days and varies with temperature. Average biting rate and amplitude of seasonality are denoted by β_0 and δ_r [22,23], respectively.

A positive dog elimination strategy reduces the source of infection and prevents more non-infected sand flies from acquiring the parasites. An infected dog prevention strategy can reduce the source of infection, but without elimination. However, preventive measures on the dog do not necessarily eliminate the parasite from the dog's organ system. Dog vaccination does not eliminate the source of infection, but it protects the remaining susceptible dogs from becoming infected. Thus, there is a reduction in the number of infected dogs by natural elimination. The use of insecticide in impregnated dog collars works (if used by all dogs) by protecting the susceptible ones (similar to the vaccine activity) and isolating the source of infection (similar to positive dog elimination).

In nature, the prevalence of *Leishmania* infection in the entire sand fly population can be very low (<0.1%), even in areas of endemicity and high transmission. As a consequence, if the replacement of parasite is accelerated, there is not enough time for the parasite to mature inside the sand fly. Therefore, we do not consider the latent status in our model [24,25].

The dynamics of the disease in humans and sand flies and two reservoir (animal) populations are described in Figure 1. From the above description, we can construct the following system of differential equations in the form given below:

$$
\begin{aligned}
\dot{S}_H &= \lambda_H - \beta\pi I_F \frac{S_H}{S_H + I_H} - \mu_H S_H, \\
\dot{I}_H &= \beta\pi I_F \frac{S_H}{S_H + I_H} - \mu_H I_H, \\
\dot{S}_A &= \lambda_A - \alpha\omega I_F \frac{S_A}{S_A + I_A} - \mu_A S_A, \\
\dot{I}_A &= \alpha\omega I_F \frac{S_A}{S_A + I_A} - \mu_A I_A, \\
\dot{S}_B &= \lambda_B - \alpha\tau I_F \frac{S_B}{S_B + I_B} - \mu_B S_B, \\
\dot{I}_B &= \alpha\tau I_F \frac{S_B}{S_B + I_B} - \mu_B I_B, \\
\dot{S}_F &= \lambda_F - \beta\gamma S_F \frac{I_H}{S_H + I_H} - \alpha\kappa S_F \frac{I_A}{S_A + I_A} - \alpha\zeta S_F \frac{I_B}{S_B + I_B} - \mu_F S_F, \\
\dot{I}_F &= \beta\gamma S_F \frac{I_H}{S_H + I_H} + \alpha\kappa S_F \frac{I_A}{S_A + I_A} + \alpha\zeta S_F \frac{I_B}{S_B + I_B} - \mu_F I_F,
\end{aligned}
\tag{1}
$$

which satisfies the conditions $S_H + I_H > 0$, $S_A + I_A > 0$, $S_B + I_B > 0$, and $S_F + I_F > 0$.

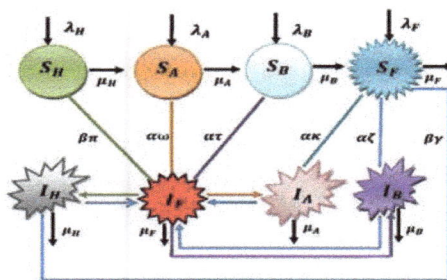

Figure 1. Schematic diagram of the model system (1) with flow of transmission.

2.1. Properties of the Model

All parameters of the model (1) are non-negative. Moreover, as the discussed model actually describes a living population, we assume the state variables to be non-negative at time $t = 0$.

Note that the total human population size, $S_H + I_H \to \frac{\lambda_H}{\mu_H}$, reservoir population size, $S_A + I_A \to \frac{\lambda_A}{\mu_A}$ and $S_B + I_B \to \frac{\lambda_B}{\mu_B}$ and $S_F + I_F \to \frac{\lambda_F}{\mu_F}$ as $t \to \infty$. It follows that the probable region is represented

by: $\mathfrak{D} = \left\{ (S_H, I_H, S_A, I_A, S_B, I_B, S_F, I_F) \in \mathbb{R}_+^8 : S_H, I_H, S_A, I_A, S_B, I_B, S_F, I_F \geq 0,\ S_H + I_H \leq \frac{\lambda_H}{\mu_H}, \right.$
$S_F + I_F \leq \frac{\lambda_F}{\mu_F},\ S_A + I_A \leq \frac{\lambda_A}{\mu_A}$ and $\left. S_B + I_B \leq \frac{\lambda_B}{\mu_B} \right\}$, a positive invariant region. Hence, the model is
mathematically efficient and proves to be adequate in estimating the dynamics of the model in the
positive invariant domain $\mathfrak{D}$. Here $\mathbb{R}_+^8$ denotes the non-negative space of $\mathbb{R}^8$, where we specify $\mathring{\mathfrak{D}}$ and
$\partial\mathfrak{D}$ to represent the boundary and the interior region of $\mathfrak{D}$, respectively.

We take the ratio between female sand fly vectors and humans as:

$$a = \frac{S_F + I_F}{S_H + I_H},\ \text{the number of female sandflies per human host.}$$

Here, a is constant because the population density of the host does not affect the number of blood
meals taken by a vector per unit time.

In either case, we take the ratio between female sand fly (vector) and reservoir as:

$$b = \frac{S_F + I_F}{S_A + I_A} \text{ and } c = \frac{S_F + I_F}{S_B + I_B}.$$

The parameters used in our model actually represent the infected cases of leishmaniasis in South
Sudan in the year 2012 [21]. In Figure 2, the data show a maximum prevalence in January–February.
After that, fewer cases occured than in previous months. Figure 3 shows that the total number of
leishmaniasis incidents for the year 2012 and the estimated model parameter values were almost
fitted with same line. The fitted model was further used to perform simulations which would
serve as a predictive tool for future cases of leishmaniasis for the forthcoming year (i.e., January
to December, 2013). Our model successfully predicted that, cumulatively, 3000 and 4770 new
cases of leishmaniasis were to be recorded during the beginning of January 2013 and the end of
December 2013, respectively. Hence, for the forthcoming year 2013, approximately 1770 new cases of
leishmaniasis were predicted. Our estimated model parameter values (Table 1) coincided with the real
data values. Thus, the initial human demographic parameters $S_H(0)$, $I_H(0)$ along with the initially
infected reservoir population $S_A(0)$, $I_A(0)$, $S_B(0)$, $I_B(0)$ and sand fly population $S_F(0)$, $I_F(0)$ were
estimated. Additionally, π, ω, and τ are disease transmission probability in humans and Type A and B
reservoirs, respectively.

From the model (1), the number of new CL cases I_{Hc} (infected human) can be written as:

$$\frac{dI_{Hc}}{dt} = \beta \pi I_F \frac{S_H}{S_H + I_H}. \tag{2}$$

This represents the rate of increase of the number of new CL occurrences, where π is the
transmission probability of the disease in humans, and β is the biting rate of sand fly on humans.

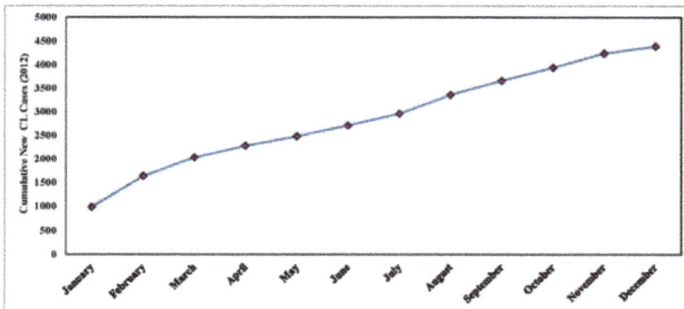

Figure 2. Cumulative number of leishmaniasis cases for the period January–December 2012. CL:
cutaneous leishmaniasis.

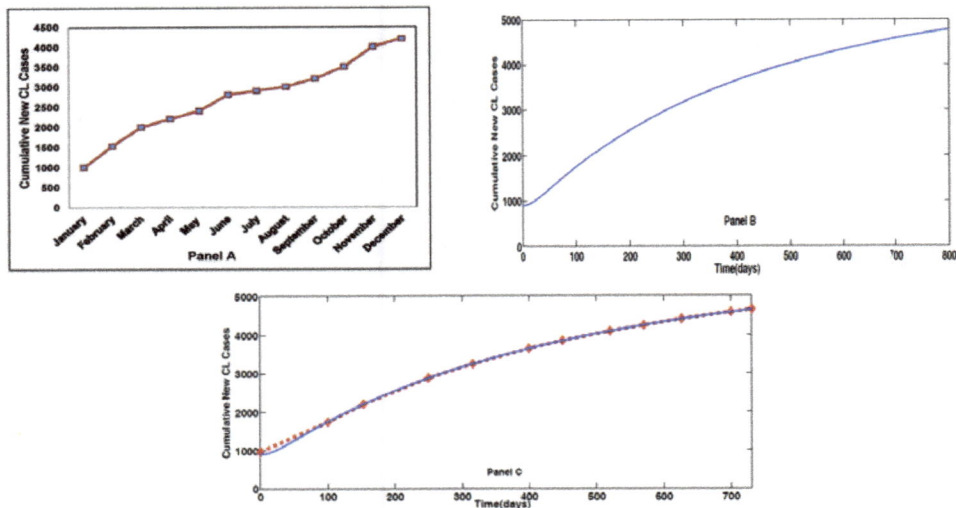

Figure 3. Panel A shows cumulative new leishmaniasis cases for the period January–December 2013 from data, and **Panel B** shows the model simulated data plotted using estimated parameter values. **Panel C** shows that the simulated parametric data values coincided with real data values, thus validating the model.

2.2. Existence Condition

The system (1) has two equilibrium points, one of which is disease-free equilibrium $E_0(\frac{\lambda_H}{\mu_H}, 0, \frac{\lambda_A}{\mu_A}, 0, \frac{\lambda_B}{\mu_B}, 0, \frac{\lambda_F}{\mu_F}, 0)$ and the other is endemic equilibrium $E^*(S_H^*, I_H^*, S_A^*, I_A^*, S_B^*, I_B^*, S_F^*, I_F^*)$, where $S_H^* = \frac{\lambda_H - \mu_H I_H^*}{\mu_H}$, $S_A^* = \frac{\lambda_A - \mu_A I_A^*}{\mu_A}$, $S_B^* = \frac{\lambda_B - \mu_B I_B^*}{\mu_B}$, $S_F^* = \frac{\lambda_F}{\frac{\beta^2 \pi \gamma I_F^*}{(\beta \pi I_F^* + \lambda_H)} + \frac{\alpha^2 \kappa \omega I_F^*}{(\alpha \omega I_F^* + \lambda_A)} + \frac{\alpha^2 \zeta \tau I_F^*}{(\alpha \tau I_F^* + \lambda_B)} + \mu_F}$,

$I_H^* = \frac{\lambda_H \beta \pi I_F^*}{(\beta \pi I_F^* + \lambda_H)\mu_H}$, $I_A^* = \frac{\alpha \omega \lambda_A I_F^*}{(\alpha \omega I_F^* + \lambda_A)\mu_A}$, $I_B^* = \frac{\lambda_B \alpha \tau I_F^*}{(\alpha \tau I_F^* + \lambda_B)\mu_B}$ and I_F^* are determined from the equation $A_1 I_F^{*3} + B_1 I_F^{*2} + C_1 I_F^* + D_1 = 0$, where

$$A_1 = \left((a_1 b_3 c_3 + b_1 a_3 c_3 + c_1 a_3 b_3)(d_1 - d_2) - d_2^2 a_3 b_3 c_3\right),$$
$$B_1 = \left((a_1(b_2 c_3 + b_3 c_2) + b_1(a_2 c_3 + a_3 c_2) + c_1(a_2 b_3 + b_3 a_2))(d_1 - d_2) - d_2^2(a_2 b_3 c_3 + b_2 c_3 a_3 + c_2 a_3 b_3)\right),$$
$$C_1 = \left((a_1 b_2 c_2 + b_1 c_2 a_2 + c_1 a_2 b_2)(d_1 - d_2) - d_2^2(a_2 b_2 c_3 + a_2 b_3 c_2 + a_3 b_2 c_2)\right),$$
$$D_1 = -d_2^2 a_2 b_2 c_2 \text{ and}$$
$$a_1 = \beta^2 \pi \gamma, a_2 = \lambda_H, a_3 = \beta \pi, b_1 = \alpha^2 \omega \kappa, b_2 = \lambda_A, b_3 = \alpha \omega,$$
$$c_1 = \alpha^2 \zeta \tau, c_2 = \lambda_B, c_3 = \alpha \tau, d_1 = \lambda_F, d_2 = \mu_F.$$

Then, the endemic equilibrium E^* exists if $\beta^2 \pi \gamma \lambda_A \lambda_B + \alpha^2 \kappa \omega \lambda_H \lambda_B + \alpha^2 \zeta \tau \lambda_A \lambda_H > \frac{\mu_F^2 \lambda_A \lambda_B \lambda_H}{\lambda_F}$.

Biological Interpretation: If the biting rate of the sand fly and transmission probabilities between infected human to vector, infected sand fly to human, infected animal to vector, and infected sand fly to animal are higher, then the system moves to its endemic state and disease persists.

2.3. Analytical Study of the Formulated Model

To find the basic reproduction ratio, four compartments $I_H{}'$, $I_A{}'$, $I_B{}'$, and $i_F{}'$ were considered here. We have

$$
\begin{bmatrix} I_H' \\ I_A' \\ I_B' \\ I_F' \end{bmatrix} = \begin{bmatrix} -\mu_H & 0 & 0 & \frac{\beta\pi S_H}{S_H+I_H} \\ 0 & -\mu_A & 0 & \frac{\alpha\omega S_A}{S_A+I_A} \\ 0 & 0 & -\mu_B & \frac{\alpha\tau S_B}{S_B+I_B} \\ \frac{\beta\gamma S_F}{S_H+I_H} & \frac{\alpha\kappa S_F}{S_A+I_A} & \frac{\alpha\zeta S_F}{S_B+I_B} & -\mu_F \end{bmatrix} \begin{bmatrix} I_H \\ I_A \\ I_B \\ I_F \end{bmatrix}. \tag{3}
$$

According to [26], the above square matrix can be re-written as the subtraction of two matrices. Thus, above matrix can be expressed as $Z' = (F - V)Z$. Here F is a non-negative matrix that contains the elements related to the generation of new infections and V is a diagonal non-negative matrix which contains the elements related to the loss of infections. F corresponds to the infectivity function of an infected population, and V^{-1} is a diagonal matrix indicating the loss of an infected population. At the disease-free equilibrium point, $(\bar{S}_H = \frac{\lambda_H}{\mu_H}, \bar{I}_H = 0, \bar{S}_A = \frac{\lambda_A}{\mu_A}, \bar{I}_A = 0, \bar{S}_B = \frac{\lambda_B}{\mu_B}, \bar{I}_B = 0, \bar{S}_F = \frac{\lambda_F}{\mu_F}, \bar{I}_F = 0)$, the matrix **NGO** (next generation operator) is $\mathbf{NGO} = FV^{-1}$, where

$$
F = \begin{pmatrix} 0 & 0 & 0 & \frac{\beta\pi \bar{S}_H}{\bar{S}_H+\bar{I}_H} \\ 0 & 0 & 0 & \frac{\alpha\omega S_A}{\bar{S}_A+\bar{I}_A} \\ 0 & 0 & 0 & \frac{\alpha\tau S_B}{\bar{S}_B+\bar{I}_B} \\ \frac{\beta\gamma \bar{S}_F}{\bar{S}_H+\bar{I}_H} & \frac{\alpha\kappa S_F}{\bar{S}_A+\bar{I}_A} & \frac{\alpha\zeta S_F}{\bar{S}_B+\bar{I}_B} & 0 \end{pmatrix}.
$$

and

$$
V = \begin{pmatrix} \mu_H & 0 & 0 & 0 \\ 0 & \mu_A & 0 & 0 \\ 0 & 0 & \mu_B & 0 \\ 0 & 0 & 0 & \mu_F \end{pmatrix}.
$$

This leads to

$$
FV^{-1} = \begin{pmatrix} 0 & 0 & 0 & \frac{\beta\pi}{\mu_F} \\ 0 & 0 & 0 & \frac{\alpha\omega}{\mu_F} \\ 0 & 0 & 0 & \frac{\alpha\tau}{\mu_F} \\ \frac{\beta\gamma\lambda_F}{\lambda_H\mu_F} & \frac{\alpha\kappa\lambda_F}{\lambda_A\mu_F} & \frac{\alpha\zeta\lambda_F}{\lambda_B\mu_F} & 0 \end{pmatrix}.
$$

From the above matrix, we can calculate the basic reproduction ratio from

$$
det(\mathbf{NGO} - \xi I) = 0.
$$

The basic reproduction ratio is the dominant eigenvalue of the matrix. It follows that the corresponding basic reproduction number is $(R_0) = \frac{\lambda_F(\alpha^2\tau\zeta\lambda_A\lambda_H+\alpha^2\kappa\omega\lambda_H\lambda_B+\beta^2\gamma\pi\lambda_B\lambda_A)}{\mu_F{}^2\lambda_H\lambda_A\lambda_B}$. Thus, if $R_0 < 1$, then the system is stable at disease-free equilibrium, while if $R_0 > 1$, the system is unstable at disease-free states and an endemic equilibrium state exists.

The calculation of disease-free equilibrium and its stability analysis, the mathematical description of basic reproduction number (R_0), and the existence and permanence of the endemic solution are discussed in Figure S1 in the Supplementary Materials.

By applying Routh–Hurwitz criteria, the system is stable around the endemic equilibrium point $E^*(S_H^*, I_H^*, S_A^*, I_A^*, S_B^*, I_B^*, S_F^*, I_F^*)$.

3. Control Theoretic Approach for the Proposed Model

To control the disease in humans, we modelled the use of the drugs sodium stibogluconate and meglumine antimoniate. To control the sand fly population, we modelled insecticide spraying. Additionally, we modelled the application of curative drugs to the infected Types A and B reservoir animals. These therapies, applied to infected human, animal, and vector populations, can limit the disease prevalence. Therefore, we prefer our control set quantifiable functions, defined on

$[t_{start}; t_{final}]$, with the constraints $0 \leq u_i(t) \leq 1$, $i = 1, 2, ..., 4$. We did not consider the side effects. We only considered the preventive case. The prevention phase is predetermined in any preventive situation. Thus, we introduce the optimal control schedule as:

(i)   The control variable u_1 acts as a prevention of human infection using drugs and the use of insecticide-treated bed nets to reduce infection.
(ii) The control variable u_2 represents the use of medicines for the prevention of an infected reservoir population of Type A.
(iii) The control variable u_3 represents the use of effective medicines for the prevention of an infected reservoir population of Type B.
(iv) The control variable u_4 corresponds to measures like spraying insecticide on residences and other places where sand flies can breed and live in order to kill them at all stages.

The parameters u_1, u_2, and u_3 reduce the transmission rate from sandflies to humans and animals of Type A and Type B. Therefore, the approach acts as a preventative method instead of as a treatment. Here the prevention of infectious humans and reservoir hosts and the reduction of vectors is possible either by taking medication that reduces the probability of the host getting infected (e.g., either a pill or a vaccine), or via the reduction of transmission by reducing the sand fly biting rate through behavior, such as less contact between sand fly vectors and people (netting, or reducing contact, application of repellent, etc.).

The most commonly used CL prevention techniques for infected host are the use of drugs, insecticide-treated bed nets, and insecticide spraying to reduce the sand fly population. Initially, each control strategy case and its effect on CL was observed separately. Figure S2 shows the effects of different cases of control strategies in comparison with no control for each population.

The aim is to reduce the rate of infection by introducing drug administration and insecticide spraying into the system. Here the levels of $u_1(t), u_2(t), u_3(t)$, and $u_4(t)$ are considered as the proper doses of drug and insecticide spraying in the system. There is a possibility of infection upon interaction between human and vector, as well as between reservoir and vector. Thus, in this circumstance, the infected human host and animal (reservoir) population are selected for drug application, and the vector population is selected for insecticide application. The control parameters $u_1(t), u_2(t), u_3(t)$, and $u_4(t)$ are introduced in the dynamical model system (1). We also consider that η_1, η_2, η_3, and η_4 are the efficacy of interventions applied in human, animal, and vector, respectively. Thus, the state system reduces to:

$$
\begin{aligned}
\dot{S}_H &= \lambda_H - \beta(1 - \eta_1 u_1(t))\pi I_F \frac{S_H}{S_H + I_H} - \mu_H S_H, \\
\dot{I}_H &= \beta(1 - \eta_1 u_1(t))\pi I_F \frac{S_H}{S_H + I_H} - \mu_H I_H, \\
\dot{S}_A &= \lambda_A - \alpha(1 - \eta_2 u_2(t))\omega I_F \frac{S_A}{S_A + I_A} - \mu_A S_A, \\
\dot{I}_A &= \alpha(1 - \eta_2 u_2(t))\omega I_F \frac{S_A}{S_A + I_A} - \mu_A I_A, \\
\dot{S}_B &= \lambda_B - \alpha(1 - \eta_3 u_3(t))\tau I_F \frac{S_B}{S_B + I_B} - \mu_B S_B, \\
\dot{I}_B &= \alpha(1 - \eta_3 u_3(t))\tau I_F \frac{S_B}{S_B + I_B} - \mu_B I_B, \\
\dot{S}_F &= \lambda_F - \beta(1 - \eta_4 u_4(t))\gamma S_F \frac{I_H}{S_H + I_H} - \alpha(1 - \eta_4 u_4(t))\kappa S_F \frac{I_A}{S_A + I_A} \\
&\quad - \alpha(1 - \eta_4 u_4(t))\zeta S_F \frac{I_B}{S_B + I_B} - (\mu_F + \mu_1)S_F,
\end{aligned}
\tag{4}
$$

$$\dot{I}_F = \beta(1-\mu_4 u_4(t))\gamma S_F \frac{I_H}{S_H+I_H} + \alpha(1-\eta_4 u_4(t))\kappa S_F \frac{I_A}{S_A+I_A}$$
$$+\alpha(1-\eta_4 u_4(t))\zeta S_F \frac{I_B}{S_B+I_B} - (\mu_F+\mu_1)I_F.$$

Note that μ_1 is the death rate of vector due to insecticide spraying, and $(\mu_F+\mu_1)$ is expressed as $\bar{\mu}_F$.

Endemic equilibrium with control are described in the following manner: $S_H^* = \frac{\lambda_H-\mu_H I_H^*}{\mu_H}$,
$S_A^* = \frac{\lambda_A-\mu_A I_A^*}{\mu_A}$, $S_B^* = \frac{\lambda_B-\mu_B I_B^*}{\mu_B}$, $S_F^* = \frac{\lambda_F}{\frac{\beta^2(1-\eta_1 u_1)(1-\eta_4 u_4)\pi\gamma I_F^*}{(\beta(1-\eta_1 u_1)\pi I_F^*+\lambda_H)} + \frac{\alpha^2(1-\eta_2 u_2)(1-\eta_4 u_4)\kappa\omega I_F^*}{(\alpha(1-\eta_2 u_2)\omega I_F^*+\lambda_A)} + \frac{\alpha^2(1-\eta_3 u_3)(1-\eta_4 u_4)\zeta\tau I_F^*}{(\alpha(1-\eta_3 u_3)\tau I_F^*+\lambda_B)} +\mu_F}$,
$I_H^* = \frac{\lambda_H\beta(1-\eta_1 u_1)\pi I_F^*}{(\beta(1-\eta_1 u_1)\pi I_F^*+\lambda_H)\mu_H}$, $I_A^* = \frac{\alpha(1-\eta_2 u_2)\omega\lambda_A I_F^*}{(\alpha(1-\eta_2 u_2)\omega I_F^*+\lambda_A)\mu_A}$, $I_B^* = \frac{\lambda_B\alpha(1-\eta_3 u_3)\tau I_F^*}{(\alpha(1-\eta_3 u_3)\tau I_F^*+\lambda_B)\mu_B}$, and I_F^* is determined from the equation $A_2 I_F^{*3}+B_2 I_F^{*2}+C_2 I_F^*+D_2=0$, where

$$A_2 = \left((a_1' b_3' c_3' + b_1' a_3' c_3' + c_1' a_3' b_3')(d_1'-d_2') - d_2'^2 a_3' b_3' c_3'\right),$$
$$B_2 = \left((a_1'(b_2' c_3'+b_3' c_2') + b_1'(a_2' c_3'+a_3' c_2') + c_1'(a_2' b_3'+b_3' a_2'))(d_1'-d_2') - d_2'^2(a_2' b_3' c_3'+b_2' c_3' a_3'+c_2' a_3' b_3')\right),$$
$$C_2 = \left((a_1' b_2' c_2' + b_1' c_2' a_2' + c_1' a_2' b_2')(d_1'-d_2') - d_2'^2(a_2' b_2' c_3'+a_2' b_3' c_2'+a_3' b_2' c_2')\right),$$
$$D_2 = -d_2'^2 a_2' b_2' c_2' \text{ and}$$
$$a_1' = \beta^2(1-\eta_1 u_1)(1-\eta_4 u_4)\pi\gamma, a_2' = \lambda_H, a_3' = \beta(1-\eta_1 u_1)\pi, b_1' = \alpha^2(1-\eta_2 u_2)(1-\eta_4 u_4)\omega\kappa,$$
$b_2' = \lambda_A, b_3' = \alpha(1-\eta_2 u_2)\omega,$
$$c_1' = \alpha^2(1-\eta_3 u_3)(1-\eta_4 u_4)\zeta\tau, c_2' = \lambda_B, c_3' = \alpha(1-\eta_3 u_3)\tau, d_1' = \lambda_F, d_2' = \bar{\mu}_F.$$

The basic reproduction number with control parameter can be expressed in the form:

$$\bar{R}_0 = \frac{\lambda_F(\alpha^2\tau\zeta\lambda_A\lambda_H(1-\eta_3 u_3)(1-\eta_4 u_4)+\alpha^2\kappa\omega\lambda_H\lambda_B(1-\eta_2 u_2)(1-\eta_4 u_4)+\beta^2\gamma\pi\lambda_B\lambda_A(1-\eta_1 u_1)(1-\eta_4 u_4))}{\mu_F^2\lambda_H\lambda_A\lambda_B}.$$

The purpose of our optimal control approach is to minimize the infected human and reservoir populations, reduce the vector population, and minimize the cost of prevention by using the possible minimal control variables $u_1(t)$, $u_2(t)$, $u_3(t)$, and $u_4(t)$. Now, we construct the objective function:

$$J(u_1,u_2,u_3,u_4) = \int_{t_{start}}^{t_{final}} [\rho_1 I_H(t) + \rho_2 I_A(t) + \rho_3 I_B(t) + \rho_4 S_F(t) + \rho_5 I_F(t)$$
$$+\frac{1}{2}(w_1 u_1^2 + w_2 u_2^2 + w_3 u_3^2 + w_4 u_4^2)]dt. \tag{5}$$

In the objective function, $\rho_1, \rho_2, \rho_3, \rho_4$, and ρ_5 represent the weight constants of the infected human, infected animal of Type A, infected animal of Type B, and vector population (susceptible and infected), respectively, and w_1, w_2, w_3, and w_4 are weight constants for the prevention of transmission to human, animals of Types A and B, and vector control, respectively. The terms $\frac{1}{2}w_1 u_1^2$, $\frac{1}{2}w_2 u_2^2$, $\frac{1}{2}w_3 u_3^2$, $\frac{1}{2}w_4 u_4^2$ describe the cost of disease prevention. The first control policy $u_1(t)$ comes from the prevention policy of the infected human class. The cost related with the second control $u_2(t)$ is the medication for reservoir Type A. The cost associated with the third control technique $u_3(t)$ is the prevention using drugs in reservoir Type B, and the cost associated with the fourth control strategy $u_4(t)$ arises from applying different types of pesticides or insecticide to kill sand flies at all life stages. Here we must consider that cost is proportional to the square of the corresponding control function. Thus, the objective function can be defined as:

$$J(u_1,u_2,u_3,u_4); (u_1,u_2,u_3,u_4) \in D.$$

Next, we consider the following minimization problem:

$$J(u_1,u_2,u_3,u_4) \to min; (u_1,u_2,u_3,u_4) \in D,$$

where the corresponding control set D is expressed in the form:

$$D = \{(u_1, u_2, u_3, u_4) : u_i(t) \text{ is the Lebesgue measurable function on}$$

$$[t_{start}; t_{final}] \text{ and } 0 \leq u_i(t) \leq 1, i = 1, 2, 3, 4\}.$$

Existence of the Optimal Control

For bounded Lebesgue measurable controls and non-negative initial conditions, there exists a non-negative bounded solution of the state system [27,28]. To find the optimal solution of the system, first, we define the Lagrangian (L) of the control system (4) as:

$$L = \rho_1 I_H + \rho_2 I_A + \rho_3 I_B + \rho_4 S_F + \rho_5 I_F + \frac{1}{2}(w_1 u_1{}^2 + w_2 u_2{}^2 + w_3 u_3{}^2 + w_4 u_4{}^2).$$

Theorem 1. *For the system* (4) *with the non-negative initial conditions, there exists an optimal control* $u^* = (u_1^*, u_2^*, u_3^*, u_4^*)$, *such that* $J(u_1^*, u_2^*, u_3^*, u_4^*) = min\{j(u_1, u_2, u_3, u_4) : (u_1, u_2, u_3, u_4) \in D\}$.

Proof. We use the result for the existence of an optimal control in [27,29] as the control variables, and the state variables are non-negative. Now, the objective function in u_1, u_2, u_3, and u_4 satisfies the condition of convexity. By definition, the control set D is convex and closed. The existence of the optimal control is confirmed by the boundedness of the solutions of the state system. Additionally, the Lagrangian, L, is convex on the control set D. This proves the existence of an optimal control. □

Now we apply Pontryagin's minimal principle [30] to the control system (4). To solve the optimal control problem, we define the Hamiltonian H with the help of the Lagrangian as follows:

$$\begin{aligned} H &= L + \lambda_1 \frac{dS_H}{dt} + \lambda_2 \frac{dI_H}{dt} + \lambda_3 \frac{dS_A}{dt} + \lambda_4 \frac{dI_A}{dt} + \lambda_5 \frac{dS_B}{dt} + \lambda_6 \frac{dI_B}{dt} + \lambda_7 \frac{dS_F}{dt} + \lambda_8 \frac{dI_F}{dt} \\ &+ \xi_1 u_1 + \xi_2(1 - u_1) + \xi_3 u_2 + \xi_4(1 - u_2) + \xi_5 u_3 + \xi_6(1 - u_3) \\ &+ \xi_7 u_4 + \xi_8(1 - u_4). \end{aligned}$$

where λ_j, $j = 1, 2, ..., 8$ are the adjoint variables and ξ_i, $i = 1, 2, ..., 8$ are the penalty multipliers:

(i) $u_1 = 0$, where $\xi_1 \neq 0$ *and* $\xi_2 = 0$ *and* $\xi_3 = \xi_4 = \xi_5 = \xi_6 = \xi_7 = \xi_8 = 0$ and
(ii) $u_1 = 1$, where $\xi_1 = 0$ *and* $\xi_2 \neq 0$ *and* $\xi_3 = \xi_4 = \xi_5 = \xi_6 = \xi_7 = \xi_8 = 0$,
(i) $u_2 = 0$, where $\xi_3 \neq 0$ *and* $\xi_4 = 0$ *and* $\xi_1 = \xi_2 = \xi_5 = \xi_6 = \xi_7 = \xi_8 = 0$ and
(ii) $u_2 = 1$, where $\xi_3 = 0$ *and* $\xi_4 \neq 0$ *and* $\xi_1 = \xi_2 = \xi_5 = \xi_6 = \xi_7 = \xi_8 = 0$,
(i) $u_3 = 0$, where $\xi_5 \neq 0$ *and* $\xi_6 = 0$ *and* $\xi_1 = \xi_2 = \xi_3 = \xi_4 = \xi_7 = \xi_8 = 0$ and
(ii) $u_3 = 1$, where $\xi_5 = 0$ *and* $\xi_6 \neq 0$ *and* $\xi_1 = \xi_2 = \xi_3 = \xi_4 = \xi_7 = \xi_8 = 0$,
(i) $u_4 = 0$, where $\xi_7 \neq 0$ *and* $\xi_8 = 0$ *and* $\xi_1 = \xi_2 = \xi_3 = \xi_4 = \xi_5 = \xi_6 = 0$ and
(ii) $u_4 = 1$, where $\xi_7 = 0$ *and* $\xi_8 \neq 0$ *and* $\xi_1 = \xi_2 = \xi_3 = \xi_4 = \xi_5 = \xi_6 = 0$.

The corresponding adjoint equations are given by,

$$\begin{aligned} \frac{d\lambda_1}{dt} &= -\frac{\partial H}{\partial S_H}, \frac{d\lambda_2}{dt} = -\frac{\partial H}{\partial I_H}, \frac{d\lambda_3}{dt} = -\frac{\partial H}{\partial S_A}, \frac{d\lambda_4}{dt} = -\frac{\partial H}{\partial I_A}, \\ \frac{d\lambda_5}{dt} &= -\frac{\partial H}{\partial S_B}, \frac{d\lambda_6}{dt} = -\frac{\partial H}{\partial I_B}, \frac{d\lambda_7}{dt} = -\frac{\partial H}{\partial S_F}, \frac{d\lambda_8}{dt} = -\frac{\partial H}{\partial I_F}, \end{aligned} \qquad (6)$$

where

$$
\begin{aligned}
\frac{\partial H}{\partial S_H} &= -\lambda_1(t)\Big(\beta(1-\eta_1 u_1)\frac{\pi I_F(t)I_H(t)}{(S_H(t)+I_H(t))^2}+\mu_H\Big)+\lambda_2(t)\Big(\beta(1-\eta_1 u_1)\frac{\pi I_F(t)I_H(t)}{(S_H(t)+I_H(t))^2}\Big)\\
&\quad +\lambda_7(t)\Big(\beta(1-\eta_4 u_4)\frac{\gamma S_F(t)I_H(t)}{(S_H(t)+I_H(t))^2}\Big)-\lambda_8(t)\Big(\beta(1-\eta_4 u_4)\frac{\gamma S_F(t)I_H(t)}{(S_H(t)+I_H(t))^2}\Big),\\[4pt]
\frac{\partial H}{\partial I_H} &= \rho_1+\lambda_1(t)\Big(\beta(1-\eta_1 u_1)\frac{\pi I_F(t)S_H(t)}{(S_H(t)+I_H(t))^2}\Big)-\lambda_2(t)\Big(\beta(1-\eta_1 u_1)\frac{\pi I_F(t)S_H(t)}{(S_H(t)+I_H(t))^2}+\mu_H\Big)\\
&\quad -\lambda_7(t)\Big(\beta(1-\eta_4 u_4)\frac{\gamma S_F(t)}{(S_H(t)+I_H(t))^2}\Big)+\lambda_8(t)\Big(\beta(1-\eta_4 u_4)\frac{\gamma S_F(t)}{(S_H(t)+I_H(t))^2}\Big),\\[4pt]
\frac{\partial H}{\partial S_A} &= -\lambda_3(t)\Big(\alpha(1-\eta_2 u_2)\frac{\omega I_F(t)I_A(t)}{(S_A(t)+I_A(t))^2}+\mu_A\Big)+\lambda_4(t)\Big(\alpha(1-\eta_2 u_2)\frac{\omega I_F(t)I_A(t)}{(S_A(t)+I_A(t))^2}\Big)\\
&\quad +\lambda_7(t)\Big(\alpha(1-\eta_4 u_4)\frac{\kappa S_F(t)I_A(t)}{(S_A(t)+I_A(t))^2}\Big)-\lambda_8(t)\Big(\beta(1-\eta_4 u_4)\frac{\kappa S_F(t)I_A(t)}{(S_A(t)+I_A(t))^2}\Big),\\[4pt]
\frac{\partial H}{\partial I_A} &= \rho_2+\lambda_3(t)\Big(\alpha(1-\eta_2 u_2)\frac{\omega I_F(t)S_A(t)}{(S_A(t)+I_A(t))^2}\Big)-\lambda_4(t)\Big(\alpha(1-\eta_2 u_2)\frac{\omega I_F(t)S_A(t)}{(S_A(t)+I_A(t))^2}+\mu_A\Big)\\
&\quad -\lambda_7(t)\Big(\alpha(1-\eta_4 u_4)\frac{\kappa S_F(t)S_A(t)}{(S_A(t)+I_A(t))^2}\Big)+\lambda_8(t)\Big(\alpha(1-\eta_4 u_4)\frac{\kappa S_F(t)S_A(t)}{(S_A(t)+I_A(t))^2}\Big),\\[4pt]
\frac{\partial H}{\partial S_B} &= -\lambda_5(t)\Big(\alpha(1-\eta_3 u_3)\frac{\tau I_F(t)I_B(t)}{(S_B(t)+I_B(t))^2}+\mu_B\Big)+\lambda_6(t)(\alpha(1-\eta_3 u_3)\frac{\tau I_F(t)I_B(t)}{(S_B(t)+I_B(t))^2}\Big)\\
&\quad +\lambda_7(t)\Big(\alpha(1-\eta_4 u_4)\frac{\zeta S_F(t)I_B(t)}{(S_B(t)+I_B(t))^2}\Big)-\lambda_8(t)\Big(\beta(1-\eta_4 u_4)\frac{\zeta S_F(t)I_B(t)}{(S_B(t)+I_B(t))^2}\Big),\qquad(7)\\[4pt]
\frac{\partial H}{\partial I_B} &= \rho_3+\lambda_5(t)\Big(\alpha(1-\eta_3 u_3)\frac{\tau I_F(t)S_B(t)}{(S_B(t)+I_B(t))^2}\Big)-\lambda_6(t)(\alpha(1-\eta_3 u_3)\frac{\tau I_F(t)S_B(t)}{(S_B(t)+I_B(t))^2}+\mu_B\Big)\\
&\quad -\lambda_7(t)\Big(\alpha(1-\eta_4 u_4)\frac{\zeta S_F(t)S_B(t)}{(S_B(t)+I_B(t))^2}\Big)+\lambda_8(t)\Big(\beta(1-\eta_4 u_4)\frac{\zeta S_F(t)S_B(t)}{(S_B(t)+I_B(t))^2}\Big),\\[4pt]
\frac{\partial H}{\partial S_F} &= \rho_4-\lambda_7(t)\Big\{\beta(1-\eta_4 u_4)\frac{\gamma I_H(t)}{(S_H(t)+I_H(t))}-\alpha(1-\eta_4 u_4)\frac{\kappa I_A(t)}{(S_A(t)+I_A(t))}\\
&\quad -\alpha(1-\eta_4 u_4)\frac{\zeta I_B(t)}{(S_B(t)+I_B(t))}-\mu_F\Big\}+\lambda_8(t)\Big\{\beta(1-\eta_4 u_4)\frac{\gamma I_H(t)}{(S_H(t)+I_H(t))}\\
&\quad +\alpha(1-\eta_4 u_4)\frac{\kappa I_A(t)}{(S_A(t)+I_A(t))}+\alpha(1-\eta_4 u_4)\frac{\zeta I_B(t)}{(S_B(t)+I_B(t))}\Big\},\\[4pt]
\frac{\partial H}{\partial I_F} &= \rho_5-\lambda_1(t)\beta(1-\eta_1 u_1)\frac{\pi S_H(t)}{(S_H(t)+I_H(t))}+\lambda_2(t)\beta(1-\eta_1 u_1)\frac{\pi S_H(t)}{(S_H(t)+I_H(t))}\\
&\quad -\lambda_3(t)\alpha(1-\eta_2 u_2)\frac{\omega S_A(t)}{(S_A(t)+I_A(t))}+\lambda_4(t)\alpha(1-\eta_2 u_2)\frac{\omega S_A(t)}{(S_A(t)+I_A(t))}\\
&\quad -\lambda_5(t)\alpha(1-\eta_3 u_3)\frac{\tau S_B(t)}{(S_B(t)+I_B(t))}+\lambda_6(t)\alpha(1-\eta_3 u_3)\frac{\tau S_B(t)}{(S_B(t)+I_B(t))}-\lambda_8(t)\bar\mu_F.
\end{aligned}
$$

Again, H can be written as

$$
\begin{aligned}
H = \ & \frac{1}{2}w_1u_1{}^2 - \lambda_1(t)\left(1 - \eta_1 u_1\right)\beta\pi I_F(t)\frac{S_H(t)}{S_H(t) + I_H(t)} + \lambda_2(t)\left(1 - \eta_1 u_1\right)\beta\pi I_F(t)\frac{S_H(t)}{S_H(t) + I_H(t)} \\
& + \frac{1}{2}w_2u_2{}^2 - \lambda_3(t)\left(1 - \eta_2 u_2\right)\alpha\omega I_F(t)\frac{S_A(t)}{S_A(t) + I_A(t)} + \lambda_4(t)\left(1 - \eta_2 u_2\right)\alpha\omega I_F(t)\frac{S_A(t)}{S_A(t) + I_A(t)} \\
& + \frac{1}{2}w_3u_3{}^2 - \lambda_5(t)\left(1 - \eta_3 u_3\right)\alpha\tau I_F(t)\frac{S_B(t)}{S_B(t) + I_B(t)} + \lambda_6(t)\left(1 - \eta_2 u_2\right)\alpha\tau I_F(t)\frac{S_B(t)}{S_B(t) + I_B(t)} \\
& + \frac{1}{2}w_4u_4{}^2 - \lambda_7(t)\left(1 - \eta_4 u_4\right)\left(\beta\gamma S_F(t)\frac{I_H(t)}{S_H(t) + I_H(t)} + \alpha\kappa S_F(t)\frac{I_A(t}{S_A(t) + I_A(t)} + \alpha\zeta S_F(t)\frac{I_B(t)}{S_B(t) + I_B(t)}\right) \\
& + \lambda_8(t)\left(1 - \eta_4 u_4\right)\left(\beta\gamma S_F(t)\frac{I_H(t)}{S_H(t) + I_H(t)} + \alpha\kappa S_F(t)\frac{I_A(t)}{S_A(t) + I_A(t)} + \alpha\zeta S_F(t)\frac{I_B(t)}{S_B(t) + I_B(t)}\right) \\
& + \xi_1 u_1 + \xi_2\left(1 - u_1\right) + \xi_3 u_2 + \xi_4\left(1 - u_2\right) \\
& + \xi_5 u_3 + \xi_6\left(1 - u_3\right) + \xi_7 u_4 + \xi_8\left(1 - u_4\right) \\
& + \text{other terms without } u_1, u_2, u_3, \text{ and } u_4.
\end{aligned}
$$

Now, differentiating H partially with respect to $u_1, u_2, u_3,$ and u_4, we get:

$$
\frac{\partial H}{\partial u_1} = w_1 u_1 + \lambda_1(t)\eta_1\beta\pi I_F(t)\frac{S_H(t)}{S_H(t) + I_H(t)} - \lambda_2(t)\eta_1\beta\pi I_F(t)\frac{S_H(t)}{S_H(t) + I_H(t)} + \xi_1 - \xi_2
$$

$$
\frac{\partial H}{\partial u_2} = w_2 u_2 + \lambda_3(t)\eta_2\alpha\omega I_F(t)\frac{S_A(t)}{S_A(t) + I_A(t)} - \lambda_4(t)\eta_2\alpha\omega I_F(t)\frac{S_A(t)}{S_A(t) + I_A(t)} + \xi_3 - \xi_4
$$

$$
\frac{\partial H}{\partial u_3} = w_3 u_3 + \lambda_5(t)\eta_3\alpha\tau I_F(t)\frac{S_B(t)}{S_B(t) + I_B(t)} - \lambda_6(t)\eta_3\alpha\tau I_F(t)\frac{S_B(t)}{S_B(t) + I_B(t)} + \xi_5 - \xi_6
$$

$$
\frac{\partial H}{\partial u_4} = w_4 u_4 + \lambda_7(t)\eta_4\left(\beta\gamma S_F(t)\frac{I_H(t)}{S_H(t) + I_H(t)} + \alpha\kappa S_F(t)\frac{I_A(t)}{S_A(t) + I_A(t)} + \alpha\zeta S_F(t)\frac{I_B(t)}{S_B(t) + I_B(t)}\right)
$$

$$
-\lambda_8(t)\eta_4\left(\beta\gamma S_F(t)\frac{I_H(t)}{S_H(t) + I_H(t)} + \alpha\kappa S_F(t)\frac{I_A(t)}{S_A(t) + I_A(t)} + \alpha\zeta S_F(t)\frac{I_B(t)}{S_B(t) + I_B(t)}\right) + \xi_7 - \xi_8.
$$

These expressions should be equal to zero at u_1^*, u_2^*, u_3^* and u_4^*. Thus,

$$
w_1 u_1 + \lambda_1(t)\eta_1\beta\pi I_F(t)\frac{S_H(t)}{S_H(t) + I_H(t)} - \lambda_2(t)\eta_1\beta\pi I_F(t)\frac{S_H(t)}{S_H(t) + I_H(t)} + \xi_1 - \xi_2 = 0 \text{ at } u_1^*,
$$

$$
w_2 u_2 + \lambda_3(t)\eta_2\alpha\omega I_F(t)\frac{S_A(t)}{S_A(t) + I_A(t)} - \lambda_4(t)\eta_2\alpha\omega I_F(t)\frac{S_A(t)}{S_A(t) + I_A(t)} + \xi_3 - \xi_4 = 0 \text{ at } u_2^*,
$$

$$
w_3 u_3 + \lambda_5(t)\eta_3\alpha\tau I_F(t)\frac{S_B(t)}{S_B(t) + I_B(t)} - \lambda_6(t)\eta_3\alpha\tau I_F(t)\frac{S_B(t)}{S_B(t) + I_B(t)} + \xi_5 - \xi_6 = 0 \text{ at } u_3^*, \text{ and}
$$

$$
w_4 u_4 + \lambda_7(t)\eta_4\left(\beta\gamma S_F(t)\frac{I_H(t)}{S_H(t) + I_H(t)} + \alpha\kappa S_F(t)\frac{I_A(t)}{S_A(t) + I_A(t)} + \alpha\zeta S_F(t)\frac{I_B(t)}{S_B(t) + I_B(t)}\right)
$$

$$
-\lambda_8(t)\eta_4\left(\beta\gamma S_F(t)\frac{I_H(t)}{S_H(t) + I_H(t)} + \alpha\kappa S_F(t)\frac{I_A(t)}{S_A(t) + I_A(t)} + \alpha\zeta S_F(t)\frac{I_B(t)}{S_B(t) + I_B(t)}\right) + \xi_7 - \xi_8 = 0
$$

at u_4^*.

Solution for the optimal control yields:

$$
u_1^* = \frac{(\lambda_2(t) - \lambda_1(t))\eta_1\beta\pi I_F(t)\frac{S_H(t)}{S_H(t) + I_H(t)} + \xi_2 - \xi_1}{w_1},
$$

$$
u_2^* = \frac{(\lambda_4(t) - \lambda_3(t))\eta_2\alpha\omega I_F(t)\frac{S_A(t)}{S_A(t) + I_A(t)} + \xi_4 - \xi_3}{w_2},
$$

$$u_3^* = \frac{(\lambda_6(t) - \lambda_5(t))\eta_3 \alpha \tau I_F(t)\frac{S_B(t)}{S_B(t)+I_B(t)} + \xi_6 - \xi_5}{w_3}, \quad \text{and}$$

$$u_4^* = \frac{(\lambda_8(t) - \lambda_7(t))\eta_4\left(\beta\gamma S_F(t)\frac{I_H(t)}{S_H(t)+I_H(t)} + \alpha\kappa S_F(t)\frac{I_A(t)}{S_A(t)+I_A(t)} + \alpha\zeta S_F(t)\frac{I_B(t)}{S_B(t)+I_B(t)}\right) + \xi_8 - \xi_7}{w_4}.$$

There are three cases to be considered for $u_1^*(t)$.

Case 1: $0 < u_1^* < 1$, subject to the condition $\xi_1 = \xi_2 = 0$:

$$u_1^* = \frac{(\lambda_2(t) - \lambda_1(t))\eta_1\beta\pi I_F(t)\frac{S_H(t)}{S_H(t)+I_H(t)}}{w_1}. \tag{8}$$

Case 2: $u_1^* = 0$, subject to the condition $\xi_1 \neq 0$ and $\xi_2 = 0$:

$$(\lambda_2(t) - \lambda_1(t))\eta_1\beta\pi I_F(t)\frac{S_H(t)}{S_H(t) + I_H(t)} = \xi_1. \tag{9}$$

Case 3: $u_1^* = 1$, subject to the condition $\xi_1 = 0$ and $\xi_2 \neq 0$:

$$(\lambda_2(t) - \lambda_1(t))\eta_1\beta\pi I_F(t)\frac{S_H(t)}{S_H(t) + I_H(t)} + \xi_2 = w_1. \tag{10}$$

Therefore, the optimal control $u_1^*(t)$ can be stated as:

$$u_1^* = max\left(min\left(\frac{(\lambda_2(t) - \lambda_1(t))\eta_1\beta\pi I_F(t)\frac{S_H(t)}{S_H(t)+I_H(t)}}{w_1}, 1\right), 0\right). \tag{11}$$

There are also three cases to be considered for $u_2^*(t)$.

Case 1: $0 < u_2^* < 1$, subject to the condition $\xi_3 = \xi_4 = 0$:

$$u_2^* = \frac{(\lambda_4(t) - \lambda_3(t))\eta_2\alpha\omega I_F(t)\frac{S_A(t)}{S_A(t)+I_A(t)}}{w_2} \tag{12}$$

Case 2: $u_2^* = 0$, subject to the condition $\xi_3 \neq 0$ and $\xi_4 = 0$:

$$(\lambda_4(t) - \lambda_3(t))\eta_2\alpha\omega I_F(t)\frac{S_A(t)}{S_A(t) + I_A(t)} = \xi_3. \tag{13}$$

Case 3: $u_2^* = 1$, subject to the condition $\xi_3 = 0$ and $\xi_4 \neq 0$:

$$(\lambda_4(t) - \lambda_3(t))\eta_2\alpha\omega I_F(t)\frac{S_A(t)}{S_A(t) + I_A(t)} + \xi_4 = w_2. \tag{14}$$

Therefore, the optimal control $u_2^*(t)$ can be stated as:

$$u_2^* = max\left(min\left(\frac{(\lambda_4(t) - \lambda_3(t))\eta_2\alpha\omega I_F(t)\frac{S_A(t)}{S_A(t)+I_A(t)}}{w_2}, 1\right), 0\right). \tag{15}$$

There are three cases to be considered for $u_3^*(t)$.

Case 1: $0 < u_3^* < 1$, subject to the condition $\xi_5 = \xi_6 = 0$:

$$u_3^* = \frac{(\lambda_6(t) - \lambda_5(t))\eta_3\alpha\tau I_F(t)\frac{S_B(t)}{S_B(t)+I_B(t)}}{w_3}. \tag{16}$$

Case 2: $u_3^* = 0$, subject to the condition $\xi_5 \neq 0$ and $\xi_6 = 0$:

$$(\lambda_6(t) - \lambda_5(t))\eta_3\alpha\tau I_F(t)\frac{S_B(t)}{S_B(t)+I_B(t)} = \xi_5. \tag{17}$$

Case 3: $u_3^* = 1$, subject to the condition $\xi_5 = 0$ and $\xi_6 \neq 0$:

$$(\lambda_6(t) - \lambda_5(t))\eta_3\alpha\tau I_F(t)\frac{S_B(t)}{S_B(t)+I_B(t)} + \xi_6 = w_3. \tag{18}$$

Therefore, the optimal control $u_3^*(t)$ can be stated as:

$$u_3^* = max\left(min\left(\frac{(\lambda_6(t) - \lambda_5(t))\eta_3\alpha\tau I_F(t)\frac{S_B(t)}{S_B(t)+I_B(t)}}{w_3}, 1\right), 0\right). \tag{19}$$

There are also three cases to be considered for $u_4^*(t)$.

Case 1: $0 < u_4^* < 1$, subject to the condition $\xi_7 = \xi_8 = 0$:

$$u_4^* = \frac{(\lambda_8(t) - \lambda_7(t))\eta_4\left(\beta\gamma S_F(t)\frac{I_H(t)}{S_H(t)+I_H(t)} + \alpha\kappa S_F(t)\frac{I_A(t)}{S_A(t)+I_A(t)} + \alpha\zeta S_F(t)\frac{I_B(t)}{S_B(t)+I_B(t)}\right)}{w_4}. \tag{20}$$

Case 2: $u_4^* = 0$, subject to the condition $\xi_7 \neq 0$ and $\xi_8 = 0$:

$$(\lambda_8(t) - \lambda_7(t))\eta_4\left(\beta\gamma S_F(t)\frac{I_H(t)}{S_H(t)+I_H(t)} + \alpha\kappa S_F(t)\frac{I_A(t)}{S_A(t)+I_A(t)} + \alpha\zeta S_F(t)\frac{I_B(t)}{S_B(t)+I_B(t)}\right) = \xi_7. \tag{21}$$

Case 3: $u_4^* = 1$, subject to the condition $\xi_7 = 0$ and $\xi_8 \neq 0$:

$$(\lambda_8(t) - \lambda_7(t))\eta_4\left(\beta\gamma S_F(t)\frac{I_H(t)}{S_H(t)+I_H(t)} + \alpha\kappa S_F(t)\frac{I_A(t)}{S_A(t)+I_A(t)} + \alpha\zeta S_F(t)\frac{I_B(t)}{S_B(t)+I_B(t)}\right) + \xi_8 = w_4. \tag{22}$$

Therefore, the optimal control $u_4^*(t)$ can be stated as:

$$u_4^* = max\left(min\left(\frac{(\lambda_8(t) - \lambda_7(t))\eta_4\left(\beta\gamma S_F(t)\frac{I_H(t)}{S_H(t)+I_H(t)} + \alpha\kappa S_F(t)\frac{I_A(t)}{S_A(t)+I_A(t)} + \alpha\zeta S_F(t)\frac{I_B(t)}{S_B(t)+I_B(t)}\right)}{w_4}, 1\right), 0\right). \tag{23}$$

Above, the optimal controls $u_1^*, u_2^*, u_3^*, u_4^*$ are functions of time t, and all the phase variables in formulae (11), (15), (19), and (23) correspond to these optimal solutions of the state system (4).

4. Numerical Simulation

It is clear from Figure 4 that the susceptible population increased faster than the infected population when $R_0 < 1$. This verifies that disease-free equilibrium existed if $R_0 < 1$. In Figure 5, we plot the time series solutions of the model variables corresponding to susceptible human $S_H(t)$ and infected human $I_H(t)$, susceptible animal Type A S_A and infected animal Type A I_A, susceptible animal Type B S_B and infected population Type B I_B, and susceptible vector population $S_F(t)$ and infected vector population $I_F(t)$ for different values of the model parameter that are depicted in Table 1. To find the stability of the non-trivial equilibrium E^*, we chose initial values as $E^*(S_H^*, I_H^*, S_A^*, I_A^*, S_B^*, I_B^*, S_F^*, I_F^*)$ = $(300, 50, 30, 20, 15, 8, 1500, 500)$. From Figure 5, we observe that the disease-free state did not exist if

$R_0 > 1$ and the system moved towards the endemic state. From the existence and stability analysis of the system, β, π, and τ seem to be important parameters. System dynamics of the model without drug application are shown in Figure 5. From Figure 5, it is also observed that when transmission factor $\pi = 0.18$ and $\tau = 0.05$ with $\beta = 0.24$ and $\mu_H = 0.1$, the system went to an infected state condition. However, if $\beta = 0.18$, $\pi = 0.16$, and $\tau = 0.05$, the system moved to an infection-free state and the disease did not persist.

From Figure 6, we considered $\eta_1 = 0.2, \eta_2 = 0.3, \eta_3 = 0.1$, and $\eta_4 = 0.4$ as the efficacy of drug applied in human, Type A animal, and Type B animal, and insecticide spray on the vector population, respectively [31]. Then, we observed the change of character for the infected human population, the infected reservoir populations (Types A and B), and the vector population due to control effects. The population of susceptible humans increased and also that of susceptible animals increased after applying control efforts, since the control (drug) effects decreased the contact rate between infected flies and humans and also decrease the contact rate between infected vector and reservoir populations. This in turn restricted the spread of leishmaniasis.

Table 1. List of parameters.

Parameter	Definition	Range	Default Value	Reference
λ_H	Recruitment rate of human population	300–318	317	[5]
λ_F	Recruitment rate of sand fly population	14,950–15,000	14,950	[5]
λ_A	Recruitment rate of animal population of Type A	70–150	73	[5]
λ_B	Recruitment rate of animal population of Type B	3–40	20	[32]
μ_H	Death rate of human population	0.000007–0.0001	0.00004	*Assumed*
μ_F	Death rate of sand fly population	0.188–0.795	0.189	[5]
μ_A	Death rate of animal population of Type A	0.06–0.21	0.19	[5]
μ_B	Death rate of animal population of Type B	0.089–0.255	0.25	[32]
β	Biting rate of sand fly on human	0.15–0.29	0.24	[5,9]
α	Biting rate of sand fly on animals of Type A and Type B	0.15–0.25	0.16	[5]
π	Transmission probability of CL in sand fly	0.12–0.24	0.18	[5,9]
ω	Transmission probability of CL on animal of Type A	0.11–0.172	0.12	[32]
τ	Transmission probability of CL on animal of Type B	0.02–0.071428	0.05	[5,9]
γ	Transmission probability of CL in sand fly from infected human	0.11–0.25	0.14	[5,9]
κ	Transmission probability of CL in sand fly from infected animal A	0.07–0.21	13	[5,9]
ζ	Transmission probability of CL in sand fly from infected animal B	0.04–0.21	0.12	[32]

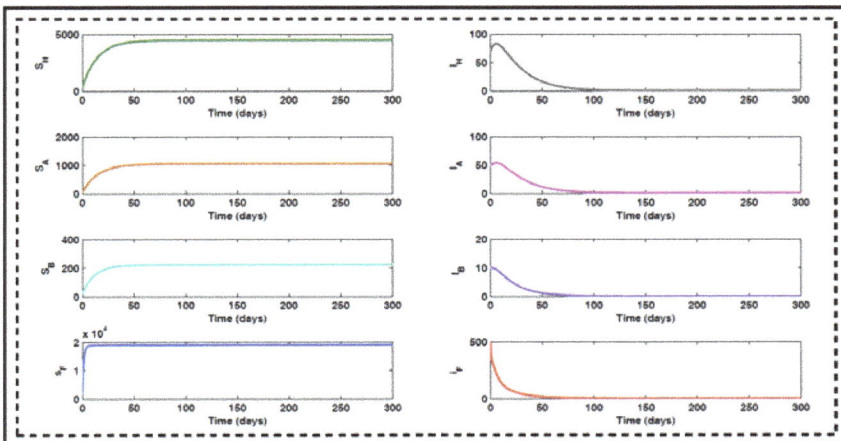

Figure 4. Population densities of the model variables for $R_0 < 1$ for disease-free state (all parameters are in Table 1).

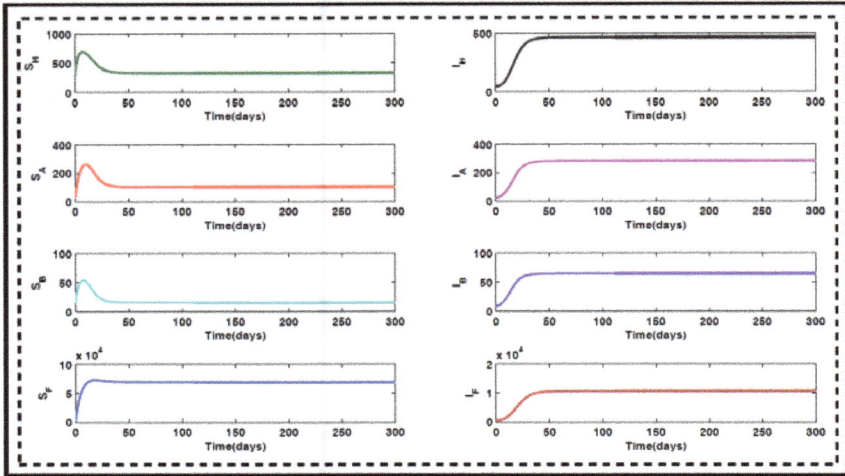

Figure 5. Population densities of the model variables for $R_0 > 1$ (other parameters are as in Table 1).

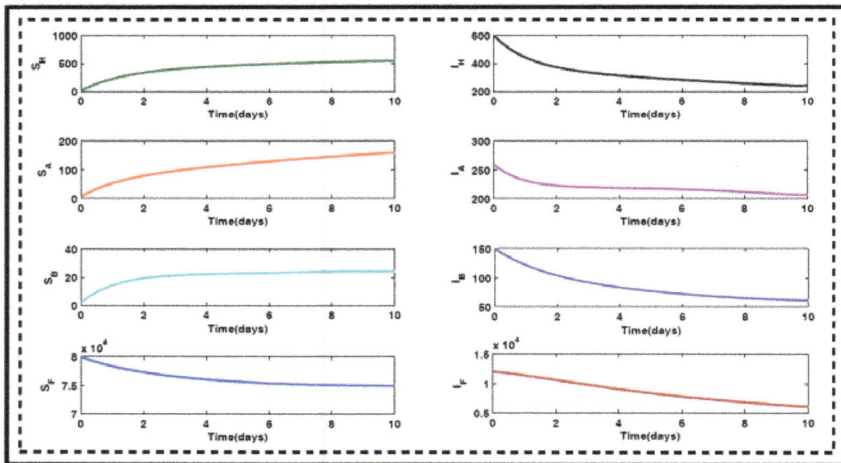

Figure 6. Behavior of the model system with optimal control schedule of the drug therapy.

4.1. Optimal Control for Different Cases

We applied all the controls continuously. We used a combination of controls for prevention of the disease, and compared all the scenarios. Actually, we compared the cases when one, two, three, and all the controls are applied separately. In fact, we do not know which combination is better to obtain the desired cost-effective result. We numerically constructed Figures A1–A8, indicating when a particular case is better than the others. Thus, we have consider all the cases for examination.

We numerically investigated the effect of the following optimal controls on the spread of leishmaniasis in a population for different cases [33,34].

- **Case I**: Prevention of the infection of animals Type A and Type B by the disease, along with spraying insecticides on the sand fly vectors.
- **Case II**: Prevention of animal Type B being infected by the disease, along with spraying insecticides on the sand fly vectors.
- **Case III**: Spraying insecticides on the sand fly vectors.
- **Case IV**: Prevention of animal Type A being infected by the disease, along with spraying insecticides on the sand fly vectors.
- **Case V**: Prevention of animal Type B being infected by the disease, along with spraying insecticides on the sand fly vectors.
- **Case VI**: Prevention of humans being infected by the disease, along with spraying insecticides on the sand fly vectors.
- **Case VII**: Prevention of humans and animal Type A being infected by the disease, along with spraying insecticides on the sand fly vectors.
- **Case VIII**: Prevention of humans and animals Type A and Type B being infected by the disease, along with spraying insecticides on the sand fly vectors.

See Appendix A.

4.2. Impact of Optimal Control on the Different Cases Proposed

We aimed to describe the long-term behavior of the disease prevalence for the future, which cannot be predicted by the application of the optimal control strategies used in our model. After effective strategies are stopped, there are often some infectious people remaining who can cause a fresh outbreak of the disease [35]. Since the basic reproduction number proved to be effective in measuring long-term endemicity in [23], the effects of our applied strategies on (R_0) were studied. Figure 7 demonstrates the results of numerical simulation of (R_0) under the various control strategy approaches. Assuming that the combinations of optimal control were implemented in the beginning of the year, it was observed that case VIII performed well in the early stages, and also kept the disease under check. Cases I, VII, V, IV, II, VI, and III performed almost similarly throughout.

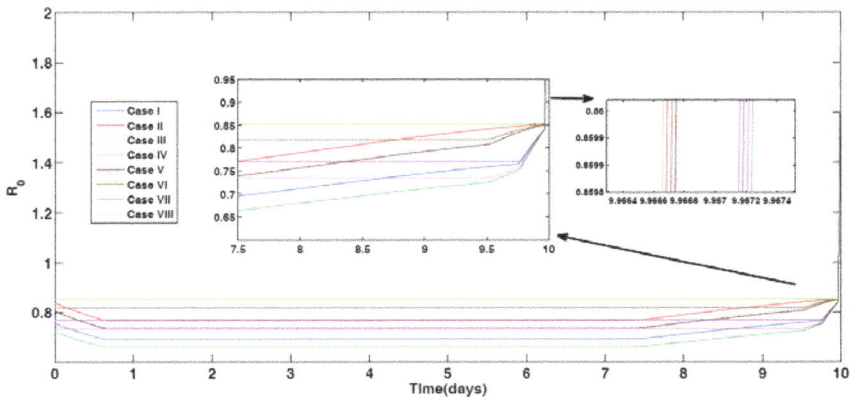

Figure 7. Effect of different control cases on R_0 and magnified for $t = 7.5$–10 days.

For longer time periods, R_0 increased, at which point further application of control policies becomes necessary. Thus, the implementation of different strategies during different points of time produces different disease dynamics. Hence, it becomes important to determine the exact point in time at which the effective strategy needs to be implemented. Further, the complete elimination of the disease by effective application of the control strategy can be possible only if these strategies are continued over long time periods, by determination of the class boundaries of the controls u_1, u_2, u_3, and u_4.

Figure S3 describes the numerical simulation outcomes of R_0 for the various cases of control strategies applied to each population.

4.3. Rescued Population and Vector Reduction for Different Cases

The useful data for finding rescued population-based strategy are in Table 2. The maximum number of humans were rescued within time in Cases VIII and VII, and the minimum number of humans were rescued in Cases III and II. The maximum number of infected animals of Type A were extricated within time in Cases VII and VIII, and the minimum number of animals of Type A were extricated in Case V. The maximum number of animals of Type B were rescued within time in Case VIII, and the minimum number of animals of Type B were rescued in Cases III and IV. The maximum vector reduction occurred within time in Cases VIII and VII, followed by Case II, and the minimum reduction occurred in Case III. Therefore, we can say empirically that the maximum number of humans were rescued in time when implementing Case VIII. Additionally, the maximum number of vectors were reduced in time compared to the other strategies. However, this cannot be the most cost-effective approach because of all of the drugs and insecticides which are applied to each of the populations to control the disease.

Table 2. Reduction in infected host population (%).

Cases	Human	Animal A	Animal B	Vector Reduction
Case I	12.67	22.31	4.67	31.10
Case II	12	11.15	4.67	29.7
Case III	11.83	11.15	4.00	29.54
Case IV	12.50	12.92	4.00	30.90
Case V	18.16	5.17	4.67	31.54
Case VI	18	11.92	4.67	31.38
Case VII	18.5	23.07	4.67	32.75
Case VIII	18.60	23.07	5.33	32.92

4.4. Cost-Effectiveness of the Different Cases

Herein, we consider the size of the population rescued from infection as well as the extent of vector reduction with the different control strategies. For this, we assumed that the cost of the controls were the square of the proportional to the number of controls deployed, and considered the cost of drug u_1 per person to be approximately 1.86\$ (in INR 118.94) [36], the cost of drug u_2 per Type A animal to be 1.2\$ (in INR 76.74), the cost of drug u_3 per Type B animal to be 1\$ (in INR 63.95), and the cost of insecticide spraying u_4 per square meter area for vector death to be 1.5\$ (in INR 95.92). The assumption was based on the understanding that the primary goal of using the drug is for the cure of those infected by the disease. Moreover, the use of insecticides is for the removal of vector. The difference between the total size of the infectious population without control and with control was used to determine the number of infections averted. We used the size of the rescued population and extent of vector reduction, as depicted in Table 3. The control strategy applied in the model gave maximum cost benefit. This was determined for each intervention strategy, illustrated in Figures 8 and 9. One can see that the most cost-effective cases in terms of number of infections averted, including the number reduction of the sand fly vector, was the spraying of insecticides (Case III).

Table 3. Cost-effectiveness for different cases.

Cases	Host Reduction	Cost (in $)	Vector Reduced	Cost (in $)
Case I	141	371.716	3729	104.878
Case II	108	70	3564	100.238
Case III	106	0	3545	99.70
Case IV	138	296.514	3709	104.316
Case V	147	1955.482	3785	106.453
Case VI	146	1868.184	3766	105.919
Case VII	178	2232.198	3930	110.531
Case VIII	180	2329.496	3950	111.094

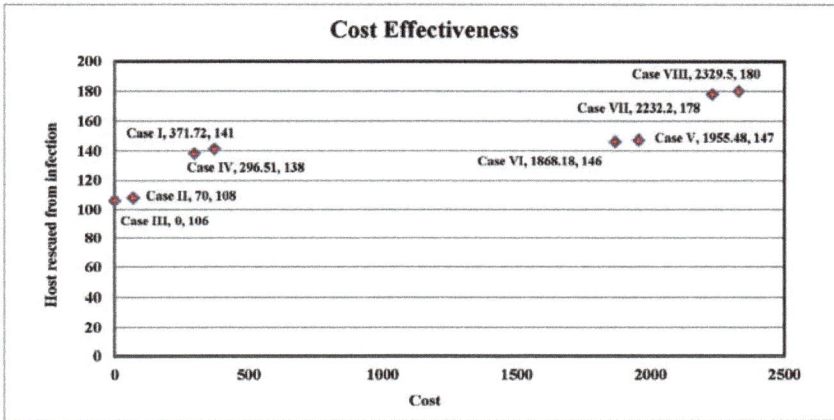

Figure 8. Cost of different cases and their corresponding rescued host population plotted as a scatter diagram.

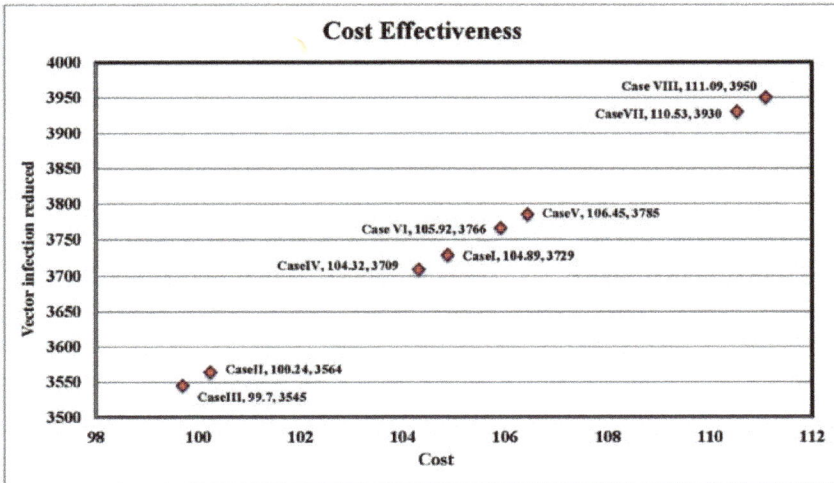

Figure 9. Cost of different cases and their corresponding vector infection reduced plotted as a scatter diagram.

4.5. Discussion

The disease leishmaniasis spreads from animal (reservoir) to human (host) via a vector. For this reason, the control of vectors and infected reservoir populations is one of the most efficient approaches to exterminating the disease cutaneous leishmaniasis. Applying a suitable drug to the reservoir population and spraying insecticide to kill the vector perform significant functions for controlling CL. We investigated the dynamics in the absence of drug applied in human hosts and animals, and without the spraying of insecticide to kill the vector. The disease-free situation existed for $R_0 < 1$. On the other hand, if $R_0 > 1$, the disease-free state lost its stability and the system tended towards the endemic condition.

From Figure 5, we observe that susceptible human host population was sharply decreased up to approximately 30 days, and the infected human host population gradually increased up to approximately 30 days. Susceptible animal of Type A sharply decreased up to approximately 25 days and susceptible animal of Type B sharply decreased up to approximately 20 days. Again, the infected animal population of Type A gradually increased up to approximately 25 days, and the population of infected animals of Type B gradually increased up to approximately 20 days. The susceptible sand fly population increased up to approximately 12 days, and the infected sand fly population gradually increased up to approximately 20 days. Thus, we can conclude that the disease can be controlled by insecticide spraying with a frequency appropriate to the size of the vector population. The change of the behavioral structure of the system dynamics depends on the biting rate and the transmission possibility of the sand fly. So, if we are able to kill the sand fly by spraying insecticide, then biting rate and transmission will automatically be reduced and then the disease can be controlled.

The inclusion of latent and recovered categories and the use of delayed terms in the model equation system may change the dynamics of the disease. However, the life cycle of the parasite is not long enough to support the incubation period. Therefore, we did not consider the latent status in our model. Additionally, the progression from susceptible to infected classes passes through a latent stage. So, we ignored the intermediate stages (latent and recovered) and considered susceptible and infected stages only. We paid no attention to the delay term, because our main aim was to control the infected population and the vector. Consequently, we considered the delay term to be in steady state.

The effectiveness of the drug dose influences the system to move towards the infection-free state. The results obtained from analytical and numerical simulations showed that the control strategies were very effective if applied at the same time in the same region. The proposed optimal control can eradicate and prevent further transmission of the disease through the vector. Although total eradication of cutaneous leishmaniasis seems complicated in a realistic environment, if our findings can be applied to an infected zone, then a pioneering insight can be achieved against cutaneous leishmaniasis in a global perspective.

To prevent vector-borne disease, different strategic cases can be applied to a finite time period. The effects of the different cases used to minimize the disease among the various populations were investigated using the analysis of optimal control, thereby depicting the real situations. Herein, we considered four controls upon three types of mammalian host and a vector population: prevention of infective individuals by using drugs and spraying of insecticides to kill vector. It was deduced that the strategy applied in Case I yielded good results for a considerable time period, however Case III was most cost effective but did not achieve as great a reduction in infected host or vector in the same time period. For the entire period of the preventive measure, the effects of our control strategies on R_0 were observed to determine the effects of the controls on the future spread of disease. The significant changes in the number of possible secondary infections from an infected individual were thus concluded to be dependent on time. Thus, it becomes important to determine the exact time interval during which the optimal control must be applied. Moreover, it is only possible to eliminate the disease entirely if the different control strategies are continued for a long duration in the future.

5. Conclusions

From the study of the effects of the various control cases on R_0, it was observed that Case VIII yielded the best results in attempting to control the disease, followed by Cases VII and V. The analysis of cost-effectiveness indicated that Case III was the most cost-effective, followed by Case II. Though the potential of Case I in trying to eliminate the disease is comparatively better than that of Case IV, it involves higher costs. It can be concluded that the cases which display low disease prevalence would require an efficient and cost-effective strategy (Cases II and IV). However, the cases where the utmost priority is to control the disease would require strategic cases that are less cost-effective but act efficiently to control the disease in a short time period (e.g., Case I). Hence, our model would suffice in assisting decisions related to the allocation of resources, where the fundamental aim is to select the best strategy to eliminate the disease in the lowest possible time.

Supplementary Materials: Supplementary materials are available online: http://www.mdpi.com/2078-2489/23/3/38/s1.

Author Contributions: Model Formulation, problem observation, and its solution were generated by professor E.V.G. and professor P.K.R. Mathematical analysis, numerical simulation, as well as most of the work in this manuscript were realistically done by the research scholar D.B. Comprehensive progression of the solution was done by the research scholars S.D. and J.C.

Acknowledgments: The research is supported by the PURSE-DST, Department of Mathematics, Jadavpur University, Government of India.

Conflicts of Interest: The authors declare no conflict of interest.

Appendix A

- *Case I: Optimal Use of Drug for Prevention From Disease for Animals Type A and Type B with Spray of Insecticides on Vector*

In this case, the objective function J is optimized by the preventive control measures u_2, u_3, and the control involving spraying of insecticides u_4, while u_1 (the control parameter for humans) is set to zero. We observed in Figure A1 that due to the control strategies, the number of infected humans I_H decreased. A similar decrease was observed in infected animals of Types A and B. Infected vector population also decreased due to the inclusion of spraying insecticide in the control strategy, while an increased number was observed for the case without control. In this case, infected human I_H was reduced by 12.67%, infected animal A was reduced by 22.13%, infected animal B was reduced by 4.67%, and vector death rate was reduced by 31.10%. Therefore, almost 100% rescued cases was achieved for humans in 78 days, for animal Type A in 45 days, and animal Type B in 200 days, and vector was removed in 32 days. From Figure A1, we find that the controls u_2 and u_4 were initially 100%, following which u_2 slowly dropped to the lower bound after 9.7 days. Moreover, the control u_4 dropped to the lower bound after 9.85 days. Here the control u_3 was 100% effective almost from the first day and reached the lower bound in almost the seventh day. Hence, the control effect of u_2 was near the optimum level but the spray of insecticide for this strategy showed the actual optimal level.

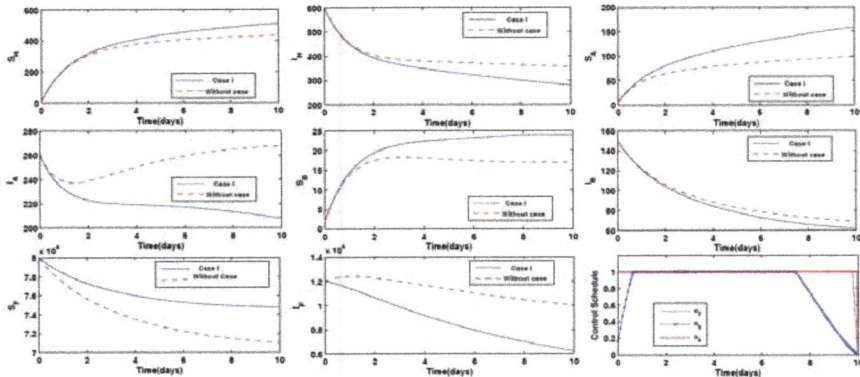

Figure A1. The behavior of the model system with control and without control for Case I for $u_1 = 0$ and $u_2 = u_3 = u_4 \neq 0$.

- *Case II: Optimal Use of Drug for Prevention of Disease for Animal Type B with Spray of Insecticides on Vector*

In this case, control parameters for preventive measure u_3 and u_4 were used to optimize the objective function J, while the other control parameters for prevention u_1, u_2 were zero. We observed in Figure A2 that due to the control strategies, the number of infected humans I_H decreased. A similar decrease was observed in infected animals of Types A and B. Additionally, the infected vector population decreased due to the inclusion of spraying insecticide in the control strategy, while an increased number was observed for the case without control. For this case, infected human I_H was reduced by 12%, infected animal A was reduced by 11.15%, infected animal B was reduced by 4.67%, and vector death was 29.7%. Therefore, almost 100% rescued cases were achieved for humans in 80 days, for animal Type A in 89 days, animal Type B in 200 days, and vector was removed in 33 days. From Figure A2, it is evident that the control u_3 was initially 100% at around 0.65 days, after which control trajectory u_3 decreased slowly in the lower level almost 7 days. Moreover, the control u_4 was initially 100% effective from the beginning and decreased to the lower bound after 9.8 days. Therefore, insecticide spray was almost effective most of the days, and for that reason the number of vectors removed was optimal compared to the other controls for this strategy.

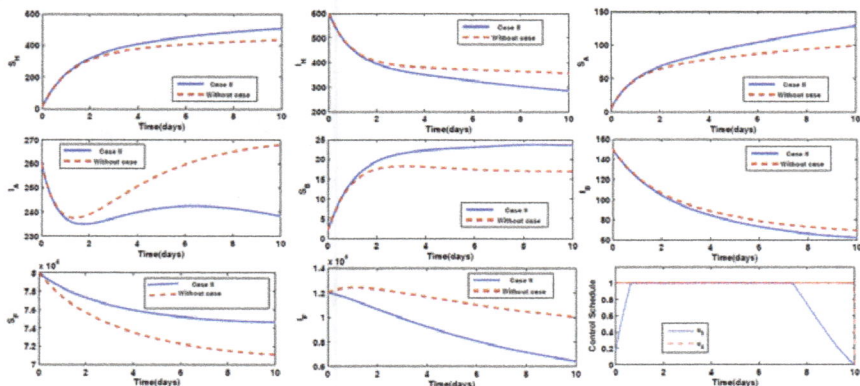

Figure A2. The behavior of the model system with control and without control for Case II for $u_1 = u_2 = 0$ and $u_3 = u_4 \neq 0$.

- *Case III: Optimal Use of Spray of Insecticides on the Sand Fly Vector*

In order to optimize the objective function *J*, this policy involves only control via spraying insecticides u_4, while we fixed the controls for preventive measure u_1, u_2, and u_3 to zero. We observed in Figure A3 that due to the control strategies, the number of infected humans I_H decreased. A similar decrease was observed in infected animals of Types A and B. Also, the infected vector population decreased due to the inclusion of spraying insecticide in the control strategy, while an increased number was observed for the case without control. In this case, infected human I_H was reduced by 11.83%, infected animal Type A was reduced by 11.15%, infected animal Type B was reduced by 4%, and vector death was 29.54%. Therefore, almost 100% rescued cases were achieved for humans in 80 days, for animal Type A in 89 days, animal Type B in 250 days, and vector was removed in 33 days. From Figure A3, we find that the control u_4 was initially 100%, following which it slowly dropped to the lower bound after 9.97 days. Therefore, effect of insecticide spray lasts longer compared to the other cases.

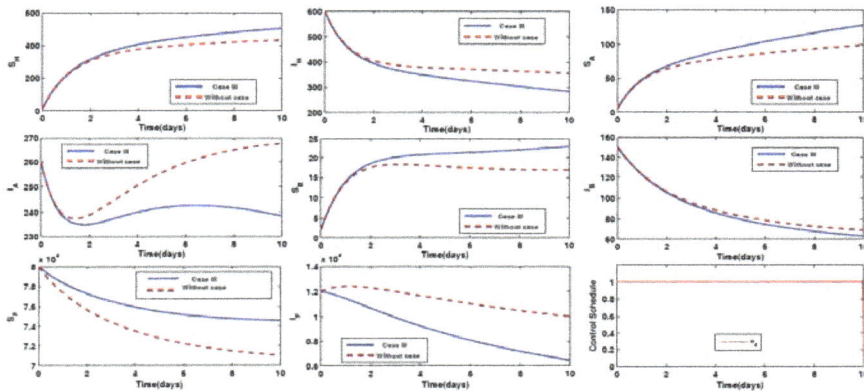

Figure A3. The behavior of the model system with control and without control for Case III for $u_1 = u_2 = u_3 = 0$ and $u_4 \neq 0$.

- *Case IV: Optimal Use of Drug for Prevention from Disease for Animal Type A with Spray of Insecticides on Vector*

In this case, the control for preventive measure u_2 is applied along with the control with the spraying of insecticides u_4, with the aim of optimizing the objective function *J*, and we fixed the controls for preventive control parameter u_1 and u_3 to zero. We observed in Figure A4 that due to the control strategies, the number of infected humans I_H decreased. A similar decrease was observed in infected animals of Types A and B. Also, the infected vector population decreased due to the inclusion of spraying insecticide in the control strategy, while an increased number was observed for the case without control. For this case, infected human I_H was reduced by 12.5%, infected animal Type A was reduced by 12.92%, infected animal Type B was reduced by 4%, and vector death was 30.90%. Therefore, almost 100% rescued cases was achieved for humans in 76 days, for animal Type A in 76 days, animal Type B in 250 days, and vector was removed in 32 days. From Figure A4, we find that the controls u_2 and u_4 were initially 100%, following which u_2 and u_4 slowly dropped to the lower bound after 9.7 days. Moreover, the control u_4 dropped to the lower bound after 9.9 days. Hence the control u_2 and u_4 were effective for almost the maximum time span. However, the control u_4 was slightly better than the other controls.

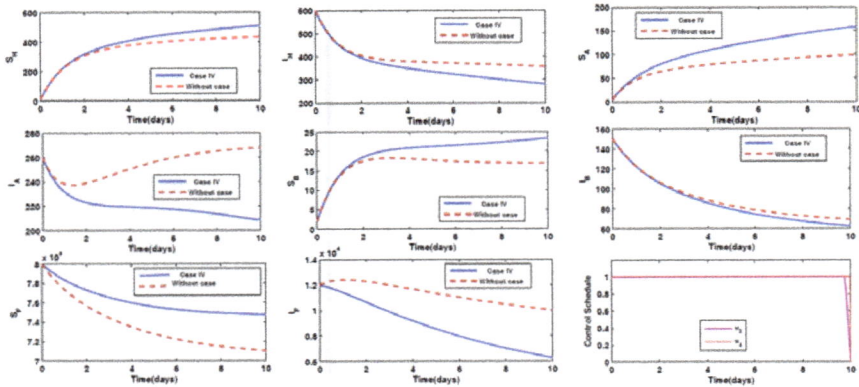

Figure A4. The behavior of the model system with control and without control for Case IV for $u_1 = u_3 = 0$ and $u_2 = u_4 \neq 0$.

- *Case V: Optimal Use of Drug for Prevention from Disease for Human and Animal Type B with Spray of Insecticides on Vector*

In this case, the preventive measure for controls u_1 and u_3, along with the spraying of insecticides u_4, are applied with the aim of optimizing the objective function J, and we fixed the control for the prevention of animal Type A u_2 to zero. We observed in Figure A5 that due to these control strategies, the number of infected humans I_H decreased. A similar decrease was observed in infected animals of Types A and B. Also, the infected vector population decreased due to the inclusion of spraying insecticide in the control strategy, while an increased number was observed for the case without control. In this case, infected human I_H was reduced by 18.16%, infected animal Type A was reduced by 5.17%, infected animal Type B was reduced by 4.67%, and vector mortality was 31.54%. Therefore, almost 100% rescue cases was achieved for humans in 55 days, for animal Type A in 153 days, animal Type B in 200 days, and vector was removed in 31 days. From Figure A5, we find that the controls u_1 and u_4 were initially 100% , following which they slowly dropped to the lower bound after 9.3 days, while control u_4 dropped to lower bound after 9.9 days. The control u_3 was at 100%, almost from 0.7 days at the beginning, and reached the lower bound after 7 days. Therefore, once spraying is complete, it stays on for the maximum time, yielding the maximum benefit under this strategy.

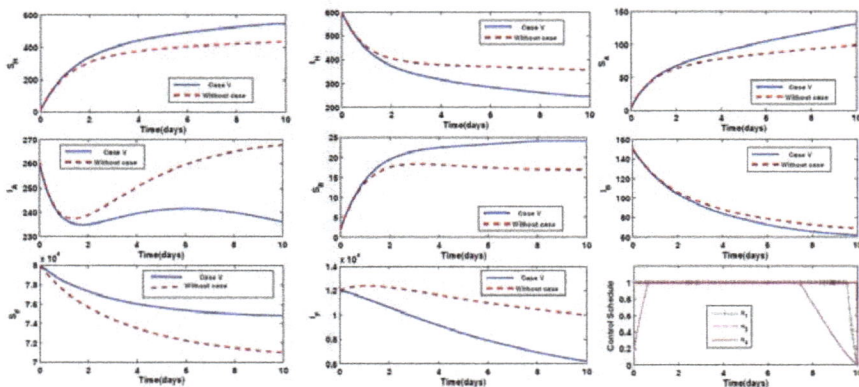

Figure A5. The behavior of the model system with control and without control for Case V for $u_2 = 0$ and $u_1 = u_3 = u_4 \neq 0$.

- *Case VI: Optimal Use of Drug for Prevention from Disease for Human and Spraying of Insecticides on Vector*

In order to optimize the objective function J, the prevention technique for controls u_1 and u_4 were applied, while the other prevention parameters u_2 and u_3 were considered as zero. We observed in Figure A6 that due to the this control case, the number of infected humans I_H decreased. A similar decrease was observed in infected animals of Types A and B. Also, the infected vector population decreased due to the inclusion of spraying insecticide in the control strategy, while an increased number was observed in the case without control. For this case, infected humans I_H was reduced by 18%, infected animal Type A was reduced by 11.92%, infected animal Type B was reduced by 4.67%, and vector death was 31.38%. Therefore, almost 100% rescue cases was achieved for humans in 56 days, for animal Type A in 83 days, animal Type B in 200 days, and vector was removed in 32 days. From Figure A6, we find that the controls u_1 and u_4 were initially 100% effective, following which u_1 slowly dropped to the lower bound after 9.3 days. Furthermore, the control u_4 dropped to the lower bound after 9.8 days. Therefore, insecticide spray is the strategy that can give us maximum benefit.

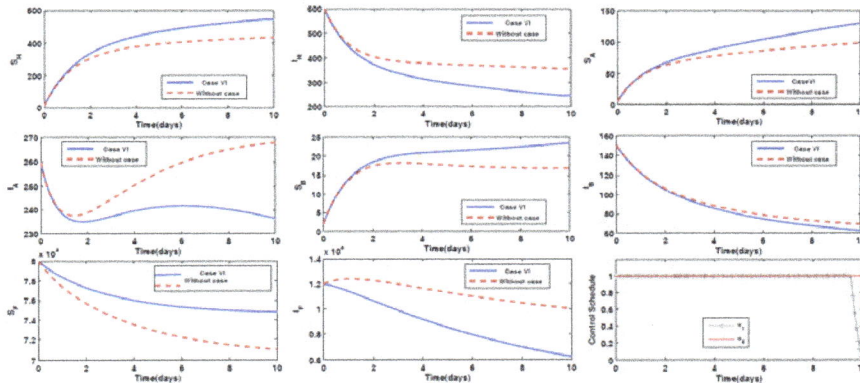

Figure A6. The behavior of the model system with control and without control for Case VI for $u_2 = u_3 = 0$ and $u_1 = u_4 \neq 0$.

- *Case VII: Optimal Use of Drug for Prevention from Disease for Human and Animal A with Spray of Insecticides on Vector*

In order to optimize the objective function J, the controls u_1 and u_2 were used for the prevention of disease in addition to the spraying of insecticide u_4, setting the control parameter for the prevention of disease in animal Type B u_3 to zero. We observed in Figure A7 that due to the different control cases, the number of infected humans I_H decreased. A similar decrease was observed in infected animals of Types A and B. Also, the infected vector population decreased due to inclusion of spraying insecticide in the control strategy, while an increased number was observed in the case without control. In this case, infected human I_H was reduced by 18.50%, infected animal Type A was reduced by 23.07%, infected animal Type B was reduced by 4.67%, and vector mortality was 32.75%. Therefore, almost 100% rescue cases was achieved for humans in 54 days, for animal Type A in 43 days, animal Type B in 200 days, and vector was removed in 30 days. From Figure A7, it is observed that the controls u_1 and u_2 were 100% effective initially, following which the control u_1 reached the lower bound after 9.4 days. At the same time, the controls u_2 and u_4 reached the lower bound after 9.6 days and 9.8 days, respectively. Here the effect of control u_4 was initially 100% and reached lower bound almost in the tenth day. Therefore, the effect of insecticide spray under this strategy stayed for maximum amount of time.

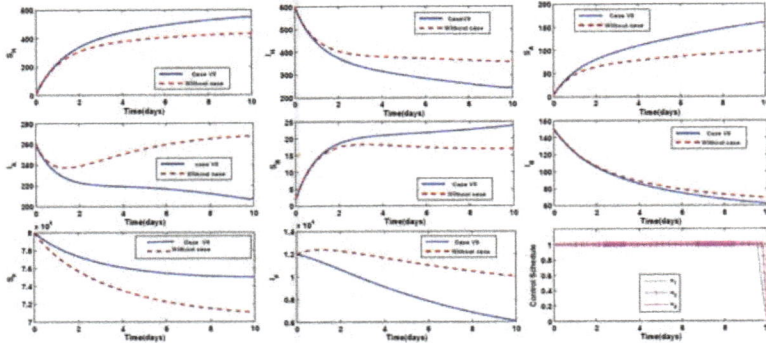

Figure A7. The behavior of the model system with control and without control for Case VII for $u_3 = 0$ and $u_1 = u_2 = u_4 \neq 0$.

- *Case VIII: Optimal Use of Drug for Prevention from Disease for Human, Animals Type A and Type B with Spraying of Insecticides on Vector*

In this case, in order to optimize the objective function J, all four controls (i.e., u_1, u_2, u_3, and u_4) were used. After comparison of Case VIII with the circumstance when no controls were applied, it could be concluded that while the human population which was susceptible increased in number, the infected human population decreased, which is illustrated in Figure A8. It was also observed in this case that in general, two types of susceptible animal populations increased in number and the infected sand fly populations decreased remarkably at an almost exponential rate. In particular, the sand fly population reduced below 3000 in around 10 days. The comparison at $t = 10$ days shows that there was an increase by 113 individuals in S_H, decrease by 60 individuals in I_A, and I_B and I_F by 8 and 3950 individuals, respectively. With the help of this strategic case, infected human I_H was reduced by 18.60%, infected animal Type A was reduced by 23.07%, infected animal Type B was reduced by 5.33%, and vector mortality was 32.92%. Therefore, almost 100% rescue cases was achieved for humans in 53 days, for animal Type A in 43 days, animal Type B in 187, days and vector was removed in 30 days. Figure A8 shows that the control u_1 was 100% effective initially, following which it dropped slowly to the lower bound after 9.5 days. At the same time, the controls u_2 and u_3 reached the lower bound after 9.3 days and 7 days, respectively. The control u_4 was initially 100% effective and reached the lower bound at approximately the tenth day. It can be concluded that a low amount of insecticide spray is necessary in this strategy.

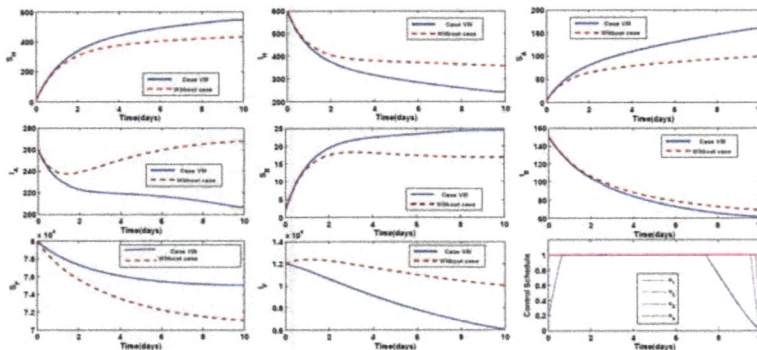

Figure A8. The behavior of the model system with control and without control for Case VIII for $u_1 = u_2 = u_3 = u_4 \neq 0$.

References

1. Sharma, U.; Singh, S. Immunobiology of leishmaniasis. *Indian J. Exp. Biol.* **2009**, *47*, 412–423. [PubMed]
2. Park, K. *Preventive and Social Medicine*; Banarsidas Bhanot Publishers: Jabalpur, India, 2005.
3. Chaves, L.F.; Hernandez, M.J. Mathematical modelling of American Cutaneous Leishmaniasis: Incidental hosts and threshold conditions for infection persistence. *Acta Trop.* **2004**, *92*, 245–252. [CrossRef] [PubMed]
4. World Health Organization. Leishmaniasis. Fact Sheet Updated April 2017. Available online: http://www.who.int/news-room/fact-sheets/detail/leishmaniasis (accessed on 18 July 2018).
5. ELmojtaba, I.M.; Mugisha, J.Y.T.; Hashim, M.H.A. Mathematical analysis of the dynamics of visceral leishmaniasis in the Sudan. *Appl. Math. Comput.* **2010**, *217*, 2567–2578. [CrossRef]
6. Kassiri, H.; Sharifinia, N.; Jalilian, M.; Shemshad, K. Epidemiological aspects of cutaneous leishmaniasis in Ilam province, west of Iran (2000–2007). *Asian Pac. J. Trop. Dis.* **2012**, *2*, S382–S386. [CrossRef]
7. Reithinger, R.; Dujardin, J.C.; Louzir, H.; Pirmez, C.; Alexander, B.; Brooker, S. Cutaneous leishmaniasis. *Lancet Infect. Dis.* **2007**, *7*, 581–596. [CrossRef]
8. Kasper, D.L.; Braunwald, E.; Fauci, A.S.; Hauser, S.L.; Longo, D.L.; Jameson, J.L.; Loscalzo, J. *Harrison's Principles of Internal Medicine*, 17th ed.; McGraw-Hill: New York, NY, USA, 2008; Volume 1, Chapter 1–216.
9. Bathena, K. A Mathematical Model of Cutaneous Leishmaniasis. Master's Thesis, Rochester Institute of Technology, Rochester, NY, USA, 2009.
10. Miller, E.; Huppert, A. The effects of Host Diversity on Vector-Borne Disease: The Conditions under Which Diversity Will Amplify or Dilute the Disease Risk. *PLoS ONE* **2013**, *8*, e80279. [CrossRef] [PubMed]
11. Jaouadi, K.; Haouas, N.; Chaara, D.; Gorcii, M.; Chargui, N.; Augot, D.; Pratlong, F.; Dedet, J.P.; Ettlijani, S.; Mezhoud, H.; et al. First detection of Leishmania killicki (Kinetoplastida, Trypanosomatidae) in Ctenodactylus gundi (Rodentia, Ctenodactylidae), a possible reservoir of human cutaneous leishmaniasis in Tunisia. *Parasites Vectors* **2011**, *4*, 159. [CrossRef] [PubMed]
12. Chatterjee, A.N.; Roy, P.K.; Mondal, J. Mathematical Model for Suppression of Sand Flies through IRS with DDT in Visceral Leishmaniasis. *Am. J. Math. Sci.* **2013**, *2*, 105–112.
13. Wu, H.J.J.; Massad, E. Mathematical modelling for Zoonotic Visceral Leishmaniasis dynamics: A new analysis considering updated parameters and notified human Brazilian data. *Infect. Dis. Model.* **2017**, *2*, 143–160.
14. Bacaer, N.; Guernaoui, S. The epidemic threshold of vector-borne diseases with seasonality the case of cutaneous leishmaniasis in Chichaoua, Morocco. *J. Math. Biol.* **2006**, *53*, 421–436. [CrossRef] [PubMed]
15. Biswas, D.; Kesh, D.K.; Datta, A.; Chatterjee, A.N.; Roy, P.K. A Mathematical Approach to Control Cutaneous Leishmaniasis Through Insecticide Spraying. *Sop Trans. Appl. Math.* **2014**, *1*, 44–54. [CrossRef]
16. Biswas, D.; Roy, P.K.; Li, X.Z.; Basir, F.A.; Pal, J. Role of macrophage in the disease dynamics of cutaneous leishmaniasis: A delay induced mathematical study. *Commun. Math. Biol. Neurosci.* **2016**, *2016*, 1–31.
17. Roy, P.K.; Li, X.Z.; Biswas, D.; Datta, A. Impulsive Application to Design Effective Therapies Against Cutaneous Leishmaniasis Under Mathematical Perceptive. *Commun. Math. Biol. Neurosci.* **2017**, *2017*, 1–17.
18. Biswas, D.; Datta, A.; Roy, P.K. Combating Leishmaniasis through Awareness Campaigning: A Mathematical Study on Media Efficiency. *Int. J. Math. Eng. Manag. Sci.* **2016**, *1*, 139–149.
19. Zamir, M.; Zaman, G.; Alshomrani, A.S. Sensitivity Analysis and Optimal Control of Anthroponotic Cutaneous Leishmania. *PLoS ONE* **2016**, *11*, e0160513. [CrossRef] [PubMed]
20. Begon, M.; Bennett, M.; Bowers, R.G.; French, N.P.; Hazel, S.M.; Turner, J. A clarication of transmission terms in host-microparasite models: numbers, densities and areas. *Epidemiol. Infect.* **2002**, *129*, 147–153. [CrossRef] [PubMed]
21. Abubakar, A.; Ruiz-Postigo, A.J.; Pita, J.; Lado, M.; Ben-Ismail, R.; Argaw, D.; Alvar, J. Visceral leishmaniasis outbreak in South Sudan 2009–2012: Epidemiological assessment and impact of a multisectoral response. *PLoS Negl. Trop Dis.* **2014**, *8*, e2720. [CrossRef] [PubMed]
22. Subramanian, A.; Singh, V.; Sarkar, R.R. Understanding Visceral Leishmaniasis Disease Transmission and its Control—A Study Based on Mathematical Modeling. *Mathematics* **2015**, *3*, 913–944. [CrossRef]
23. Biswas, S.; Subramanian, A.; ELMojtaba, I.M.; Chattopadhyay, J.; Sarkar, R.R. Optimal combinations of control strategies and cost-effective analysis for visceral leishmaniasis disease transmission. *PLoS ONE* **2017**, *12*, e0172465. [CrossRef] [PubMed]

24. Brett-Major, D.M.; Claborn, D.M. sandfly Fever: What Have We Learned in One Hundred Years? *Mil. Med.* **2009**, *174*, 426–431. [CrossRef] [PubMed]
25. Rogers, M.E.; Bates, P.A. Leishmania Manipulation of sandfly Feeding Behavior Results in Enhanced Transmission. *PLoS Pathog.* **2007**, *3*, e91. [CrossRef] [PubMed]
26. Lopez, L.F.; Coutinho, F.A.B.; Burattini, M.N.; Massad, E. Threshold conditions for infection persistence in complex host-vectors interactions. *Comptes Rendus Biol.* **2002**, *325*, 1073–1084. [CrossRef]
27. Zaman, G.; Kang, Y.H.; Jung, I.H. Stability analysis and optimal vaccination of an SIR epidemic model. *BioSystems* **2008**, *93*, 240–249. [CrossRef] [PubMed]
28. Birkhoff, G.; Rota, G.C. *Ordinary Differential Equations*, 4th ed.; John Wiley and Sons: New York, NY, USA, 1989.
29. Lukes, D.L. *Differential Equations: Classical to Controlled in Mathematics in Science and Engineering*; Academic Press: New York, NY, USA, 1982; Volume 162.
30. Kirschner, D.; Lenhart, S.; Serbin, S. Optimal control of the chemotherapy of HIV. *J. Math. Biol.* **1997**, *35*, 775–792. [CrossRef] [PubMed]
31. Shimozako, H.J.; Wu, J.; Massad, E. The Preventive Control of Zoonotic Visceral Leishmaniasis: Efficacy and Economic Evaluation. *Comput. Math. Methods Med.* **2017**, *2017*, 4797051. [CrossRef] [PubMed]
32. Kaabi, B.; Ahmed, S.B. Assessing the effect of zooprophylaxis on zoonotic cutaneous leishmaniasis transmission: A system dynamics approach. *Biosystems* **2013**, *114*, 253–260. [CrossRef] [PubMed]
33. Okosuna, K.O.; Rachid, O.; Marcus, N. Optimal control strategies and cost-effectiveness analysis of a malaria model. *BioSystems* **2013**, *111*, 83–101. [CrossRef] [PubMed]
34. Sardar, T.; Mukhopadhyay, S.; Bhowmick, A.R.; Chattopadhyay, J. An optimal cost effectiveness study on Zimbabwe cholera seasonal data from 2008–2011. *PLoS ONE* **2013**, *8*, e81231. [CrossRef] [PubMed]
35. Stauch, A.; Sarkar, R.R.; Picado, A.; Ostyn, B.; Sundar, S.; Rijal, S.; Boelaert, M.; Dujardin, J.; Duerr, H. Visceral Leishmaniasis in the Indian Subcontinent: Modelling Epidemiology and Control. *PLoS Negl. Trop. Dis.* **2011**, *5*, e1405. [CrossRef] [PubMed]
36. Costs of Medicines in Current Use for the Treatment of Leishmaniasis (Annex6), Drug Prices (January 2010). Available online: http://www.who.int/leishmaniasis/research/978_92_4_12_949_6_Annex6.pdf (accessed on 18 July 2018).

![Mathematical and Computational Applications logo]

Mathematical and Computational Applications

MDPI

Article

Optimal Control and Computational Method for the Resolution of Isoperimetric Problem in a Discrete-Time SIRS System

Fadwa El Kihal , **Imane Abouelkheir, Mostafa Rachik and Ilias Elmouki** *

Department of Mathematics and Computer Sciences, Faculty of Sciences Ben M'Sik,
Hassan II University of Casablanca, Casablanca 20000, Morocco; fadwa.elkihal@gmail.com (F.E.K.);
abouelkheir88@gmail.com (I.A.); m_rachik@yahoo.fr (M.R.)
* Correspondence: i.elmouki@gmail.com

Received: 7 September 2018; Accepted: 22 September 2018; Published: 24 September 2018

Abstract: We consider a discrete-time susceptible-infected-removed-susceptible "again" (SIRS) epidemic model, and we introduce an optimal control function to seek the best control policy for preventing the spread of an infection to the susceptible population. In addition, we define a new compartment, which models the dynamics of the number of controlled individuals and who are supposed not to be able to reach a long-term immunity due to the limited effect of control. Furthermore, we treat the resolution of this optimal control problem when there is a restriction on the number of susceptible people who have been controlled along the time of the control strategy. Further, we provide sufficient and necessary conditions for the existence of the sought optimal control, whose characterization is also given in accordance with an isoperimetric constraint. Finally, we present the numerical results obtained, using a computational method, which combines the secant method with discrete progressive-regressive schemes for the resolution of the discrete two-point boundary value problem.

Keywords: discrete-time model; SIRS model; optimal control; isoperimetric problem

1. Introduction

Many mathematical models in epidemiology are used to assist in finding the most appropriate control strategies for a given group of individuals who belong to different classes. These classes are often represented in epidemic systems, using compartments that are usually named susceptible (S), exposed (E), infectious or infected (I) and removed or recovered (R) [1]. In this paper, we are interested in the study of a population infected by an epidemic and whose dynamics are described using a discrete-time SIRS system. The SIRS models in the continuous-time case have been widely studied by many researchers as in [2], where Acedo et al. proposed an analytical approach to find the exact global solution of the classical SIRS epidemic system. Furthermore, there are Alexander and Moghadas in [3] and Hu et al. in [4], who all provided bifurcation analysis of the SIRS model with different incidence rates. The authors who contributed with Teng in [5] and in [6] found significant results from the study of the persistence and extinction of disease using SIRS models. As for Jin et al. in [7], Liu and Zhou in [8] and Chen in [9], they obtained stability conditions for other SIRS systems. A stability analysis of the SIRS model in the discrete-time case is not often available, but there exist interesting analyses done for some classes of this type of model; see for example, the work of Hu et al. in [10]. As an application of such models in a particular case of disease, Mukhopadhyay and Tapaswi published their paper about Japanese encephalitis in [11]. Other authors studied SIRS dynamics when the model framework was in the form of a discrete metapopulation-like system [12].

In parallel, there are many researchers who have benefited from modeling approaches in epidemics, in order to determine the best prevention strategies against the spread of infection to susceptibles, using different optimization techniques such as optimal control methods; see examples in [13–20]. On the other hand, some models as in [21–23] discussed the impact of limited public health resources in the propagation of infectious diseases, but there are very few optimization problems that have been adapted to such subjects. Here, we try to resolve this issue by exploiting studies published in [24–26], where medical constraints have been modeled differently with a constraint called "isoperimetric". More precisely, we propose an anti-epidemic control strategy that targets susceptible people, under the isoperimetric condition that we could not control all individuals of this category due to restricted health resources.

We consider a simple discrete-time epidemic compartmental model devised in the form of difference equations, which describe the dynamics of a discrete-time SIRS model with a temporary controlled class, meaning that the controlled people cannot acquire long-lived immunity to move towards the removed compartment due to the temporary effect of the control parameter. Thereafter, we characterize the sought optimal control, and we show the effectiveness of this limited control policy. This optimal control problem leads to the execution of two numerical methods all combined together at the same time, namely the forward-backward sweep method to generate the optimal state and control functions and the secant method adapted to the isoperimetric restriction.

2. Materials

Let us define a discrete-time model with the four following main compartments:

- S: the number of susceptible people to infection or who are not yet infected,
- C_S: the number of susceptible people who are temporarily controlled, so they cannot move to the removed class due to the limited effect of control. It can represent the compartment of vaccinated people in case a vaccination is not 100% effective due to the difficulty of producing a perfect vaccine, the heterogeneity of the population or a vaccine not conferring a lifelong immunity [17,27],
- I: the number of infected people who are capable of spreading the epidemic to those in the susceptible and temporarily controlled categories,
- R: the number of removed people from the epidemic, but can return to the susceptible class because of the short-term removal individuals' immunity.

In our modeling approach, we aim to describe the dynamics of variables S, C_S, I and R at time i based on the following difference equations:

$$
\begin{cases}
S_{i+1} & = S_i - \beta S_i I_i - a\theta_i S_i + \Pi_i - \mu S_i + \sigma R_i \\
C_{S_{i+1}} & = C_{S_i} + a\theta_i S_i - b\beta C_{S_i} I_i - \mu C_{S_i} \\
I_{i+1} & = I_i + \beta(S_i + bC_{S_i})I_i - \gamma I_i - \mu I_i \\
R_{i+1} & = R_i + \gamma I_i - \mu R_i - \sigma R_i
\end{cases}
\tag{1}
$$

with initial conditions $S_0 > 0$, $C_{S_0} \geq 0$, $I_0 \geq 0$ and $R_0 \geq 0$ and where $\Pi_i = \mu N_i$ with $N_i = S_i + C_{S_i} + I_i + R_i$, gives the newborn people, $a\theta$ ($0 \leq a \leq 1$) is the recruitment rate of susceptibles to the controlled class with θ defining the control parameter as a constant between 0 and 1 (see such consideration in the case of vaccination in [27]) and "a" modeling the reduced chances of a susceptible individual to be controlled, $\beta = \frac{\delta}{N_i}$ with δ the infection transmission rate, μ the natural death rate, $b\theta$ ($0 \leq b \leq 1$) the recruitment rate of controlled people to the infected class even in the presence of θ with "b" modeling the reduced chances of a temporarily controlled individual to be infected, γ the recovery rate and σ the losing removal individuals' immunity rate. We note that the population size N_i is constant at any time i because $N_{i+1} = S_{i+1} + C_{S_{i+1}} + I_{i+1} + R_{i+1} = N_i$. Hence, $\Pi_i = \Pi = constant$.

3. Methods

Now, we consider the mathematical model (1) with θ as a discrete control function.

Motivated by the desire to reduce the number of infected people as much as possible while minimizing the value of the control θ over N times, our objective is to seek an optimal control θ^* such that:

$$J(\theta^*) = \min_{\theta \in \Theta} J(\theta) \tag{2}$$

where J is the functional defined by:

$$J(\theta) = \sum_{i=0}^{N-1} \left(A I_i + \frac{B}{2} \theta_i^2 \right) + A I_N \tag{3}$$

and where the control space Θ is defined by the set:

$$\Theta = \{ \theta \in \mathbb{R}^N | \theta_{\min} \leq \theta_i \leq \theta_{\max}, \ \theta_{\max} \leq 1 \ \theta_{\min} \geq 0, \ i = 0, ..., N-1 \} \}$$

A and B represent constant severity weights associated with functions I and θ, respectively.

Managers of the anti-epidemic resources cannot well predict whether their control strategy will reach all the susceptible population over N times. To model the situation in which a restricted resource of control is available, we consider that the number of susceptible people we can control is equal to a constant $C > 0$ for N days. Hence, we try to find θ^* under the definition of the following isoperimetric restriction:

$$C = \sum_{i=0}^{N-1} a \theta_i S_i \tag{4}$$

In [25,26], the authors defined an isoperimetric constraint on the control variable only, to model the total tolerable dosage amount of a therapy along the treatment period. In their conferences talks [28,29], Kornienko et al. and de Pinho et al. introduced state constraints in an optimal control problem that is subject to an S-exposed-I-Rdifferential system to model the situation of the limited supply of vaccine based on the work in [24] and where the isoperimetric constraint is defined on the product of the control and state variables. Our study aims to highlight more the importance of such optimal control approaches by considering a discrete model rather than a continuous one. This would be interesting since data are often collected at discrete times, as noted in [30].

In our case, to take into account the constraint (4) for the resolution of the optimal control problem (2), we consider a new variable Z defined as:

$$Z_{i+1} = Z_i + a \theta_i S_i \tag{5}$$

with $Z_0 = 0$ and $Z_N = C$.

The discrete-time system of (1) becomes:

$$\begin{cases} S_{i+1} &= S_i + \Pi - \beta S_i I_i - a \theta_i S_i - \mu S_i + \sigma R_i \\ C_{S_{i+1}} &= C_{S_i} + a \theta_i S_i - b \beta C_{S_i} I_i - \mu C_{S_i} \\ I_{i+1} &= I_i + \beta (S_i + b C_{S_i}) I_i - \gamma I_i - \mu I_i \\ R_{i+1} &= R_i + \gamma I_i - \mu R_i - \sigma R_i \\ Z_{i+1} &= Z_i + a \theta_i S_i \end{cases} \tag{6}$$

In the following, we announce two theorems for proving the existence and the characterization of the sought optimal control θ^*.

Theorem 1. *(Sufficient conditions) For the isoperimetric optimal control problem given by (2) along with the discrete state equations in (6), there exists a control $\theta^* \in \Theta$ such that $J(\theta^*) = \min_{\theta \in \Theta} J(\theta)$.*

Proof. In order to prove the existence of a solution θ^* in Θ, we try to prove that $\min_{\theta \in \Theta} J(\theta)$ exists.

We have a finite number of time steps N and discrete state equations in System (6) with bounded coefficients $\gamma, \mu, b, \sigma, \Pi, a$ and β, then for all θ in the control set Θ, the N-component state variables:

$$S = (S_0, S_1, ..., S_i, ..., S_{N-1}),$$
$$C_S = (C_{S_0}, C_{S_1}, ..., C_{S_i}, ..., C_{S_{N-1}}),$$
$$I = (I_0, I_1, ..., I_i, ..., I_{N-1})$$
$$\text{and } R = (R_0, R_1, ..., R_i, ..., R_{N-1}) \; \forall i = 0, ..., N-1$$

are uniformly bounded, which implies that $\forall \theta \in \Theta$, $J(\theta)$ is uniformly bounded.

We can deduce then that $inf_{\theta \in \Theta} J(\theta)$ is finite since $J(\theta)$ is bounded, and there exists a finite number j of uniformly bounded sequences $\theta^j \in \Theta$ such that $lim_{j \to \infty} J(\theta^j) = inf_{\theta \in \Theta} J(\theta)$ and corresponding sequences of states S^j, C_S^j, I^j and R^j.

Thus, there exists $\theta^* \in \Theta$ and $S^*, C_S^*, I^*, R^* \in \mathbb{R}^N$ such that on a subsequence,

$$\theta^j \to \theta^*,$$
$$S^j \to S^*,$$
$$C_S^j \to C_S^*,$$
$$I^j \to I^*$$
$$\text{and } R^j \to R^*.$$

Finally, due to the finite dimensional structure of the system (6) and the objective function $J(\theta)$, θ^* is an optimal control with corresponding states S^*, C_S^*, I^* and R^* [26]. Therefore, taking into account the structure of J being a convex function, $inf_{\theta \in \Theta} J(\theta)$ is achieved. $\square$

In order to derive the necessary conditions of optimality, we employ the discrete version of Pontryagin's maximum principle stated in Theorem A1 in Appendix A.

Theorem 2. *(Necessary conditions) Given the optimal control θ^* and solutions S^*, C_S^*, I^* and R^*, there exist $\lambda_{l,i}$, $l = 1, ..., 5$, $i = 0, ..., N$, the adjoint variables satisfying the following equations:*

$$\triangle\lambda_{1,i} = \lambda_{1,i+1}(-1 + \beta I_i^* + \mu + a\theta_i^*) - a\lambda_{2,i+1}\theta_i^* - \beta\lambda_{3,i+1}I_i^* - a\theta_i^*\lambda_{5,i+1} \tag{7}$$

$$\triangle\lambda_{2,i} = \lambda_{2,i+1}(-1 + b\beta I_i^* + \mu) - b\lambda_{3,i+1}\beta I_i^* \tag{8}$$

$$\triangle\lambda_{3,i} = -A + \lambda_{1,i+1}\beta S_i^* + b\lambda_{2,i+1}\beta C_{S_i}^* - \lambda_{3,i+1}(-1 + \beta(S_i^* + bC_{S_i}^*) - \mu - \gamma) \tag{9}$$
$$\quad - \lambda_{4,i+1}\gamma$$

$$\triangle\lambda_{4,i} = \lambda_{4,i+1}(\mu + \sigma - 1) - \sigma\lambda_{1,i+1} \tag{10}$$

$$\triangle\lambda_{5,i} = -\lambda_{5,i+1} \tag{11}$$

with $\triangle\lambda_{l,i} = \lambda_{l,i+1} - \lambda_{l,i}$, $l = 1, ..., 5$, $i = 0, ..., N-1$ the difference operator and $\lambda_{1,N} = \lambda_{2,N} = \lambda_{4,N} = 0$, $\lambda_{3,N} = A$ and $\lambda_{5,N} = constant$ to be determined, as the transversality conditions associated with adjoint Equations (7)–(11).

In addition, the optimal control θ^* is characterized at each iteration i by:

$$\theta_i^* = \min\left(\max\left(\theta_{min}, \frac{aS_i^*(\lambda_{1,i+1} - \lambda_{2,i+1} - \lambda_{5,i+1})}{B}\right), \theta_{max}\right) \; \forall i = 0, ..., N-1 \tag{12}$$

Proof. With the application of a discrete version of Pontryagin's maximum principle in Appendix A and as done in [26,31,32], we can determine the discrete optimal control θ^* for the problem (6) and its associated trajectories S^*, C_S^*, I^* and R^*.

We define a discrete Hamiltonian H_i as a brief notation of the function H defined for $i = 0, ..., N - 1$. as follows:

$$H(i, S_i, C_{S_i}, I_i, R_i, \lambda_{1,i+1}, \lambda_{2,i+1}, \lambda_{3,i+1}, \lambda_{4,i+1}, \lambda_{5,i+1}, \theta_i)$$
$$= AI_i + \frac{B}{2}\theta_i^2 + \lambda_{1,i+1}S_{i+1} + \lambda_{2,i+1}C_{S_{i+1}} + \lambda_{3,i+1}I_{i+1} + \lambda_{4,i+1}R_{i+1} + \lambda_{5,i+1}Z_{i+1}$$

The discrete-time adjoint system is resolved using the following formulations:

$$\triangle\lambda_{1,i} = -\frac{\partial H_i}{\partial S_i}, \triangle\lambda_{2,i} = -\frac{\partial H_i}{\partial C_{S_i}}, \triangle\lambda_{3,i} = -\frac{\partial H_i}{\partial I_i}, \triangle\lambda_{4,i} = -\frac{\partial H_i}{\partial R_i}$$
$$\text{and } \triangle\lambda_{5,i} = -\frac{\partial H_i}{\partial Z_i}$$

that we associate with the following transversality conditions:

$$\lambda_{1,N} = \frac{\partial\phi_N}{\partial S_N}, \lambda_{2,N} = \frac{\partial\phi_N}{\partial C_{S_N}}, \lambda_{3,N} = \frac{\partial\phi_N}{\partial I_N}, \lambda_{4,N} = \frac{\partial\phi_N}{\partial R_N}$$

with ϕ_N representing the payoff term function in (3), namely AI_N.

Then, we obtain the following discrete-time adjoint system:

$$\begin{cases} \triangle\lambda_{1,i} &= \lambda_{1,i+1}(-1 + \beta I_i + \mu + a\theta_i) - a\lambda_{2,i+1}\theta_i - \beta\lambda_{3,i+1}I_i - a\theta_i\lambda_{5,i+1} \\ \triangle\lambda_{2,i} &= \lambda_{2,i+1}(-1 + b\beta I_i + \mu) - b\lambda_{3,i+1}\beta I_i \\ \triangle\lambda_{3,i} &= -A + \lambda_{1,i+1}\beta S_i + b\lambda_{2,i+1}\beta C_{S_i} - \lambda_{3,i+1}(-1 + \beta(S_i + bC_{S_i}) - \mu - \gamma) - \lambda_{4,i+1}\gamma \\ \triangle\lambda_{4,i} &= \lambda_{4,i+1}(\mu + \sigma - 1) - \sigma\lambda_{1,i+1} \\ \triangle\lambda_{5,i} &= -\lambda_{5,i+1} \end{cases}$$

with the transversality conditions $\lambda_{1,N} = 0, \lambda_{2,N} = 0, \lambda_{4,N} = 0, \lambda_{3,N} = A$ and $\lambda_{5,N}$ is unknown.

In order to find the transversality condition $\lambda_{5,N} = constant$, we use the secant-method as the appropriate numerical technique for finding the zero of the function $\lambda_{5,N} \to V(\lambda_{5,N}) = \tilde{Z}_N - Z_N$ where $\tilde{Z}_N$ is the value of Z at final iteration N for various values of $\lambda_{5,N}$ and Z_N is the value fixed by C [25,33].

Since θ_i is a bounded control, we can then define a Lagrangian L as follows:

$$L((i, S_i, C_{S_i}, I_i, R_i, \lambda_{1,i+1}, \lambda_{2,i+1}, \lambda_{3,i+1}, \lambda_{4,i+1}, \lambda_{5,i+1}, \theta_i, \omega_{1,i}, \omega_{2,i})$$
$$= H_i + \omega_{1,i}(\theta_{max} - \theta_i) + \omega_{2,i}(\theta_i - \theta_{min})$$

where $\omega_{1,i}, \omega_{2,i} \geq 0 \; \forall i$ verifying at $\theta_i = \theta_i^*$, the two conditions $\omega_{1,i}(\theta_{max} - \theta_i^*) = 0$ and $\omega_{2,i}(\theta_i^* - \theta_{min}) = 0$.

Let L_i be the brief notation of L and L_i^* be the brief notation of L at S^*, C_S^*, I^*, R^* and θ^*. The condition of minimization is defined as:

$$L_i^* = \min_{\theta_i \in \Theta} L_i(**).$$

In order to find the solution θ_i^* of (**), we differentiate the Lagrangian L_i with respect to θ_i on the set Θ to obtain the optimality equation:

$$\frac{\partial L_i}{\partial \theta_i} = B\theta_i + aS_i(\lambda_{2,i+1} - \lambda_{1,i+1}) + aS_i\lambda_{5,i+1} - \omega_{1,i} + \omega_{2,i} = 0 \text{ at } \theta_i^*.$$

Furthermore, we find

$$\theta_i^* = \frac{aS_i^*(\lambda_{1,i+1} - \lambda_{2,i+1} - \lambda_{5,i+1}) - \omega_{2,i} + \omega_{1,i}}{B}.$$

If:

$$\theta_{\min} < \theta_i^* < \theta_{\max},$$

then:

$$\omega_{1,i} = \omega_{2,i} = 0,$$

therefore:

$$\theta_i^* = \frac{aS_i^*\left(\lambda_{1,i+1} - \lambda_{2,i+1} - \lambda_{5,i+1}\right)}{B}.$$

If:

$$\theta_i^* = \theta_{\min},$$

then:

$$\omega_{1,i} = 0,$$

therefore:

$$\theta_{\min} = \frac{aS_i^*\left(\lambda_{1,i+1} - \lambda_{2,i+1} - \lambda_{5,i+1}\right) - \omega_{2,i}}{B},$$

implying that:

$$\omega_{2,i} = a\left(S_i^*(\lambda_{1,i+1} - \lambda_{2,i+1}) - S_i^*\lambda_{5,i+1}\right) - B\theta_{\min}.$$

Knowing that $\omega_{2,i} \geq 0$ and $B > 0$, we obtain $\theta_i^* \leq \dfrac{aS_i^*\left(\lambda_{1,i+1} - \lambda_{2,i+1} - \lambda_{5,i+1}\right)}{B}.$
If:

$$\theta_i^* = \theta_{\max},$$

then:

$$\omega_{2,i} = 0,$$

thus:

$$\theta_{\max} = \frac{aS_i^*\left(\lambda_{1,i+1} - \lambda_{2,i+1} - \lambda_{5,i+1}\right) + \omega_{1,i}}{B},$$

implying that $\omega_{1,k} = B\theta_{\max} - a\left(S_i^*(\lambda_{1,i+1} - \lambda_{2,i+1}) - S_i^*\lambda_{5,i+1}\right).$

Knowing that $\omega_{1,i} \geq 0$ and $B > 0$, we obtain $\theta_i^* \geq \dfrac{aS_i^*\left(\lambda_{1,i+1} - \lambda_{2,i+1} - \lambda_{5,i+1}\right)}{B}.$
Using these standard optimality arguments, we characterize the control u_k^* by:

$$\theta_i^* = \begin{cases} \frac{aS_i^*(\lambda_{1,i+1} - \lambda_{2,i+1} - \lambda_{5,i+1})}{B} & \text{if } \theta_{\min} < \frac{aS_i^*(\lambda_{1,i+1} - \lambda_{2,i+1} - \lambda_{5,i+1})}{B} < \theta_{\max} \\[2ex] \theta_{\min} & \text{if } \frac{aS_i^*(\lambda_{1,i+1} - \lambda_{2,i+1} - \lambda_{5,i+1})}{B} \leq \theta_{\min} \\[2ex] \theta_{\max} & \text{if } \frac{aS_i^*(\lambda_{1,i+1} - \lambda_{2,i+1} - \lambda_{5,i+1})}{B} \geq \theta_{\max} \end{cases}$$

or by a more reduced form, we can write

$$\theta_i^* = \min\left(\max\left(\theta_{\min}, \frac{aS_i^*\left(\lambda_{1,i+1} - \lambda_{2,i+1} - \lambda_{5,i+1}\right)}{B}\right), \theta_{\max}\right) \qquad \square$$

4. Numerical Results and Discussion

In this section, we resolve the discrete two-point value problem defined by System (6) with initial conditions along with Equations (7)–(11) with final conditions, using a discrete version of the forward-backward sweep method (FBSM) [26,33] with the incorporation of a discrete progressive iterative scheme to stock at each iteration i, the values of the state variables corresponding to the forward discrete-time system (6), to use them in a second discrete regressive iterative scheme incorporated for stocking at each time i, the values of the adjoint state variables corresponding to the backward discrete-time adjoint system (7)–(11). In fact, at each time i, the values stocked of both state and adjoint state variables were utilized in the characterization of the optimal control θ^*. In brief, our algorithm is defined by the following four steps of numerical calculus (Algorithm 1).

Algorithm 1: Resolution steps of the discrete two-point boundary value optimal control problem (6)–(11).

Step 0:

  Guess an initial estimation of θ.

Step 1:

  Use the initial condition $S(0)$, $C_S(0)$, $I(0)$, $R(0)$ and $Z(0)$ and the stocked values by θ.
  Find the optimal states S^*, C_S^*, I^*, R^* and Z^*, which iterate forward in the discrete two-point boundary value problem (6).

Step 2:

  Use the stocked values by θ and the transversality conditions $\lambda_{l,N+1}$ for $l = 1,2,3,4$ while searching the constant $\lambda_{5,N+1}$ using the secant-method. More precisely, the secant method is used to obtain the zero of the function $\lambda_{5,N} \to V(\lambda_{5,N}) = \tilde{Z}_N - Z_N$ where $\tilde{Z}_N$ is the value of Z at final iteration N for various values of $\lambda_{5,N}$, and Z_N is the value fixed by C. In addition, due its structure in (4), we choose the constant C in a way that it cannot exceed an upper bound $N_0 \times N$ where N_0 is the initial population size and N is the number of iterations.
  Find the adjoint variables λ_l for $l = 1,2,3,4,5$, which iterate backward in the discrete two-point boundary value problem (6).

Step 3:

  Update the control utilizing new S, C_S, I, R, Z and λ_l for $l = 1,2,3,4,5$ in the characterization of θ^* as presented in (12).

Step 4:

  Test the convergence. If the values of the sought variables in this iteration and the final iteration are sufficiently small, check out the recent values as solutions. If the values are not small, go back to Step 1.

Figure 1 depicts the behavior of the number of susceptible people in the absence and presence of the control, and we can see that the number of susceptible people had decreased from its initial condition once the control had been introduced, while there was no significant decrease of the S function compared to the case when there was yet no control. With these parameters used, it reached only three people because of the maximal value of one taken by the optimal control θ^* in almost alltimes of the control strategy, as seen in the last figure.

In Figure 2, we can well understand the increase of the number of the removed people because of a natural recovery, but it cannot represent a significant recovery because it has not reached even 14 people, and this means that only a very small number of people have been removed based on the initial condition $I(0)$ considered. In the presence of the control, the R function increased towards a much higher number of removed people and showed it can even reach more than 31 individuals recovered from the disease in the first 17 days; this number decreased thereafter because of the results in the next figure, which will show that infection will disappear as we move forward in time.

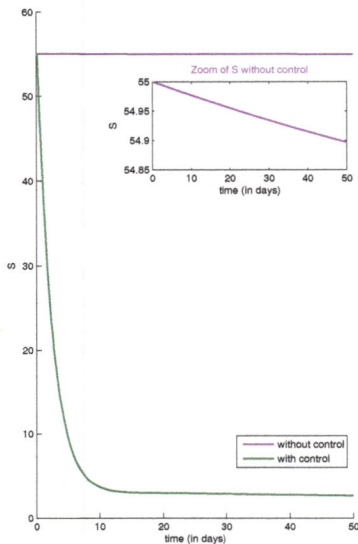

Figure 1. Number of susceptible people in the absence and presence of the control in the two cases $\theta = 0$ and $\theta \neq 0$. Parameter values: $\Pi = \mu N_0$, $a = 0.5$, $b = 0.1$, $\beta = 10^{-5}$, $\mu = 0.00045$, $\gamma = 0.75 \times 10^{-2}$, $\sigma = 5 \times 10^{-4}$. Initial conditions: $S(0) = 55$, $C_S(0) = 0$, $I(0) = 42$, $R(0) = 0$. Severity weight constants: $A = 1$ and $B = 4 \times 10^5$.

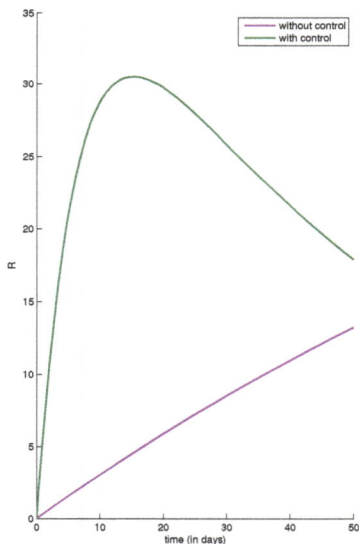

Figure 2. Number of removed people in the absence and presence of the control in the two cases $\theta = 0$ and $\theta \neq 0$, with the same parameter values, initial conditions and severity weights as in Figure 1.

In Figure 3, the simulation shows that the number of infected people could decrease only because of a natural recovery or death, while the infection was still serious and remained present in more than 28 individuals. After the introduction of the control, the I function started to decrease once the anti-epidemic was followed, and it tended to zero values after 37 days; this means most people would recover from the disease at the end of the control strategy.

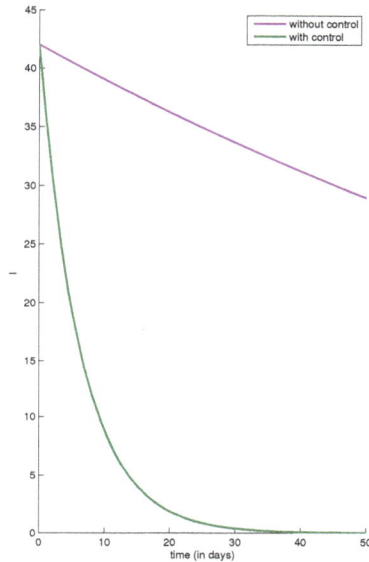

Figure 3. Number of infected people in the absence and presence of the control in the two cases $\theta = 0$ and $\theta \neq 0$, with the same parameter values, initial conditions and severity weights as in Figure 1.

As regards Figure 4, we sought to verify the condition $C = 110$, which meets the value $2 \times S_0$, and we presented associated simulations of the number of controlled people, which increased to 53 individuals once the optimal control had been introduced and increased thereafter, again showing that it could even exceed that number, approximately towards 80 people when we go forward towards the end. In the same figure, we show the values of the optimal control θ^*, which take the value of one as the maximal peak for almost alldays, and we can also see that the imposed isoperimetric constant has been verified in the final instant with some error $\epsilon = 3.6589$. In fact, it is not evident that it reached any imposed value while verifying convergence tests of both methods used. Sometimes, the program did not stop iterating or could not show the plot because of a NANvalue, and then, the only solution was to fix the number of iterations of the secant method in which the imposed initial guess of C was approximately reached.

In Figure 5, we exhibit the value of the sought constant missing transversality condition $\lambda_{5,N}$, which will be essential to verify the necessary conditions announced in Theorem 2. As we can observe from this figure, the value obtained equals -1.3830×10^8.

Figure 6 presents a numerical simulation of the Z function when we did not seek the verification of the condition $Z_N = C$, and we let Z_N free, so we could prove that our algorithm in the case of the isoperimetric constraint helped to approximate Z_N to C or even verify the equality between them, far from the value that could reach Z_N when it was free; as we can see from the mentioned figure, $Z_N = 105.5729$, which led to an important error of about 4.4271 from C that was sought in Figure 4.

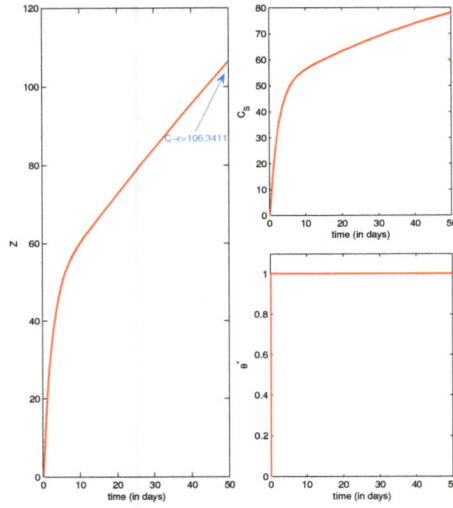

Figure 4. Number of controlled people, the optimal control θ^* and the variable Z with an imposed constant $C = 110$, with the same parameter values, initial conditions and severity weights as in Figure 1.

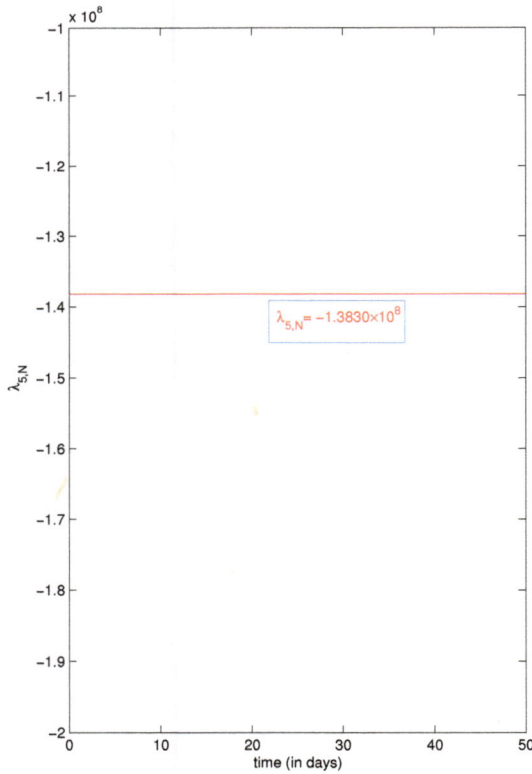

Figure 5. Value of the sought transversality condition $\lambda_{5,N}$.

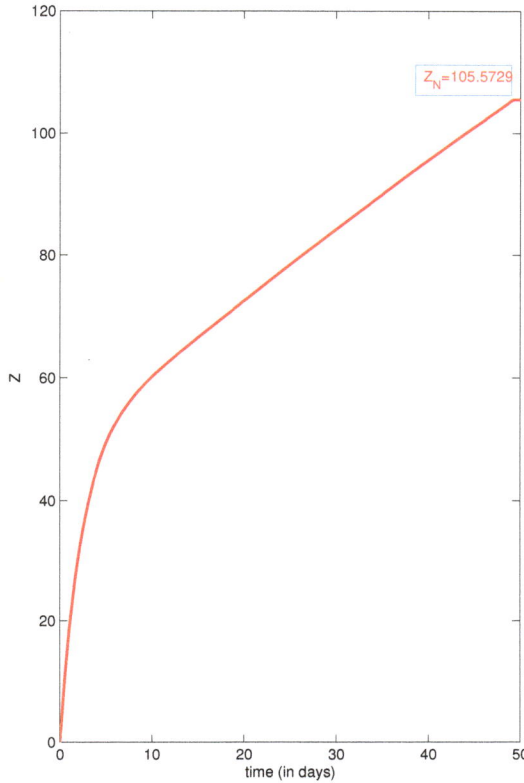

Figure 6. Z function when Z_N is free and not forced to equal C.

5. Conclusions

In this paper, an optimal control approach with an isoperimetric constraint has been applied to a discrete-time SIRS model, which was in the form of a four-compartmental epidemic model where it was supposed that the controlled population did not reach the removed class due to the temporary effect of the control. The isoperimetric restriction, which has been proposed to define the number of susceptible people who receive the control along the anti-epidemic measures period, allowed us to find the optimal control needed to fight against a disease when there were limited resources.

Author Contributions: All authors contributed equally to this work. All authors read and approved the final manuscript.

Acknowledgments: The authors would like to thank all the members of the Editorial Board who were responsible for dealing with this paper and the anonymous referees for their valuable comments and suggestions, improving the content of this paper.

Conflicts of Interest: The authors declare no conflict of interest.

Appendix A

Let us define the following:

- I: the set $\{0, ..., N-1\}$
- X_i: the real n-component column vector; $i = 0, ..., N$

- θ_i: the real m-component column control vector; $i = 0, ..., N-1$
- $\Theta = \{\theta \in \mathbb{R}^N | \theta_{min} \leq \theta_i \leq \theta_{max}, \; \theta_{max} \leq 1 \; \theta_{min} \geq 0, \; i = 0, .., N-1\}\}$: the set of admissible controls
- $f : \mathbb{R}^n \times \mathbb{R}^m \times \Theta \to \mathbb{R}^n$, $F : \mathbb{R}^n \times \mathbb{R}^m \times I \to \mathbb{R}$, $\phi : \mathbb{R}^m \times I \cup \{N\} \to \mathbb{R}$, continuously differentiable functions.

Let us consider a discrete-time optimal control problem over times $0, ..., N$, defined by:

$$\min\{J\} = \sum_{i=0}^{N-1} F(X_i, \theta_i, i) + \phi(X_N, N) \tag{A1}$$

subject to the discrete-time system:

$$X_{i+1} = X_i + f(X_i, \theta_i, i), \; i = 0, ..., N-1 \tag{A2}$$

$$X_0 \; given \tag{A3}$$

$$\theta_i \in \Theta \tag{A4}$$

We now define the Hamiltonian function H_i to be:

$$H_i = H(X_i, \theta_i, i) = F(X_i, \theta_i, i) + \lambda_{i+1} f(X_i, \theta_i, i)$$

and in the optimal control and state by $H_i^* = H(X_i^*, \theta_i^*, i)$.

Then, based on results of the discrete version of the maximum principle discussed in [34], we can derive the following necessary conditions for our problem (A1) based on the following theorem.

Theorem A1. *(A discrete version of the maximum principle) Given a discrete optimal control θ_i^* in the sense of sufficient conditions and given solutions X_i^*, then the necessary conditions for θ_i^* to be optimal for Problem (A1)–(A4) are:*

$$X_{i+1}^* = X_i^* + f(X_i^*, \theta_i^*, i), \; X_0 \; given$$

$$\Delta\lambda_i = -\frac{\partial H_i^*}{\partial X_i}, \; \lambda_N = \frac{\partial \phi(X_N^*, N)}{\partial X_N}$$

$$H_i^* \leq H(X_i^*, \theta_i, i)$$

$$for \; all \; \theta_i^* \in \Theta$$

References

1. Kermack, W.O.; McKendrick, A.G. A Contribution to the Mathematical Theory of Epidemics. *Proc. R. Soc. A* **1927**, *115*, 700–721. [CrossRef]
2. Acedo, L.; González-Parra, G.; Arenas, A.J. An exact global solution for the classical SIRS epidemic model. *Nonlinear Anal. Real World Appl.* **2010**, *11*, 1819–1825. [CrossRef]
3. Alexander, M.E.; Moghadas, S.M. Bifurcation analysis of an SIRS epidemic model with generalized incidence. *SIAM J. Appl. Math.* **2005**, *65*, 1794–1816. [CrossRef]
4. Hu, Z.; Bi, P.; Ma, W.; Ruan, S. Bifurcations of an SIRS epidemic model with nonlinear incidence rate. *Discret. Contin. Dyn. Syst. Ser. B* **2011**, *15*, 93–112.
5. Teng, Z.; Liu, Y.; Zhang, L. Persistence and extinction of disease in non-autonomous SIRS epidemic models with disease-induced mortality. *Nonlinear Anal. Theory Methods Appl.* **2008**, *69*, 2599–2614. [CrossRef]
6. Xamxinur, A.; Teng, Z. On the persistence and extinction for a non–autonomous SIRS epidemic model. *J. Biomath.* **2006**, *21*, 167–176.
7. Jin, Y.; Wang, W.; Xiao, S. An SIRS model with a nonlinear incidence rate. *Chaos Solitons Fractals* **2007**, *34*, 1482–1497.

8. Liu, J.; Zhou, Y. Global stability of an SIRS epidemic model with transport-related infection. *Chaos Solitons Fractals* **2009**, *40*, 145–158.

9. Chen, J. An SIRS epidemic model. *Appl. Math. J. Chin. Univ.* **2004**, *19*, 101–108.

10. Hu, Z.; Teng, Z.; Jiang, H. Stability analysis in a class of discrete SIRS epidemic models. *Nonlinear Anal. Real World Appl.* **2012**, *13*, 2017–2033. [CrossRef]

11. Mukhopadhyay, B.B.; Tapaswi, P.K. An SIRS epidemic model of Japanese encephalitis. *Int. J. Math. Math. Sci.* **1994**, *17*, 347–355. [CrossRef]

12. Abouelkheir, I.; El Kihal, F.; Rachik, M.; Zakary, O.; Elmouki, I. A multi-regions SIRS discrete epidemic model with a travel-blocking vicinity optimal control approach on cells. *Br. J. Math. Comput. Sci.* **2017**, *20*, 1–16. [CrossRef] [PubMed]

13. Zakary, O.; Rachik, M.; Elmouki, I. On the impact of awareness programs in HIV/AIDS prevention: An SIR model with optimal control. *Int. J. Comput. Appl.* **2016**, *133*. [CrossRef]

14. Zakary, O.; Larrache, A.; Rachik, M.; Elmouki, I. Effect of awareness programs and travel-blocking operations in the control of HIV/AIDS outbreaks: A multi-domains SIR model. *Adv. Differ. Equ.* **2016**, *2016*, 169. [CrossRef]

15. Shim, E. A note on epidemic models with infective immigrants and vaccination. *Math. Biosci. Eng.* **2006**, *3*, 557. [CrossRef] [PubMed]

16. Roy, P.K.; Saha, S.; Al Basir, F. Effect of awareness programs in controlling the disease HIV/AIDS: An optimal control theoretic approach. *Adv. Differ. Equ.* **2015**, *2015*, 217. [CrossRef]

17. Rodrigues, H.S.; Monteiro, M.T.T.; Torres, D.F. Vaccination models and optimal control strategies to dengue. *Math. Biosci.* **2014**, *247*, 1–12. [CrossRef] [PubMed]

18. Kumar, A.; Srivastava, P.K. Vaccination and treatment as control interventions in an infectious disease model with their cost optimization. *Commun. Nonlinear Sci. Numer. Simul.* **2017**, *44*, 334–343. [CrossRef]

19. Liu, X.; Takeuchi, Y.; Iwami, S. SVIR epidemic models with vaccination strategies. *J. Theor. Biol.* **2008**, *253*, 1–11. [CrossRef] [PubMed]

20. Nainggolan, J.; Supian, S.; Supriatna, A.K.; Anggriani, N. Mathematical model of tuberculosis transmission with reccurent infection and vaccination. In *Journal of Physics: Conference Series*; IOP Publishing: Bristol, UK, 2013; Volume 423, No. 1, p. 012059.

21. Zhou, L.; Fan, M. Dynamics of an SIR epidemic model with limited medical resources revisited. *Nonlinear Anal. Real World Appl.* **2012**, *13*, 312–324. [CrossRef]

22. Abdelrazec, A.; Bélair, J.; Shan, C.; Zhu, H. Modeling the spread and control of dengue with limited public health resources. *Math. Biosci.* **2016**, *271*, 136–145. [CrossRef] [PubMed]

23. Yu, T.; Cao, D.; Liu, S. Epidemic model with group mixing: Stability and optimal control based on limited vaccination resources. *Commun. Nonlinear Sci. Numer. Simul.* **2018**, *61*, 54–70. [CrossRef]

24. Neilan, R.M.; Lenhart, S. An Introduction to Optimal Control with an Application in Disease Modeling. In *Modeling Paradigms and Analysis of Disease Trasmission Models*; American Mathematical Society: Providence, RI, USA, 2010; pp. 67–82.

25. Elmouki, I.; Saadi, S. BCG immunotherapy optimization on an isoperimetric optimal control problem for the treatment of superficial bladder cancer. *Int. J. Dyn. Control* **2016**, *4*, 339–345. [CrossRef]

26. Alkama, M.; Rachik, M.; Elmouki, I. A discrete isoperimetric optimal control approach for BCG immunotherapy in superficial bladder cancer: Discussions on results of different optimal doses. *Int. J. Appl. Comput. Math.* **2017**, *3*, 1–18. [CrossRef]

27. Sharomi, O.; Malik, T. Optimal control in epidemiology. *Ann. Oper. Res.* **2017**, *251*, 55–71. [CrossRef]

28. Kornienko, I.; Paiva, L.T.; De Pinho, M.D.R. Introducing state constraints in optimal control for health problems. *Procedia Technol.* **2014**, *17*, 415–422. [CrossRef]

29. De Pinho, M.D.R.; Kornienko, I.; Maurer, H. Optimal control of a SEIR model with mixed constraints and L 1 cost. In Proceedings of the 11th Portuguese Conference on Automatic Control, CONTROLO'2014, Porto, Portugal, 21–23 July 2014; Springer: Cham, Switzerland, 2015; pp. 135–145.

30. Zakary, O.; Rachik, M.; Elmouki, I. On the analysis of a multi-regions discrete SIR epidemic model: An optimal control approach. *Int. J. Dyn. Control* **2017**, *5*, 917–930. [CrossRef]

31. Zakary, O.; Rachik, M.; Elmouki, I.; Lazaiz, S. A multi-regions discrete-time epidemic model with a travel-blocking vicinity optimal control approach on patches. *Adv. Differ. Equ.* **2017**, *2017*, 120. [CrossRef]

32. Zakary, O.; Rachik, M.; Elmouki, I. A new epidemic modeling approach: Multi-regions discrete-time model with travel-blocking vicinity optimal control strategy. *Infect. Dis. Model.* **2017**, *2*, 304–322. [CrossRef] [PubMed]
33. Lenhart, S.; Workman, J.T. *Optimal Control Applied to Biological Models*; CRC Press: Boca Raton, FL, USA, 2007.
34. Sethi, S.P.; Thompson, G.L. *What Is Optimal Control Theory?* Springer: New York, NY, USA, 2000; pp. 1–22.

*Mathematical
and Computational
Applications*

MDPI

Article

Solution of Optimal Harvesting Problem by Finite Difference Approximations of Size-Structured Population Model

Johanna Pyy [1,*] , **Anssi Ahtikoski** [2], **Alexander Lapin** [3] and **Erkki Laitinen** [1]

1   Faculty of Sciences, University of Oulu, FI-90014 Oulu, Finland; erkki.laitinen@oulu.fi
2   Natural Resources Institute Finland (Luke) Oulu, FI-90014 Oulu, Finland; anssi.ahtikoski@oulu.fi
3   Institute of Computational Mathematics & Information Technology, Kazan Federal University,
     420008 Kazan, Russia; avlapine@mail.ru
*   Correspondence: pyyj@student.oulu.fi

Received: 11 April 2018; Accepted: 24 April 2018 ; Published: 26 April 2018

Abstract: We solve numerically a forest management optimization problem governed by a nonlinear partial differential equation (PDE), which is a size-structured population model. The formulated problem is supplemented with a natural constraint for a solution to be non-negative. PDE is approximated by an explicit or implicit in time finite difference scheme, whereas the cost function is taken from the very beginning in the finite-dimensional form used in practice. We prove the stability of the constructed nonlinear finite difference schemes on the set of non-negative vectors and the solvability of the formulated discrete optimal control problems. The gradient information is derived by constructing discrete adjoint state equations. The projected gradient method is used for finding the extremal points. The results of numerical testing for several real problems show good agreement with the known results and confirm the theoretical statements.

Keywords: size-structured population model; nonlinear partial differential equation; finite difference approximation; optimization; gradient method

---

1. Introduction

The well-posedness of the continuous size-structured model has been studied in several papers (e.g., [1–4]). In Ref. [1], authors proved the local existence and uniqueness of a solution of the continuous model, where birth and mortality functions depend on total population. In Ref. [2], the authors established the local existence and uniqueness of a solution of the size-structured nonlinear population model, where also growth rate depends on total population. In the papers [3,4], the authors proved global existence and uniqueness of a solution of the continuous nonlinear population model, where all vital rates depend on total population. The total population can be described by e.g., total number of individuals (e.g., [3]), total biomass (e.g., [5]) or basal area.

A continuous nonlinear size-structured population model has been used in a forest management optimization problem (e.g., [6–8]). In a continuous formulation, this nonlinear optimization problem cannot be solved by analytic methods. A natural approach is to solve this problem by approximating a continuous model by a discrete one and further solving a discrete optimization problem by iterative algorithms. In this paper, we focus on development of finite difference schemes to approximate the solution of a continuous nonlinear population model. Efficient schemes are essential for solving optimal control problems or parameter estimation problems as such problems require solving the model numerous times before an optimal solution is obtained.

When continuous population model is approximated by a finite difference scheme, it becomes a matrix population model [9]. In matrix models, trees are divided into classes with respect to their size—for instance, diameter. The matrix describes how the class division changes at one time step. Matrix population models have also been used for forest management optimizations (e.g., [10,11]).

In optimization, using iterative algorithms is inevitable. Higher-order algorithms are usually sensitive to the regularity of the solution, and, therefore, they usually yield a convergence rate of first order as soon as the compatibility conditions are not satisfied. Moreover, in practice, the vital rates are determined on a statistical basis and the compatibility conditions required for high-order convergence are hardly valid with real-life data. These suggest that, in most cases, a first-order method should be the most adequate. Hence, it is desirable to have a robust scheme that can produce many useful qualitative and quantitative properties of the solutions of the differential problem but requires minimum regularity of the solution [12]. Unfortunately, one could not derive the explicit formula for the optimal strategy since the strategy, the state and the costate are coupled into a complex system. The results at this stage may be regarded as a middle step to real world applications and serve as a starting point for numerical computations [13].

In our knowledge, the comprehensive theoretical investigation of the forest management optimization problem with a continuous nonlinear population model as the state equation is still lacking, and, in that sense, the problem is an open problem. Hence, in this work, the investigation of the problem in its differential form is omitted, whereas, we consider the finite dimensional counterpart of the problem constructed by finite difference approximations of the state problem and taking for cost function a finite dimensional form used in practice. We prove the stability of constructed finite difference schemes on the set of non-negative solutions and solvability of the optimization problem, and deduce the necessary gradient information for iterative solution methods. We solve several applied problems, where different approximations schemes are used, and compare the computed results. The rest of the article is organized as follows. In Section 2, a mathematical model of optimal harvesting problem for the size-structured forest is formulated. In Section 3, we construct and investigate two finite difference approximation schemes for a nonlinear boundary value problem that simulates the growth and the harvesting of a forest. A gradient method for minimizing the cost function is constructed in Section 4. The theoretical details of this method are set out in the Appendix A to the article. Section 5 is devoted to the numerical solution of a real-life problem and comparative analysis of the computing results. Finally, in Section 6, we present discussions.

2. Formulation of the Optimal Control Problem

In order to formulate the mathematical model for the optimal harvesting problem for the size-structured forest, we define the following notations. In space, we denote by $x \in \Omega := (L_0, L]$ the thickness of the tree, where L_0 and L are the lower and upper bounds of the space domain, respectively. Moreover, $t \in (0, T]$ is the time, where T is the upper limit. By Q, we denote the product space $\Omega \times (0, T]$. We denote by $y(x, t)$ and $h(x, t)$ the number of trees per unit area (state) and the number of removed trees per unit area (control), respectively. Now, the optimal harvesting problem where the cost functional $J(y, h)$ characterizes a net present value (NPV) of ongoing rotation, and $d(x, t)$ is the discounted price function, is formulated as follows:

$$\max_{(y,h) \in \mathcal{K}} J(y, h) := \int_Q d(x, t) h(x, t). \tag{1}$$

Above $\mathcal{K} = \mathcal{Y}_{ad} \times \mathcal{H}_{ad}$ is the set of constraints for the state and the control, where

$$\mathcal{Y}_{ad} = \{y \mid \text{for all } (x, t) \in Q : y(x, t) \geq 0; \ y \text{ is a solution for Equations (4)–(6)}\}, \tag{2}$$

$$\mathcal{H}_{ad} = \{h \mid \text{for all } (x, t) \in Q : 0 \leq h(x, t) \leq h_{max}, \text{ for all } t \in (0, T] : \int_\Omega h(x, t) \geq B \text{ or } \int_\Omega h(x, t) = 0\} \tag{3}$$

From the point of real-life problems, it is obvious that there exist constants $h_{max} > 0$ and B, which denote the upper limit for harvesting and lower limit for making profitable thinning of trees at time event t; otherwise, the thinning is not done. Notice that harvesting h depends on the state y (via constraint sets), which is defined by the population model

$$\frac{\partial y(x,t)}{\partial t} + \frac{\partial (g(x,P(t))y(x,t))}{\partial x} + m(x,P(t))y(x,t) + h(x,t) = 0, \text{ in } Q, \tag{4}$$

$$g(0,P(t))y(0,t) = 0, \text{ in } (0,T], \tag{5}$$

$$y(x,0) = y^0(x), \text{ in } \Omega, \tag{6}$$

where $g(x,P(t))$ is growth rate, $m(x,P(t))$ mortality rate and $y^0(x) \geqslant 0$ is initial diameter distribution of the trees. Growth and mortality rates depend on diameter x of a tree and on the basal area, $P(t)$, of the forest stand, where

$$P(t) = \pi \int_0^L \left(\frac{x}{2}\right)^2 y(x,t)dx.$$

In the case $h = 0$, the problems (4)–(6) are a particular case of the problem that have been investigated in [1–4]. In these articles, the existence of a non-negative continuous solution of this problem has been proved under some "natural" assumptions for input data. They are:

1. $g(x,P)$ is continuous and strictly positive for all x and P and continuously differentiable with respect to x;
2. $m(x,P)$ is non-negative for x and P and integrable in x;
3. $g(x,P)$ and $m(x,P)$ are Lipschitzian with respect to P;
4. $\sup\limits_{x,P} m(x,P) < \infty$.

We also assume that these assumptions are satisfied. We use growth rate g and mortality m in a bilinear form

$$g(x,P(t)) = g_{11} + g_{12}x + (g_{21} + g_{22}x)P(t),$$
$$m(x,P(t)) = m_{11} + \frac{m_{12}}{x} + \frac{m_{13}}{x^2} + \left(m_{21} + \frac{m_{22}}{x} + \frac{m_{23}}{x^2}\right)P(t),$$

where the constants g_{ij} and m_{ij} are such that $g(x,P) > 0$ and $m(x,P) \geqslant 0$ for all $x \in \Omega$ and $P \geqslant 0$. Obviously, because of suppositions $g(0,P(t)) > 0$, the boundary condition (5) reads as $y(0,t) = 0$.

The optimal harvesting problem has been investigated in [6–8]. The authors of these publications considered the case where the harvesting function has the form $h(x,t) = c(x,t)y(x,t)$, where $c(x,t)$ is the control. Thus, they investigated a coefficient identification problem while we solve an optimal control problem with distributed (on the right-hand side) control.

3. Finite Difference Approximations

In this chapter, we derive explicit and (semi)implicit finite difference approximations for the state problems (4)–(6) and prove their stability estimates on non-negative solutions. The investigation of existence, uniqueness and convergence of approximations is beyond the scope of our article. For the size-structured population model with recruitment, the existence, uniqueness and convergence of explicit approximations is investigated in [14] and implicit approximation in [5,15].

The following notations are used throughout the paper: $\Delta t = \frac{T}{M}$ and $\Delta x = \frac{L-L_0}{N}$ denote the temporal and spatial mesh size, respectively. The non-overlapping mesh intervals are $(t^{k-1}, t^k]$, $k = 1,\ldots,M$, and $(x_{i-1}, x_i]$, $i = 1,\ldots,N$, where $t^0 = 0, t^M = T, x_0 = L_0, x_N = L$.

Let us denote by y_i^k and h_i^k the finite difference approximations of $y(x_i, t^k)$ and $h(x_i, t^k)$, respectively. Moreover, we denote $g_i^k := g(x_i, P^k)$ and $m_i^k := m(x_i, P^k)$ the discrete values of the growth rate and mortality rate, respectively, in size class $[x_{i-1}, x_i]$. The discretized value of the basal area at time t^k is $P^k := \pi \sum\limits_{i=1}^{N} (\frac{x_i}{2})^2 y(x_i, t^k)$.

3.1. Explicit Approximation of the State Equation

For all meshpoints $i = 1, \ldots, N$; $k = 1, \ldots, M$, the explicit finite difference approximation of the size-structured population model (4)–(6) reads

$$\frac{y_i^k - y_i^{k-1}}{\Delta t} + \frac{g_i^{k-1} y_i^{k-1} - g_{i-1}^{k-1} y_{i-1}^{k-1}}{\Delta x} + m_i^{k-1} y_i^{k-1} + h_i^k = 0,$$

$$y_0^k = 0, \tag{7}$$

$$y_i^0 \geqslant 0 \text{ constant.}$$

Note that we use so-called upwind approximation for the first order derivative in space (variable x) using the positivity of coefficient $g(x, P)$ on the set of non-negative mesh functions y. The explicit scheme (7) can be written in the form:

$$y_i^k - \left(1 - \frac{\Delta t}{\Delta x} g_i^{k-1} - \Delta t m_i^{k-1}\right) y_i^{k-1} - \left(\frac{\Delta t}{\Delta x} g_{i-1}^{k-1}\right) y_{i-1}^{k-1} + \Delta t h_i^k = 0.$$

Later on, we denote by $a_i^k = 1 - \frac{\Delta t}{\Delta x} g_i^k - \Delta t m_i^k$, and, $b_i^k = \frac{\Delta t}{\Delta x} g_i^k$. Moreover, we denote by $\mathbf{y}^k := (y_1^k, \ldots, y_N^k)$, $\mathbf{h}^k := (h_i^k, \ldots, h_N^k)$ the vectors of the nodal values and by

$$\mathbf{A}^k = \begin{bmatrix} a_1^k & 0 & 0 & \cdots & 0 & 0 \\ b_1^k & a_2^k & 0 & \cdots & 0 & 0 \\ 0 & b_2^k & a_3^k & \cdots & 0 & 0 \\ \vdots & \vdots & \ddots & \ddots & \vdots & \vdots \\ 0 & 0 & 0 & \cdots & b_{N-1}^k & a_N^k \end{bmatrix}$$

the matrix of coefficients. Now, we can write explicit difference scheme (7) in the following algebraic form:

$$\mathbf{y}^k - \mathbf{A}^{k-1} \mathbf{y}^{k-1} + \Delta t \mathbf{h}^k = 0, \quad k = 1, \ldots, M. \tag{8}$$

Note that this scheme is just the forest growth model studied in [11]. Moreover, the numerical calculation of the next temporal state involves only matrix to vector calculations. The drawback of the explicit scheme is that the following stability condition (9) must be satisfied.

Lemma 1. *Let the condition*

$$\Delta x \geqslant \Delta t \sup_{x,t} g(x, P(t)) \tag{9}$$

be satisfied. Then, on the set of non-negative mesh functions y, the finite difference scheme (7) is stable

$$\max_k \|\mathbf{y}^k\|_1 \leqslant C(T) \left(\|\mathbf{y}^0\|_1 + \sum_{k=1}^M \Delta t \|\mathbf{h}^k\|_1\right), \tag{10}$$

where $\|v\|_1 = \sum\limits_{i=1}^N |v_i|$.

Proof. On the non-negative mesh functions y, the coefficients $g_i(P)$ are positive and $m_i(P) \geqslant 0$. For the mesh steps satisfying condition (9), the diagonal entries of matrix $\mathbf{A}^k$ satisfy the inequality

$$|a_i^k| \leqslant 1 - \frac{\Delta t}{\Delta x} g_i(P^k) + \Delta t m_i(P^k).$$

Because of this inequality, we have the following estimate for $\|.\|_1$-norm of matrices, connected with $\|.\|_1$-norm of vectors:

$$\|\mathbf{A}^k\|_1 = \sum_{i=1}^{N}(|a_i^k| + |b_i^k|) \leqslant 1 + C_m \Delta t$$

with $C_m = \sup_{x,t} m(x, P(t))$. Due to this estimate and condition (9), we obtain from Equation (8) the inequality

$$\|\mathbf{y}^k\|_1 \leqslant (1 + C_m \Delta t)\|\mathbf{y}^{k-1}\|_1 + \Delta t\|\mathbf{h}^k\|_1 \text{ for all } k = 1, 2, \ldots, M,$$

whence stability estimate (10) follows. $\square$

The condition (9) means that the length of the time step Δt and width of the size class Δx have to be chosen so that a tree cannot grow over one size class during one time step Δt (compare with [16]).

3.2. Implicit Approximation of the State Equation

For all meshpoints $i = 1, \ldots, N$; $k = 1, \ldots, M$, the implicit finite difference approximation of the models (4)–(6) is the following linearized problem, with nonlinear coefficients calculated on the previous time level:

$$\frac{y_i^k - y_i^{k-1}}{\Delta t} + \frac{g_i^{k-1} y_i^k - g_{i-1}^{k-1} y_{i-1}^k}{\Delta x} + m_i^{k-1} y_i^k + h_i^k = 0,$$
$$y_0^k = 0, \tag{11}$$
$$y_i^0 \geqslant 0 \text{ constant.}$$

Equation (11) can be rewritten as

$$\left(1 + \frac{\Delta t}{\Delta x} g_i^{k-1} + \Delta t m_i^{k-1}\right) y_i^k - \frac{\Delta t}{\Delta x} g_{i-1}^{k-1} y_{i-1}^k - y_i^{k-1} + \Delta t h_i^k = 0,$$
$$\text{for all } i = 1, \ldots, N, \ k = 1, \ldots, M.$$

Using the notations $a_i^k = 1 + \frac{\Delta t}{\Delta x} g_i^k + \Delta t m_i^k$ and $b_i^k = -\frac{\Delta t}{\Delta x} g_i^k$, we rewrite Equation (11) in a form of linear algebraic equations

$$\mathbf{B}^{k-1}\mathbf{y}^k - \mathbf{y}^{k-1} + \Delta t\mathbf{h}^k = 0, \ k = 1, \ldots, M, \tag{12}$$

where

$$\mathbf{B}^k := \begin{bmatrix} a_1^k & 0 & 0 & \ldots & 0 & 0 \\ b_1^k & a_2^k & 0 & \ldots & 0 & 0 \\ 0 & b_2^k & a_3^k & \ldots & 0 & 0 \\ \vdots & \vdots & \ddots & \ddots & \vdots & \vdots \\ 0 & 0 & 0 & \ldots & b_{N-1}^k & a_N^k \end{bmatrix}$$

is a matrix of nonlinear coefficients.

Lemma 2. *Finite difference scheme* (11) *is unconditionally stable on the set of non-negative mesh functions y: for any Δt and Δx the following stability estimate holds:*

$$\max_k \|\mathbf{y}^k\|_1 \leqslant \|\mathbf{y}^0\|_1 + \sum_{k=1}^{M} \Delta t\|\mathbf{h}^k\|_1. \tag{13}$$

Proof. By direct calculations, we obtain from Equation (12) the equality

$$\sum_{i=1}^{N} a_i^k y_i^k + \sum_{i=1}^{N-1} b_i^k y_i^k - \sum_{i=1}^{N} y_i^{k-1} + \Delta t \sum_{i=1}^{N} h_i^k = 0.$$

Since $a_i^k + b_i^k \geqslant 1$ and $a_N^k \geqslant 1$, then, from this equality, we get

$$\sum_{i=1}^{N} y_i^k \leqslant \sum_{i=1}^{N} y_i^{k-1} + \Delta t \sum_{i=1}^{N} |h_i^k|.$$

Because of positivity of vectors $\mathbf{y}^k$ and $\mathbf{y}^{k-1}$, the last inequality can be written in the form

$$\|\mathbf{y}^k\|_1 \leqslant \|\mathbf{y}^{k-1}\|_1 + \Delta t \|\mathbf{h}^k\|_1 \ \forall k,$$

whence stability estimate (13) follows. □

Notice, contrary to the explicit scheme the time step Δt and class width Δx has no mutual dependence, hence the growth of a tree during a time step is not restricted less than one size class. This characteristic of the implicit scheme is useful in the optimal harvesting problem, covered by the models (4)–(6) or parameter identification problem because such problems require solving the model many times before an optimal solution is obtained.

3.3. Approximation of the Optimal Control Problem

We denote $d_i^k := d(x_i, t^k)$ the discounted price for size class $(x_{i-1}, x_i]$ at time t^k, and $\mathbf{d}^k = (d_1^k, \dots, d_N^k)$. Moreover, $(\mathbf{u}, \mathbf{v}) := \sum_1^N u_i v_i$ is the vector product of vectors $\mathbf{u}, \mathbf{v} \in \mathbb{R}^N$. Approximating the cost function (1) by the right-hand Riemann sum, we get the following approximation for the harvesting problem:

$$\max_{(\mathbf{y},\mathbf{h}) \in K} \left\{ \tilde{J}(\mathbf{y}, \mathbf{h}) := \sum_{k=1}^{M} (\mathbf{d}^k, \mathbf{h}^k) = \sum_{k=1}^{M} \sum_{i=1}^{N} d_i^k h_i^k \right\}. \tag{14}$$

Above, we denote by $K = Y_{ad} \times H_{ad}$, where

$$Y_{ad} = \{(\mathbf{y}, \mathbf{h}) \,|\, \mathbf{y} \geqslant 0, \ \mathbf{y} \text{ is a solution for Equations (7) or (11)}\}, \tag{15}$$

$$H_{ad} = \{\mathbf{h} \,|\, 0 \leqslant \mathbf{h}^k \leqslant \mathbf{h_{max}}, \ \|\mathbf{h}^k\|_1 \geqslant B \text{ or } \mathbf{h}^k = 0, \ k = 1, \dots, M\}. \tag{16}$$

Moreover, $\mathbf{y} = (\mathbf{y}^1, \dots, \mathbf{y}^M)$ and $\mathbf{h} = (\mathbf{h}^1, \dots, \mathbf{h}^M)$.

The following propositions show that the discrete optimal harvesting problem (14) has at least one solution in both cases, i.e., if models (4)–(6) is approximated explicitly or implicitly.

Proposition 1. *Let the mesh steps Δt and Δx satisfy the inequality*

$$1 - \frac{\Delta t}{\Delta x} \sup_{x,t} g(x, P(t)) - \Delta t \sup_{x,t} m(x, P(t)) \geqslant 0. \tag{17}$$

Then, Problem (14) has at least one solution if $\mathbf{y}$ satisfies Equation (7).

Proof. The set K is non-empty. In fact, due to assumption (17), the solution $\mathbf{y}$ of finite difference scheme (7) with $\mathbf{y}^0 \geqslant 0$ is non-negative if $\mathbf{h} = 0$. This statement can be easily verified using form (8) of the difference scheme and noting that all entries a_i^k and b_i^k of the matrices $\mathbf{A}^k$ are non-negative.

Obviously, assumption (9) follows from inequality (17), so stability estimate (10) holds. Since vector $\mathbf{h} \in K$ is bounded, then, due to inequality (10), there exists a constant Y such that $\|\mathbf{y}\|_1 \leqslant Y$, i.e., the set K is bounded. It is closed because of the continuity of functions $g(P)$ and $m(P)$ with respect to P, while P is obviously continuous with respect to $\mathbf{y}$. Thus, K is compact. At last, cost function $\bar{J}$ of Problem (14) is continuous, whence the existence of a solution to Problem (14) follows from Weierstrass's theorem. $\square$

Proposition 2. *Problem* (14) *has at least one solution if* $\mathbf{y}$ *satisfies Equation* (11).

Proof. Proof is very similar to the proof of Proposition 1. Namely, the set K is non-empty because, for $\mathbf{h} = \mathbf{0}$, the solution $\mathbf{y}$ of finite difference scheme (11) is non-negative for all Δx and Δt. Since $\mathbf{h}$ is bounded, then, due to stability estimate (13), $\mathbf{y}$ is also bounded, so the set K is bounded. It is closed because of the continuity of functions $g(P)$ and $m(P)$ with respect to P, and continuity of P with respect to $\mathbf{y}$. Thus, K is compact. At last, cost function $\bar{J}$ of Problem (14) is continuous, whence the existence of a solution to Problem (14) follows. $\square$

Remark 1. *Since neither the function* $\bar{J}$ *is strictly concave nor the set* K *is strictly convex, the optimization problem can have a non-unique solution.*

4. Realization of the Optimal Strategies

In this section, a first order method to approximate the optimal harvesting problem (14) is constructed. In real-life applications, the growth rate g and mortality rate m are determined on a statistical basis and the compatibility conditions required for high-order methods can be hardly validated. Hence, a first order method, which is desirable to have a robust scheme but requires minimum regularity of the solution should be the most adequate. The first order methods require computing of the Fréchet derivatives (Jacobian matrix), which can be computationally expensive. However, when we consider the nonlinear optimization problem, only the gradient of the object function is needed, and the gradient can be computed without the Fréchet derivatives. In this work, the adjoint approach developed in the 1970s in [17] is applied for calculation of the functional gradient. The adjoint method has a great advantage against the direct method because only one linear state problem, so called adjoint state, need to be solved for obtaining the gradient information. Today, it is a well-known method for computing the gradient of a functional with respect to model parameters when this functional depends on those model parameters trough state variables, which are solutions of the state problem. However, this method is less well understood in the control of population models, and, as far as we know, no applications to distributed optimal control of harvesting is presented in literature. Duality and adjoint equations are essential tools in studying existence of the optimal pair $(\mathbf{y}, \mathbf{h})$, and, for a periodic age-dependent harvesting problem and for age-spatial structured harvesting problem, it is applied for proving the existence of the bang-bang control in [18] and in [19,20], respectively. For continuous size-structured harvesting, problem duality and adjoint equations are applied for proving the existence of the bang-bang control in [6,8].

In this work, we apply the Lagrange method and give a recipe to systematically define the adjoint state equations and gradient information. We formulate the Lagrangian of the problem (14) with respect to the state constraint (15) only, and use the projection method regarding the control constraint (16). In the projection method, if solution goes outside the constraint set (16), it is projected back to there. Let us generalize and denote by $A(\mathbf{y}, \mathbf{h}) = 0$ the operator Equation (8) (or Equation (12)). Moreover, $A^k(\mathbf{y}, \mathbf{h}) = 0$ is the operator equation at the time level k, $k = 1, \dots, M$.

Suppose the functional $\bar{J}$ and operator A to be differentiable in the sense that there exist the following partial derivatives:

$$\bar{J}_y \delta \mathbf{y} = \lim_{t \to 0} \frac{\bar{J}(\mathbf{y} + t\delta\mathbf{y}, \mathbf{h}) - \bar{J}(\mathbf{y}, \mathbf{h})}{t}, \ \bar{J}_h \delta \mathbf{h} = \lim_{t \to 0} \frac{\bar{J}(\mathbf{y}, \mathbf{h} + t\delta\mathbf{h}) - \bar{J}(\mathbf{y}, \mathbf{h})}{t},$$

$$A_y \delta \mathbf{y} = \lim_{t \to 0} \frac{A(\mathbf{y} + t\delta\mathbf{y}, \mathbf{h}) - A(\mathbf{y}, \mathbf{h})}{t}, \ A_h \delta \mathbf{h} = \lim_{t \to 0} \frac{A(\mathbf{y}, \mathbf{h} + t\delta\mathbf{h}) - A(\mathbf{y}, \mathbf{h})}{t}$$

for all vectors $\delta\mathbf{y}$ and $\delta\mathbf{h}$ (or at least for the vectors such that $\mathbf{y}, \mathbf{y} + \delta\mathbf{y} \in Y_{ad}$ and $\mathbf{h}, \mathbf{h} + \delta\mathbf{h} \in H_{ad}$). Note that for the fixed $\mathbf{h}$ and $\mathbf{y}$, $\bar{J}_y \equiv \bar{J}_y(\mathbf{y}, \mathbf{h})$ and $\bar{J}_h \equiv \bar{J}_h(\mathbf{y}, \mathbf{h})$ are vectors while $A_y = A_y(\mathbf{y}, \mathbf{h})$ and $A_h = A_h(\mathbf{y}, \mathbf{h})$ are matrices. By A_y^* and A_h^*, we denote the corresponding transpose matrices.

Let us define Lagrange function, $\mathcal{L}$, of the problem (14) by

$$\mathcal{L}(\mathbf{y}, \mathbf{h}, \lambda) = \bar{J}(\mathbf{y}(\mathbf{h}), \mathbf{h}) - \sum_{k=1}^{M} (\lambda^k, A^k(\mathbf{y}, \mathbf{h})),$$

where $\lambda^k \in \mathbb{R}^N$. Now, for all feasible pair $(\mathbf{y}, \mathbf{h})$ holds $A(\mathbf{y}, \mathbf{h}) = 0$, and, for any λ, we have:

$$\mathcal{L}(\mathbf{y}, \mathbf{h}, \lambda) = \bar{J}(\mathbf{y}(\mathbf{h}), \mathbf{h}),$$

and, since λ does not depend on $\mathbf{h}$, we have

$$\frac{\partial \bar{J}}{\partial \mathbf{h}} = \frac{\partial \mathcal{L}(\mathbf{y}, \mathbf{h}, \lambda)}{\partial \mathbf{y}} \frac{\partial \mathbf{y}}{\partial \mathbf{h}} + \frac{\partial \mathcal{L}(\mathbf{y}, \mathbf{h}, \lambda)}{\partial \mathbf{h}}. \tag{18}$$

Above, one method to approximate $\dfrac{\partial \mathbf{y}}{\partial \mathbf{h}}$ is to compute N finite differences over control variable $\mathbf{h}$. However, each computation requires solving the equation $A(\mathbf{y}, \mathbf{h}) = 0$, and, for large N, this method is computationally expensive. In the *adjoint method*, we can avoid to compute $\dfrac{\partial \mathbf{y}}{\partial \mathbf{h}}$ by solving the linear adjoint state equation only once.

The theory of constrained optimization, see [21], says that $(\mathbf{y}, \mathbf{h})$ is the optimal pair for the problem (14) if $(\mathbf{y}, \mathbf{h}, \lambda)$ is a saddle point of $\mathcal{L}$. The derivatives of $\mathcal{L}$ with respect to $\mathbf{y}$, $\mathbf{h}$ and λ are:

$$\frac{\partial \mathcal{L}(\mathbf{y}, \mathbf{h}, \lambda)}{\partial \mathbf{y}} = \frac{\partial \bar{J}(\mathbf{y}, \mathbf{h})}{\partial \mathbf{y}} - \left(\frac{\partial A(\mathbf{y}, \mathbf{h})}{\partial \mathbf{y}}\right)^* \lambda,$$

$$\frac{\partial \mathcal{L}(\mathbf{y}, \mathbf{h}, \lambda)}{\partial \mathbf{h}} = \frac{\partial \bar{J}(\mathbf{y}, \mathbf{h})}{\partial \mathbf{h}} - \left(\frac{\partial A(\mathbf{y}, \mathbf{h})}{\partial \mathbf{h}}\right)^* \lambda,$$

$$\frac{\partial \mathcal{L}(\mathbf{y}, \mathbf{h}, \lambda)}{\partial \lambda} = -A(\mathbf{y}, \mathbf{h}).$$

Now, $\dfrac{\partial \mathcal{L}(\mathbf{y}, \mathbf{h}, \lambda)}{\partial \lambda} = 0$ gives the state equation, $\dfrac{\partial \mathcal{L}(\mathbf{y}, \mathbf{h}, \lambda)}{\partial \mathbf{y}} = 0$ gives the adjoint state equation and $\dfrac{\partial \bar{J}}{\partial \mathbf{h}} = \dfrac{\partial \mathcal{L}(\mathbf{y}, \mathbf{h}, \lambda)}{\partial \mathbf{h}}$ gives the gradient.

Now, the calculation of the gradient can be summarized by the following steps when the Lagrangian $\mathcal{L}(\mathbf{y}, \mathbf{h}, \lambda) = \bar{J}(\mathbf{y}, \mathbf{h}) - (\lambda, A(\mathbf{y}, \mathbf{h}))$ is first formulated:

I   Solve the state equation $A(\mathbf{y}, \mathbf{h}) = 0$;

II   Solve the adjoint state equation

$$\frac{\partial \mathcal{L}(\mathbf{y}, \mathbf{h}, \lambda)}{\partial \mathbf{y}} = \frac{\partial \bar{J}(\mathbf{y}, \mathbf{h})}{\partial \mathbf{y}} - \left(\frac{\partial A(\mathbf{y}, \mathbf{h})}{\partial \mathbf{y}}\right)^* \lambda = 0;$$

III   Compute the gradient

$$\frac{\partial \bar{J}}{\partial \mathbf{h}} = \frac{\partial \bar{J}(\mathbf{y}, \mathbf{h})}{\partial \mathbf{h}} - \left(\frac{\partial A(\mathbf{y}, \mathbf{h})}{\partial \mathbf{h}}\right)^* \lambda.$$

Partial derivatives of $\bar{J}(\mathbf{y}, \mathbf{h})$ and $A(\mathbf{y}, \mathbf{h})$ are presented in Appendix A. Gradient $\dfrac{\partial \bar{J}}{\partial \mathbf{h}}$ we used in projected gradient method [22], which we applied for iteration of a solution of the optimal harvesting problem.

5. Numerical Example

In this section, we study numerical examples of problem (14). We compared two cases where the state constraint (4) was approximated with explicit approximation (7) and implicit approximation (11). As the discounted price for size class $(x_{i-1}, x_i]$ at time t^k, we used $d_i^k = \dfrac{c^p v_i^p + c^s v_i^s}{(1+r)^{t^k}}$, where r is the interest rate, c^p and c^s are the prices of the pulpwood and sawlog, respectively, and v_i^p and v_i^s are the volumes of pulpwood and sawlog of a tree in size class $(x_{i-1}, x_i]$, respectively. In the optimizations, we used the following values for parameters: price of pulpwood $c^p = 16.56 \ \text{€m}^{-3}$ and sawlog $c^s = 58.44 \ \text{€m}^{-3}$, interest rate $r = 3\%$ and lower bound for harvested trees $B = 50 \ \text{m}^3 \ \text{ha}^{-1}$. The pulpwood and sawlog volumes v_i^p and v_i^s we got from [10]. The optimization results of problem (14) are presented in Tables 1 and 2, and in Figures 1 and 2.

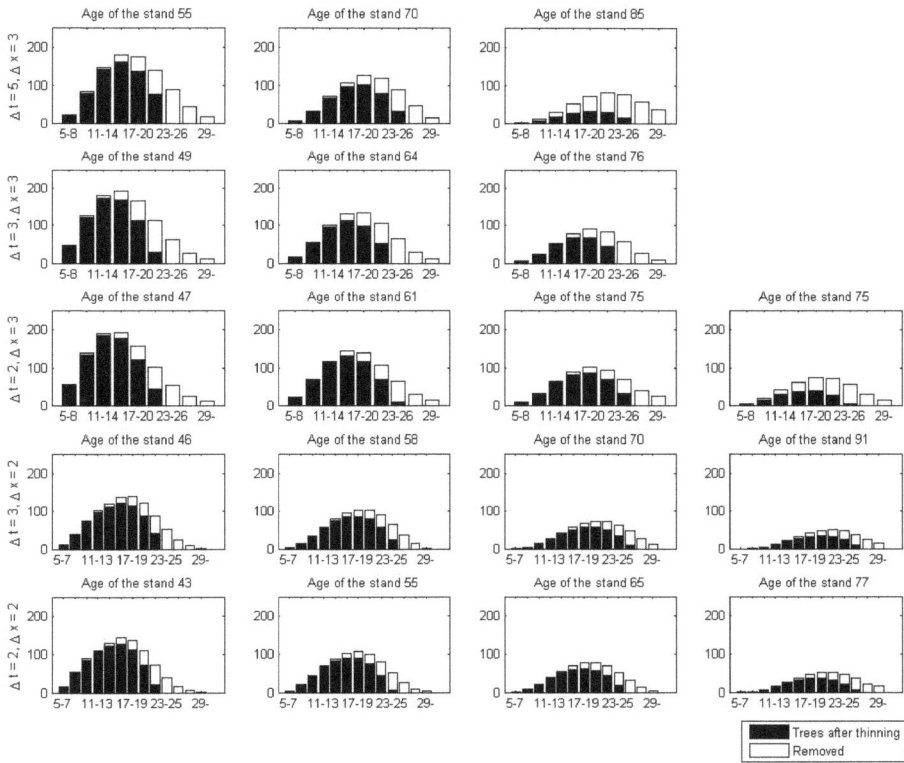

Figure 1. Diameter distributions associated with optimal thinnings of problem (14) with state equation approximated by explicit scheme (7). Numbers (e.g., 5–8, 11–14) represent diameter in centimetres.

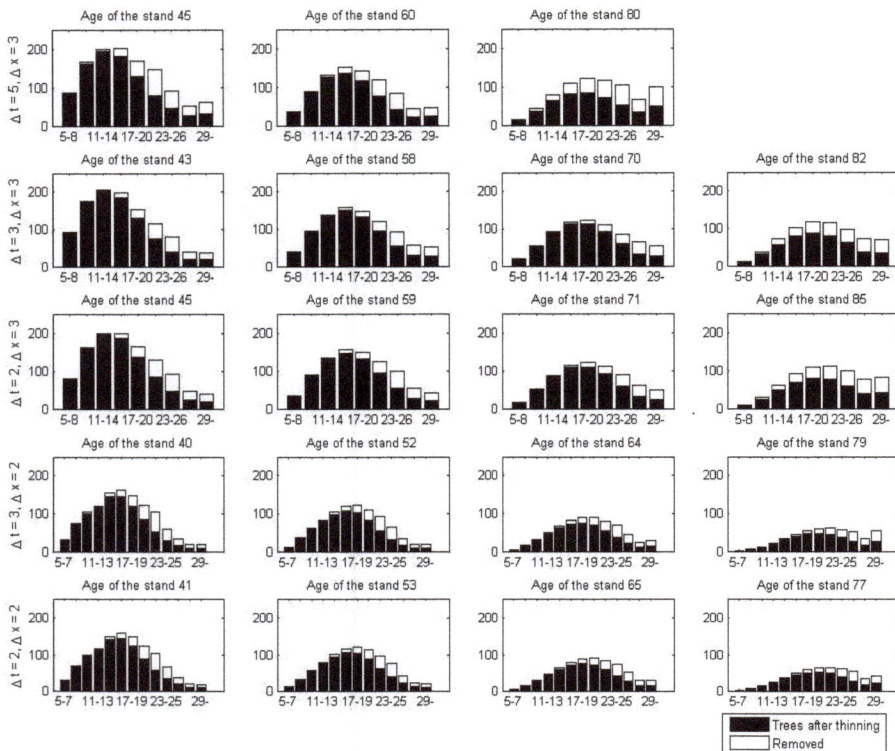

Figure 2. Diameter distributions associated with optimal thinnings of problem (14) with state equation approximated by implicit scheme (11). Numbers (e.g., 5–8, 11–14) represent diameter in centimetres.

Table 1. Maximum net present values, MaxNPVs, (i.e., optimal cost function values of the problem (14)) and mean annual increments (MAI) associated with optimal stand-level managements. Initial density is 1000 stems ha^{-1}.

	MaxNPV (€ha^{-1})		MAI (m^3 ha^{-1} a^{-1})	
	Explicit	Implicit	Explicit	Implicit
Time step $\Delta t = 5$ years and class width $\Delta x = 3$ cm	4095	3996	2.94	2.83
Time step $\Delta t = 3$ years and class width $\Delta x = 3$ cm	4301	4231	2.94	3.11
Time step $\Delta t = 3$ years and class width $\Delta x = 2$ cm	4676	4648	2.96	3.00
Time step $\Delta t = 2$ years and class width $\Delta x = 3$ cm	4384	4311	2.93	3.04
Time step $\Delta t = 2$ years and class width $\Delta x = 2$ cm	4776	4726	3.09	3.09

Table 2. Optimal stand-level managements generated by explicit (7) and implicit (11) approximation of the forest growth model (4). Initial density is 1000 stems ha^{-1}.

	Stand Age (a)		Removal (m^3 ha^{-1})		Thinning Intensity (% of Basal Area Removed)		Saw Log Proportion (%)	
	Explicit	Implicit	Explicit	Implicit	Explicit	Implicit	Explicit	Implicit
Time step Δt = 5 years and class width Δx = 3 cm								
1st thinning	55	45	91.0	77.2	48	48	87	85
2nd thinning	70	60	68.9	59.3	46	46	88	87
3rd thinning	85	80	98.6	90.6	79	72	87	86
final thinning	105	105	50.0	69.9	100	100	90	86
total			308.5	297.0			88	86
Time step Δt = 3 years and class width Δx = 3 cm								
1st thinning	49	43	78.9	52.4	48	35	83	86
2nd thinning	64	58	68.2	54.3	50	35	86	89
3rd thinning	76	70	55.4	50.0	56	35	87	91
4th thinning	-	82	-	72.4	-	59	-	86
final thinning	100	100	91.5	82.2	100	100	88	85
total			293.9	311.3			86	87
Time step Δt = 3 years and class width Δx = 2 cm								
1st thinning	46	40	60.6	60.5	36	40	83	81
2nd thinning	58	52	56.5	51.8	38	40	85	83
3rd thinning	70	64	59.7	52.1	45	46	88	86
4th thinning	79	79	50.0	55.0	52	56	89	90
final thinning	94	97	72.6	71.4	100	100	89	86
total			299.5	290.7			87	85
Time step Δt = 2 years and class width Δx = 3 cm								
1st thinning	47	45	62.2	59.1	39	38	84	86
2nd thinning	61	59	59.9	54.8	41	37	87	88
3rd thinning	75	71	58.9	50.0	44	37	90	90
4th thinning	83	85	64.4	78.0	70	64	84	87
final thinning	101	103	50.0	71.0	100	100	86	86
total			295.5	313.3			86	87
Time step Δt = 2 years and class width Δx = 2 cm								
1st thinning	43	41	55.3	58.5	36	38	80	82
2nd thinning	55	53	55.6	53.3	39	39	84	84
3rd thinning	65	65	50.0	52.0	42	44	86	87
4th thinning	77	77	53.0	50.2	53	51	89	89
final thinning	95	95	79.5	79.2	100	100	89	87
total			293.4	293.1			86	86

The results show that the maximum net present value (NPV) associated with the explicit approximation (7) was higher than the corresponding of the implicit approximation (Table 1). When class width or time step decreased, maximum NPV increased in both cases. The difference of maximum NPVs between the two cases decreased when class width or time step decreased. Only when time step decreased from three to two years, the difference of maximum NPVs increased. The difference was biggest (99 €) when time step Δt = 5 years and class width Δx = 3 cm and smallest (28 €) when time step Δt = 3 years and class width Δx = 2 cm.

With both approximations, three or four intermediate thinnings were made (Table 2). Number of thinnings increased when time step and class width decreased. When implicit approximation (11) was used, first thinnings were made 1–2 time steps earlier, while the last few thinnings were made 0–2 time steps later than when explicit approximation (7) was used. The thinning intensities were almost identical between the two approximations. If there was some difference, intensity was usually bigger when explicit approximation (7) was used (Table 2). The thinning pattern was in all optimal managements quite similar: in each thinning, more big trees than small ones were removed indicating a thinning from-above method (for different thinning types, see e.g., [23], pp. 727, 733). Thinning from above has proven to be the best thinning type in stand-level optimizations of even-aged boreal forests (e.g., [24]). When explicit approximation (7) was used, all trees from two or three of the biggest size classes were removed (Figure 1). On the other hand, when implicit approximation (11) was applied, only part of the trees from those size classes were removed (Figure 2).

6. Discussion

This study contributes to existing literature on forest management by providing a theoretically sound framework to solve nonlinear optimization problem of even-aged stands. We compared the results of forest management optimizations, when the explicit and implicit approximations of the forest growth model was used. The optimization results show that the differences of the results between approximations are diminutive. This was expected as solutions of both approximation equations are proved to converge to solutions of continuous equation [5,14].

In numerical examples, we used data from the Scots pine (*Pinus sylvestris* L.) stands that were located in Northern Ostrobothnia, Finland, on nutrient-poor soil type. The data was the same as in [11]. The difference is that, in [11], data was fitted directly to the matrix model, as, in this study, we first fitted data to the continuous model and then approximated it with a matrix model. In [11], the time step was five years and class width 3 cm. The results are in line with each other. Both methods gave four thinnings in optimal management and thinning from above dominated as the thinning type. In [11], the optimal net present value was slightly higher and, in the optimal management, the thinnings were made slightly earlier than in this study.

The optimal harvesting problem with a continuous size-structured population model was studied in [6–8]. In those papers, harvesting was defined as a proportion of removed trees. The maximum principle for the problem was proved in [6,8]. Moreover, in [7], the strong bang-bang principle under some additional (but realistic) conditions was proved. This means that the optimal solution has the structure, where all trees bigger than some certain size are removed. In our results, the solution of the optimization problem, where state constraint was approximated with explicit approximation, was nearer that structure. In addition, the optimization results were a little better then. However, when explicit approximation is used, the time step and class width have to be chosen so that a tree cannot grow over one size class during one time step [16]. We proved that only then is the explicit approximation scheme stable. For the implicit approximation scheme, we proved that it is unconditionally stable. Thus, in implicit approximation, the time step and class width can be chosen freely. In general, explicit approximation of the population model is more commonly used as a forest growth model [9,16].

Author Contributions: J.P. made the approximations and conducted the optimizations under supervision of E.L.; A.L. proved the theoretical results; J.P. and A.A. analyzed the numerical results; all authors contributed to the writing of the manuscript.

Acknowledgments: We want to acknowledge the Jenny and Antti Wihuri foundation for financial support.

Conflicts of Interest: The authors declare no conflict of interest.

Appendix A

We used the adjoint method to solve the optimization problem (14). For that, we needed the partial derivatives of the Lagrangian

$$\mathcal{L}(\mathbf{y}, \mathbf{h}, \lambda) = \bar{J}(\mathbf{y}, \mathbf{h}) - \sum_{k=1}^{M} (\lambda^k, A^k(\mathbf{y}, \mathbf{h})),$$

where $\bar{J}(\mathbf{h}) = \sum_{k=1}^{M} (\mathbf{d}^k, \mathbf{h}^k)$ is cost function of problem (14) and $A^k(\mathbf{y}, \mathbf{h})$ is constraint (8) or (12).

First, we calculate the partial derivatives of the cost function $\bar{J}(\mathbf{y}, \mathbf{h})$. Since it depends only on $\mathbf{h}$, obviously $\dfrac{\partial \bar{J}}{\partial \mathbf{y}} = 0$. The partial derivative of the cost function $\bar{J}$ with respect to $\mathbf{h}$ is

$$\frac{\partial \bar{J}(\mathbf{h})}{\partial \mathbf{h}} = \mathbf{d}.$$

Next, we calculate the partial derivatives of constraint function $A(\mathbf{y}, \mathbf{h})$. In both forms of A (constraints (8) or (12)), the partial derivative with respect to $\mathbf{h}$ is $\frac{\partial A(\mathbf{y}, \mathbf{h})}{\partial \mathbf{h}} = \Delta t$.

Let us calculate the partial derivative of constraint function (8) (explicit approximation of the state Equation (4)) with respect to $\mathbf{y}$

$$\frac{\partial (\lambda, A(\mathbf{y}, \mathbf{h}))}{\partial y} = \frac{\partial}{\partial \mathbf{y}} \sum_{k=1}^{M} (\lambda^k, \mathbf{y}^k - \mathbf{A}^{k-1} \mathbf{y}^{k-1} + \Delta t \mathbf{h}^k). \tag{A1}$$

Let us denote

$$\mathbf{H}^k := \mathbf{A}^k \mathbf{y}^k = \begin{bmatrix} a_1^k y_1^k \\ a_2^k y_2^k + b_1^k y_1^k \\ \vdots \\ a_N^k y_N^k + b_{N-1}^k y_{N-1}^k \end{bmatrix}.$$

Then, the partial derivative (A1) can be written in the form

$$\frac{\partial (\lambda, A(\mathbf{y}, \mathbf{h}))}{\partial y} = \frac{\partial}{\partial \mathbf{y}} \left(\sum_{k=1}^{M} (\lambda^k, \mathbf{y}^k - \mathbf{H}^{k-1}) \right).$$

By rearranging the terms and defining $\lambda^{M+1} = 0$, we get

$$\frac{\partial (\lambda, A(\mathbf{y}, \mathbf{h}))}{\partial y} = \frac{\partial}{\partial \mathbf{y}} \left((\lambda^1, \mathbf{H}^0) + \sum_{k=1}^{M} ((\lambda^k, \mathbf{y}^k) - (\lambda^{k+1}, \mathbf{H}^k)) \right).$$

Now, $\frac{\partial \mathbf{H}^0}{\partial y} = 0$ by definition of $\mathbf{y}^0$ and $\frac{\partial \mathbf{H}^k}{\partial y} = \mathbf{A}^k + \mathbf{H}_1^k$, $k = 1, \ldots, M-1$, where

$$\mathbf{H}_1^k = \begin{bmatrix} a_{11}^k y_1^k & \cdots & a_{1N}^k y_1^k \\ a_{21}^k y_2^k + b_{11}^k y_1^{k+1} & \cdots & a_{2N}^k y_2^k + b_{1N}^k y_1^k \\ \vdots & \ddots & \vdots \\ a_{N1}^k y_N^k + b_{N-1,1}^k y_{N-1}^k & \cdots & a_{NN}^k y_N^k + b_{N-1,N}^k y_{N-1}^k \end{bmatrix},$$

and

$$a_{ij}^k = \frac{\partial a_i^k}{\partial y_j^k} = -\frac{\Delta t}{\Delta x} (g_{21} + g_{22} x_i) \left(\frac{x_j}{2} \right)^2 \pi - \Delta t \left(m_{21} + \frac{m_{22}}{x_i} + \frac{m_{23}}{x_i^2} \right) \left(\frac{x_j}{2} \right)^2 \pi, \tag{A2}$$

$$b_{ij}^k = \frac{\partial b_i^k}{\partial y_j^k} = \frac{\Delta t}{\Delta x} (g_{21} + g_{22} x_i) \left(\frac{x_j}{2} \right)^2 \pi. \tag{A3}$$

Thus, we can define

$$\frac{\partial (\lambda, A(\mathbf{y}, \mathbf{h}))}{\partial y} = (\mathbf{w}^1, \mathbf{w}^2, \ldots, \mathbf{w}^M),$$

where

$$\mathbf{w}^k = \lambda^k - ((\mathbf{A}^k)^* + (\mathbf{H}_1^k)^*) \lambda^{k+1}, \ k = 1, \ldots, M-1$$
$$\mathbf{w}^M = \lambda^M.$$

Then, we calculate the partial derivative of constraint (12) (implicit approximation of the state Equation (4)) with respect to **y**

$$\frac{\partial(\lambda, A(\mathbf{y}, \mathbf{h}))}{\partial \mathbf{y}} = \frac{\partial}{\partial \mathbf{y}} \sum_{k=1}^{M} (\lambda^k, \mathbf{B}^{k-1}\mathbf{y}^k - \mathbf{y}^{k-1} + \Delta t \mathbf{h}^k). \tag{A4}$$

Let us denote

$$\mathbf{G}^k = \mathbf{B}^{k-1}\mathbf{y}^k = \begin{bmatrix} a_1^{k-1}y_1^k \\ a_2^{k-1}y_2^k - b_1^{k-1}y_1^k \\ \vdots \\ a_N^{k-1}y_N^k - b_{N-1}^{k-1}y_{N-1}^k \end{bmatrix}.$$

Then, the partial derivative (A4) can be written in the form

$$\frac{\partial(\lambda, A(\mathbf{y}, \mathbf{h}))}{\partial \mathbf{y}} = \frac{\partial}{\partial \mathbf{y}} \sum_{k=1}^{M} (\lambda^k, \mathbf{G}^k - \mathbf{y}^{k-1}).$$

Note that function $\mathbf{G}^k$ depends on $\mathbf{y}^{k-1}$ and $\mathbf{y}^k$ so the partial derivative $\dfrac{\partial \mathbf{G}^k}{\partial \mathbf{y}} = \dfrac{\partial \mathbf{G}^k}{\partial \mathbf{y}^{k-1}} + \dfrac{\partial \mathbf{G}^k}{\partial \mathbf{y}^k}$. By definition of $\mathbf{y}^0$, derivative $\dfrac{\partial(\lambda^1, \mathbf{y}^0)}{\partial \mathbf{y}} = 0$ and derivative $\dfrac{\partial(\lambda^1, \mathbf{G}^1)}{\partial \mathbf{y}} = \dfrac{\partial(\lambda^1, \mathbf{G}^1)}{\partial \mathbf{y}^1}$. By rearranging the terms and defining $\lambda^{M+1} = 0$, we get

$$\frac{\partial(\lambda, A(\mathbf{y}, \mathbf{h}))}{\partial \mathbf{y}} = \sum_{k=1}^{M} \left(\frac{\partial(\lambda^k, \mathbf{G}^k)}{\partial \mathbf{y}^k} + \frac{\partial(\lambda^{k+1}, \mathbf{G}^{k+1} - \mathbf{y}^k)}{\partial \mathbf{y}^k} \right).$$

Derivative $\dfrac{\partial \mathbf{G}^k}{\partial \mathbf{y}^k} = \mathbf{B}^{k-1}$ and derivative

$$\frac{\partial \mathbf{G}^{k+1}}{\partial \mathbf{y}^k} = \begin{bmatrix} a_{11}^k y_1^{k+1} & \cdots & a_{1N}^k y_1^{k+1} \\ a_{21}^k y_2^{k+1} - b_{11}^k y_1^{k+1} & \cdots & a_{2N}^k y_2^{k+1} - b_{1N}^k y_1^{k+1} \\ \vdots & \ddots & \vdots \\ a_{N1}^k y_N^{k+1} - b_{N-1,1}^k y_{N-1}^{k+1} & \cdots & a_{NN}^k y_N^{k+1} - b_{N-1,N}^k y_{N-1}^{k+1} \end{bmatrix},$$

where a_{ij}^k and b_{ij}^k are derivatives of coefficients a_i^k and b_i^k with respect to y_j^k defined in Equations (A2) and (A3), respectively. Thus, we can define

$$\frac{\partial(\lambda, A(\mathbf{y}, \mathbf{h}))}{\partial \mathbf{y}} = (\mathbf{q}^1, \mathbf{q}^2, \dots, \mathbf{q}^M),$$

where

$$\mathbf{q}^k = (\mathbf{B}^{k-1})^* \lambda^k + \left(\left(\frac{\partial \mathbf{G}^{k+1}}{\partial \mathbf{y}^k} \right)^* - \mathbf{1}_N \right) \lambda^{k+1}, \quad k = 1, \dots, M-1,$$

$$\mathbf{q}^M = (\mathbf{B}^{M-1})^* \lambda^M,$$

and $\mathbf{1}_N$ is $N \times N$ identity matrix.

References

1. Kato, N.; Torikata, H. Local existence for a general model of size-dependent population dynamics. *Abstr. Appl. Anal.* **1997**, *2*, 207–226. [CrossRef]

2. Kato, N. A general model of size-dependent population dynamics with nonlinear growth rate. *J. Math. Anal. Appl.* **2004**, *297*, 234–256. [CrossRef]

3. Calsina, A.; Saldana, J. A model of physiologically structured population dynamics with a nonlinear individual growth rate. *J. Math. Biol.* **1995**, *33*, 335–364. [CrossRef]

4. Calsina, A.; Saldana, J. Basic theory for a class of models of hierarchically structured population dynamics with distributed states in the recruitment. *Math. Model. Meth. Appl. Sci.* **2006**, *16*, 16951722. [CrossRef]

5. Ackleh, A.S.; Deng, K.; Hu, S.A. Quasilinear Hierarchical Size-Structured Model: Well-Posedness and Approximation. *Appl. Math. Optim.* **2005**, *51*, 3559. [CrossRef]

6. Hritonenko, N.; Yatsenko, Y.; Goetz, R.-U.; Xabadia, A. Maximum principle for size-structured model of forest and carbon sequestration management. *Appl. Math. Lett.* **2008**, *21*, 1090–1094 [CrossRef]

7. Hritonenko, N.; Yatsenko, Y.; Goetz, R.-U.; Xabadia, A. A bang-bang regime in optimal harvesting of size-structured populations. *Nonlinear Anal. Theory Methods Appl.* **2009**, *71*, e2331–e2336. [CrossRef]

8. Hritonenko, N.; Yatsenko, Y.; Goetz, R.-U.; Xabadia, A. Optimal harvesting in forestry: Steady-state analysis and climate change impact. *J. Biol. Dyn.* **2013**, *7*, 41–58. [CrossRef] [PubMed]

9. Liang, J.; Picard, N. Matrix model of Forest Dynamics: An Overview and Outlook. *For. Sci.* **2013**, *59*, 359–378. [CrossRef]

10. Rämö, J.; Tahvonen, O. Economics of harvesting uneven-aged forest stands in Fennoscandia. *Scand. J. For. Res.* **2014**, *29*, 777–792. [CrossRef]

11. Pyy, J.; Ahtikoski, A.; Laitinen, E.; Siipilehto, J. Introducing a non-stationary matrix model for stand-level optimization, an even aged Pine (*Pinus sylvestris* L.) stand in Finland. *Forests* **2017**, *8*, 163. [CrossRef]

12. Anita, S.; Ianneli, M.; Kim, M.-Y.; Park, E.-J. Optimal harvesting for periodic age-dependent population dynamics. *SIAM J. Appl. Math.* **1998**, *58*, 1648–1666.

13. Xie, Q.; He, Z.-R.; Wang, X. Optimal harvesting in diffusive population models with size random growth and distributed recruitment. *Electron. J. Differ. Eq.* **2016**, *214*, 1–13.

14. Ackleh, A.S.; Farkas, J.Z.; Li, X.; Ma, B. Finite difference approximations for a size-structured population model with distributed states in the recruitment. *J. Biol. Dyn.* **2015**, *9*, 2–31. [CrossRef] [PubMed]

15. Ackleh, A.S.; Ito, K. An implicit finite difference scheme for the nonlinear size-structured population model. *Numer. Func. Anal. Opt.* **1997**, *18*, 865–884. [CrossRef]

16. Picard, N.; Liang, J. Matrix models for size structured populations: Unrealistic fast growth or simply diffusion? *PLoS ONE* **2014**, *9*, e98254. [CrossRef] [PubMed]

17. Lions, J. *Nonhomogeneous Boundary Value Problems and Applications*; Springer Verlag: Berlin, German, 1972.

18. Anita, S.; Arnautu, V.; Stefanescu, R. Numerical optimal harvesting for periodic age-structured population dynamics with logistic term. *Numer. Func. Anal. Opt.* **2009**, *30*, 183–198. [CrossRef]

19. Kang, Y.H.; Lee, M.J.; Jung, I.H. Optimal Harvesting for an Age-Spatial-Structured Population Dynamic Model with External Mortality. *Abstr. Appl. Anal.* **2012**. [CrossRef]

20. Kim, Y.K.; Lee, M.J.; Jung, I.H. Duality in an Optimal Harvesting Problem by a Nonlinear Age-Spatial Structured Population Dynamic System. *KYUNGPOOK Math. J.* **2011**, *51*, 353–364. [CrossRef]

21. Ciarlet, P.G. *Introduction to Numerical Linear Algebra and Optimization*; Cambridge University Press: New York, United States, 1989.

22. Anita, S.; Arnautu, V.; Capasso, V. *An Introduction to Optimal Control Problems in Life Sciences and Economics: From Mathematical Models to Numerical Simulation with MATLAB*; Springer: Dordrecht, The Netherlands, 2011.

23. Kuuluvainen, T.; Tahvonen, O.; Aakala, T. Even-aged and uneven-aged forest management in boreal Fennoscandia: A review. *AMBIO* **2012**, *41*, 720–737. [CrossRef] [PubMed]

24. Tahvonen, O.; Pihlainen, S.; Niinimäki, S. On the economics of optimal timber production in boreal Scots pine stand. *Can. J. For. Res.* **2013**, *43*, 719–730. [CrossRef]

Mathematical and Computational Applications

MDPI

Article

Solution of Fuzzy Differential Equations Using Fuzzy Sumudu Transforms

Raheleh Jafari [1],* [iD] and Sina Razvarz [2] [iD]

[1]   Department of Information and Communication Technology, University of Agder, 4879 Grimstad, Norway
[2]   Departamento de Control Automático, CINVESTAV-IPN (National Polytechnic Institute),
     07360 Mexico City, Mexico; srazvarz@yahoo.com
*   Correspondence: jafari3339@yahoo.com; Tel.: +47-3723-3000

Received: 15 December 2017; Accepted: 16 January 2018; Published: 17 January 2018

Abstract: The uncertain nonlinear systems can be modeled with fuzzy differential equations (FDEs) and the solutions of these equations are applied to analyze many engineering problems. However, it is very difficult to obtain solutions of FDEs. In this paper, the solutions of FDEs are approximated by utilizing the fuzzy Sumudu transform (FST) method. Significant theorems are suggested in order to explain the properties of FST. The proposed method is validated with three real examples.

Keywords: uncertain nonlinear system; modeling; fuzzy Sumudu transform

---

1. Introduction

In many physical and dynamical processes, mathematical modeling leads to the deterministic initial and boundary value problems. In practice, the boundary values may be different from crisp and displays in the form of unknown parameters [1]. The qualitative behavior of solutions of the equations is associated with the errors. If the errors are random, in this case, we have a stochastic differential equation along with the random boundary value. Moreover, if the errors are not probabilistic, the fuzzy numbers are substituted by random variables [1,2]. The fuzzy derivative, as well as fuzzy differential equations (FDE), have been discussed in [3,4]. The Peano-like theorems for FDEs, and system of FDE on R (Real line) is investigated in [5]. The first-order fuzzy initial value problem, and the fuzzy partial differential equation, have been studied in [5]. The simulation of the fuzzy system is discussed in [6–11]. The application of numerical techniques for resolving FDEs has been illustrated in [12]. The Lipschitz condition and the theorem for existence and uniqueness of the solution related to FDEs, are discussed in [13–15]. The fractional fuzzy Laplace transformation has been mentioned in [13].

An advanced method to solve FDEs is laid down based on the Sumudu transform. Sumudu transform along with broad applications has been utilized in the area of system engineering and applied physics. Recently, Sumudu transform is popularized in order to solve fractional local differential equations [16–20]. In [21], Sumudu transform is suggested in order to solve fuzzy partial differential equations. Some fundamental theorems along with some properties for Sumudu transform are mentioned in [22]. In [23] the variational iteration technique is proposed utilizing Sumudu transform for solving ordinary equations.

In this paper, we use FST to approximate the solutions of the FDEs. We extend our previous work [24] by generating more theorems for describing the properties of FST. Moreover, the comparison between our method with other numerical methods has been carried out. The FST reduces the FDE to an algebraic equation. A very important property of the FST is that it can solve the equation without resorting to a new frequency domain. By utilizing the proposed technique, the fuzzy boundary value problem can be resolved directly without determining a general solution.

This paper is organized as follows: in Section 2 some definitions which have been used in this paper are given. Section 3 demonstrates the properties of FST. In Section 4 solving FDEs by utilizing

FST approach is described. Three real examples are used to demonstrate the efficiency of the proposed method in Section 5. Section 6 provides the conclusion to the paper.

2. Preliminaries

Some concepts related to the fuzzy calculations are laid down in this section [25,26].

Definition 1. *A fuzzy number B is a function of $B \in E : R \to [0, 1]$, in such a manner, (1) B is normal, (there exists $a_0 \in R$ in such a manner $B(a_0) = 1$); (2) B is convex, $B(\gamma a + (1 - \gamma)c) \geq \min\{B(a), B(c)\}$, $\forall a, c \in R, \forall \gamma \in [0, 1]$; (3) B is upper semi-continuous on R, i.e., $B(a) \leq B(a_0) + \epsilon, \forall a \in N(a_0), \forall a_0 \in R$, $\forall \epsilon > 0$, $N(a_0)$ is a neighborhood; (4) The set $B^+ = \{a \in R, B(a) > 0\}$ is compact.*

Definition 2. *The r-level of the fuzzy number B is defined as follows*

$$[B]^r = \{a \in R : B(a) \geq r\} \tag{1}$$

where $0 < r \leq 1$, $B \in E$.

Definition 3. *Let $B_1, B_2 \in E$ and $\xi \in R$, the operations addition, subtraction, multiplication and scalar multiplication are defined as*

$$[B_1 \oplus B_2]^r = [B_1]^r + [B_2]^r = [\underline{B}_1^r + \underline{B}_2^r, \overline{B}_1^r + \overline{B}_2^r] \tag{2}$$

$$[B_1 \ominus B_2]^r = [B_1]^r - [B_2]^r = [\underline{B}_1^r - \underline{B}_2^r, \overline{B}_1^r - \overline{B}_2^r] \tag{3}$$

$$[B_1 \odot B_2]^r = \begin{pmatrix} \min\{\underline{B}_1^r\underline{B}_2^r, \underline{B}_1^r\overline{B}_2^r, \overline{B}_1^r\underline{B}_2^r, \overline{B}_1^r\overline{B}_2^r\} \\ \max\{\underline{B}_1^r\underline{B}_2^r, \underline{B}_1^r\overline{B}_2^r, \overline{B}_1^r\underline{B}_2^r, \overline{B}_1^r\overline{B}_2^r\} \end{pmatrix} \tag{4}$$

$$[\xi B_1]^r = \xi[B_1]^r = \begin{cases} (\xi\underline{B}_1^r, \xi\overline{B}_1^r), & \xi \geq 0 \\ (\xi\overline{B}_1^r, \xi\underline{B}_1^r), & \xi \leq 0 \end{cases} \tag{5}$$

Definition 4. *The Hausdroff distance between two fuzzy numbers B_1 and B_2 is defined as [27,28]*

$$D(B_1, B_2) = \sup_{0 \leq r \leq 1} \{\max(|\underline{B}_1^r - \underline{B}_2^r|, |\overline{B}_1^r - \overline{B}_2^r|)\} \tag{6}$$

$D(B_1, B_2)$ *has the following properties*

(i) $D(B_1 \oplus u, B_2 \oplus u) = D(B_1, B_2), \forall B_1, B_2, u \in E$
(ii) $D(\xi B_1, \xi B_2) = |\xi| D(B_1, B_2), \forall \xi \in R, B_1, B_2 \in E$
(iii) $D(B_1 \oplus B_2, u \oplus v) \leq D(B_1, u) + D(B_2, v), \forall B_1, B_2, u, v \in E$
(iv) (D, E) *is stated as complete metric space.*

Definition 5. *The function $\psi : [a_1, a_2] \longrightarrow E$ is integrable on $[a_1, a_2]$, if it satisfies in the below mentioned relation*

$$\int_{a_1}^{\infty} \psi(x)dx = (\int_{a_1}^{\infty} \underline{\psi}(x, r)dx, \int_{a_1}^{\infty} \overline{\psi}(x, r)dx) \tag{7}$$

If $\psi(x)$ be a fuzzy value function, as well as $q(x)$ be a fuzzy Riemann integrable on $[a_1, \infty]$ so $\psi(x) \oplus q(x)$ can be a fuzzy Riemann integrable on $[a_1, \infty]$. Therefore,

$$\int_{a_1}^{\infty} (\psi(x) \oplus q(x))dx = \int_{a}^{\infty} \psi(x)dx \oplus \int_{a}^{\infty} q(x)dx \tag{8}$$

According to fuzzy concept or in the case of interval arithmetic, equation $B_1 = B_2 \oplus s$ is not equivalent with $s = B_1 \ominus B_2 = B_1 \oplus (-1)B_2$ or to $B_2 = B_1 \ominus s = B_1 \oplus (-1)s$ and this is the main reason in introducing the following Hukuhara difference (H-difference).

Definition 6. *The definition of H-difference [29,30], is proposed by* $B_1 \ominus_H B_2 = s \iff B_1 = B_2 \oplus s$. *If* $B_1 \ominus_H B_2$ *prevails, its r-level is* $[B_1 \ominus_H B_2]^r = [\underline{B}_1^r - \underline{B}_2^r, \overline{B}_1^r - \overline{B}_2^r]$. *Precisely,* $B_1 \ominus_H B_1 = 0$ *but* $B_1 \ominus B_1 \neq 0$.

Definition 7. *Suppose* $\psi : [a_1, a_2] \longrightarrow E$ *and* $x_0 = [a_1, a_2]$. *ψ is strongly generalized differentiable at x_0, if for all $k > 0$ adequately minute, $\psi'(x_0) \in E$ exists in such a manner that*

(i) $\exists \, \psi(x_0 + k) \ominus_H \psi(x_0), \, \psi(x_0) \ominus_H \psi(x_0 - k)$ *and*

$$\lim_{k \to 0+} \frac{\psi(x_0+k) \ominus_H \psi(x_0)}{k} = \lim_{k \to 0+} \frac{\psi(x_0) \ominus_H \psi(x_0-k)}{k} = \psi'(x_0)$$

(ii) $\exists \, \psi(x_0) \ominus_H \psi(x_0 + k), \, \psi(x_0 - k) \ominus_H \psi(x_0)$ *and*

$$\lim_{k \to 0+} \frac{\psi(x_0) \ominus_H \psi(x_0+k)}{(-k)} = \lim_{k \to 0+} \frac{\psi(x_0-k) \ominus_H \psi(x_0)}{(-k)} = \psi'(x_0),$$

(iii) $\exists \, \psi(x_0 + k) \ominus_H \psi(x_0), \, \psi(x_0 - k) \ominus_H \psi(x_0)$ *and*

$$\lim_{k \to 0+} \frac{\psi(x_0+k) \ominus_H \psi(x_0)}{k} = \lim_{k \to 0+} \frac{\psi(x_0-k) \ominus_H \psi(x_0)}{(-k)} = \psi'(x_0)$$

(iv) $\exists \, \psi(x_0) \ominus_H \psi(x_0 + k), \, \psi(x_0) \ominus_H \psi(x_0 - k)$ *and*

$$\lim_{k \to 0+} \frac{\psi(x_0) \ominus_H \psi(x_0+k)}{(-k)} = \lim_{k \to 0+} \frac{\psi(x_0) \ominus_H \psi(x_0-k)}{k} = \psi'(x_0)$$

Remark 1. *It is clear that case (i) is H-derivative. Furthermore, a function is (i)-differentiable only when it is H-derivative.*

Remark 2. *It can be concluded from [29] that, the definition of differentiability is non contradictory [31].*

Let us consider $\psi : R \to E$ where $\psi(t)$ has a parametric form as $[\psi(t, r)] = [\underline{\psi}(t, r), \overline{\psi}(t, r)]$, for all $0 \leq r \leq 1$, thus [31]

(i)   If ψ be (i)-differentiable, so $\underline{\psi}(t, r)$ and $\overline{\psi}(t, r)$ are differentiable functions, moreover $\psi'(t) = (\underline{\psi}'(t, r), \overline{\psi}'(t, r))$.

(ii)   If ψ be (ii)-differentiable, so $\underline{\psi}(t, r)$ and $\overline{\psi}(t, r)$ are differentiable functions, moreover $\psi'(t) = (\overline{\psi}'(t, r), \underline{\psi}'(t, r))$.

Suppose $f : (a_1, a_2) \to R$ is differentiable on (a_1, a_2), furthermore ψ' has finite root in (a_1, a_2), and $m \in E$, therefore, $\psi(x) = mf(x)$ is strongly generalized differentiable on (a_1, a_2) along with $\psi'(x) = mf'(x), \forall x \in (a_1, a_2)$.

Theorem 1. *In [30] Assume $\psi : R \times E \to E$ is taken to be a continuous fuzzy function. If $x_0 \in R$, the fuzzy initial value constraint*

$$\begin{cases} \phi'(t) = \psi(x, \phi) \\ \phi(x_0) = \phi_0 \end{cases} \tag{9}$$

is incorporated with two solutions: (i)-differentiable, also (ii)-differentiable. Hence the successive iterations

$$\phi_{n+1}(x) = \phi_0 + \int_{x_0}^{x} \psi(t, \phi_n(t)) dt, \quad \forall x \in [x_0, x_1] \tag{10}$$

and

$$\phi_{n+1}(x) = \phi_0 \ominus_H (-1) \int_{x_0}^{x} \psi(t, \phi_n(t)) dt, \quad \forall x \in [x_0, x_1] \tag{11}$$

approaches towards the two solutions sequentially.

Theorem 2. *[29] The FDE is equivalent to a system of ordinary differential equations under generalized differentiability.*

3. Fuzzy Sumudu Transform

Fuzzy initial and boundary value problems can be resolved by utilizing fuzzy Laplace transform [13]. In this paper, the FST methodology is illustrated; furthermore, the properties of this methodology are stated. By applying the FST methodology, the FDE is reduced to an algebraic equation. The main advantageous of the FST is that it can resolve the equation without resorting to a new frequency domain. The methodology of converting FDEs to an algebraic equation is expressed in [13].

Definition 8. *Suppose $\psi(t)$ be a continuous fuzzy value function, also, $\psi(Bt) \odot e^{-t}$ be an improper fuzzy Riemann integrable on $[0, \infty)$. Accordingly, $\int_0^\infty \psi(Bt) \odot e^{-t} dt$ is expressed as FST and it is defined by $\Omega(B) = S[\psi(t)] = \int_0^\infty \psi(Bt) \odot e^{-t} dt$, where $0 \leq B < K$, $K \geq 0$, also e^{-t} is real valued function. Based on the Theorem 4 we have the following relation*

$$\int_0^\infty \psi(Bt) \odot e^{-t} dt = (\int_0^\infty \underline{\psi}(Bt, r) e^{-t} dt, \int_0^\infty \overline{\psi}(Bt, r) e^{-t} dt) \tag{12}$$

Let

$$\begin{aligned} S[\underline{\psi}(t, r)] &= \int_0^\infty \underline{\psi}(Bt, r) e^{-t} dt \\ S[\overline{\psi}(t, r)] &= \int_0^\infty \overline{\psi}(Bt, r) e^{-t} dt \end{aligned} \tag{13}$$

hence we obtain the following relation

$$S[\psi(t)] = (S[\underline{\psi}(t, r), S\overline{\psi}(t, r)]) \tag{14}$$

Theorem 3. *Suppose $\psi'(t)$ be a fuzzy value integrable function, as well as $\psi(t)$ be the primitive of $\psi'(t)$ on $[0, \infty)$. Therefore,*

$$S[\psi'(t)] = \frac{1}{B} \odot S[\psi(t)] \ominus (\frac{1}{B} \odot [\psi(0)]) \tag{15}$$

where ψ is considered to be (i)-differentiable, or

$$S[\psi'(t)] = \frac{-1}{B} \odot [\psi(0)] \ominus (\frac{-1}{B} \odot S[\psi(t)]) \tag{16}$$

where ψ is considered to be (ii)-differentiable.

Proof. For arbitrary fixed $r \in [0, 1]$ we have

$$\begin{aligned} &\frac{1}{B} \odot S[\psi(t)] \ominus (\frac{1}{B} \odot \psi(0)) \\ &= (\frac{1}{B} S[\underline{\psi}(t, r)] - \frac{1}{B} S[\underline{\psi}(0, r)], \frac{1}{B} S[\overline{\psi}(t, r)] - \frac{1}{B} S[\overline{\psi}(0, r)]) \end{aligned} \tag{17}$$

We have the following relations

$$\begin{aligned} S[\overline{\psi}'(t, r)] &= \frac{1}{B} S[\overline{\psi}(t, r)] - \frac{1}{B} [\overline{\psi}(0, r)] \\ S[\underline{\psi}'(t, r)] &= \frac{1}{B} S[\underline{\psi}(t, r)] - \frac{1}{B} [\underline{\psi}(0, r)] \end{aligned} \tag{18}$$

Hence, we obtain

$$\frac{1}{B} \odot S[\psi(t)] \ominus (\frac{1}{B} \odot \psi(0)) = (S[\underline{\psi}'(t, r)], S[\overline{\psi}'(t, r)]) \tag{19}$$

If ψ is cosidered to be (i)-differentiable, so

$$\frac{1}{B} \odot S[\psi(t)] \ominus (\frac{1}{B} \odot \psi(0)) = S[\psi'(t)] \tag{20}$$

Let ψ is (ii)-differentiable. For arbitrary fixed $\alpha \in [0,1]$ we obtain

$$
\begin{aligned}
&\frac{-1}{B} \odot [\psi(0)] \ominus (\frac{-1}{B} \odot S[\psi(t)]) \\
&= (\frac{-1}{B}\overline{\psi}(0,r) + \frac{1}{B}S[\overline{\psi}(t,r)], \frac{-1}{B}\underline{\psi}(0,r) + \frac{1}{B}S[\underline{\psi}(t,r)])
\end{aligned} \tag{21}
$$

The above equation can be written as the following relation

$$
\begin{aligned}
&\frac{-1}{B} \odot [\psi(0)] \ominus (\frac{-1}{B} \odot S[\psi(t)]) \\
&= (\frac{1}{B}S[\overline{\psi}(t,r)] - \frac{1}{B}\overline{\psi}(0,r), \frac{1}{B}S[\underline{\psi}(t,r)] - \frac{1}{B}\underline{\psi}(0,r))
\end{aligned} \tag{22}
$$

We obtain

$$
\begin{aligned}
S[\overline{\psi}'(t,r)] &= \frac{1}{B}S[\overline{\psi}(t,r)] - \frac{1}{B}\overline{\psi}(0,r) \\
S[\underline{\psi}'(t,r)] &= \frac{1}{B}S[\underline{\psi}(t,r)] - \frac{1}{B}\underline{\psi}(0,r)
\end{aligned} \tag{23}
$$

So, we have

$$(\frac{-1}{B}\psi(0)) \ominus (\frac{-1}{B} \odot S[\psi(t)]) = (S[\overline{\psi}'(t,r)], S[\underline{\psi}'(t,r)]) \tag{24}$$

Hence

$$(\frac{-1}{B}\psi(0)) \ominus (\frac{-1}{B} \odot S[\psi(t)]) = S([\overline{\psi}'(t,r)], [\underline{\psi}'(t,r)]) \tag{25}$$

Since ψ is (ii)-differentiable, therefore,

$$(\frac{-1}{B}\psi(0)) \ominus (\frac{-1}{B} \odot S[\psi(t)]) = S[\psi'(t)] \tag{26}$$

□

Theorem 4. *Taking into consideration that Sumudu transform is a linear transformation, so if $\psi(t)$ and $\vartheta(t)$ be continuous fuzzy valued functions, moreover k_1 as well as k_2 be constant, therefore the following relation can be obtained*

$$S[(k_1 \odot \psi(t)) \oplus (k_2 \odot \vartheta(t))] = (k_1 \odot S[\psi(t)]) \oplus (k_2 \odot S[\vartheta(t)]) \tag{27}$$

Proof. We have

$$
\begin{aligned}
S[(k_1 \odot \psi(t)) \oplus (k_2 \odot \vartheta(t))] &= \int_0^\infty (k_1 \odot \varphi(Bt) \oplus k_2 \odot \vartheta(Bt)) \odot e^{-t}dt \\
&= \int_0^\infty k_1 \odot \psi(Bt) \odot e^{-t}dt \oplus \int_0^\infty k_2 \odot \vartheta(Bt) \odot e^{-t}dt \\
&= k_1 \odot (\int_0^\infty \psi(Bt) \odot e^{-t}dt) \oplus k_2 \odot (\int_0^\infty \vartheta(Bt) \odot e^{-t}dt) \\
&= k_1 \odot S[\psi(t)] \oplus k_2 \odot S[\vartheta(t)]
\end{aligned} \tag{28}
$$

Therefore, we conclude

$$S[(k_1 \odot \psi(t)) \oplus (k_2 \odot \vartheta(t))] = (k_1 \odot S[\psi(t)]) \oplus (k_2 \odot S[\vartheta(t)]) \tag{29}$$

□

Lemma 1. *Assume that the $\psi(t)$ is a continuous fuzzy value function on $[0, \infty)$, also $\gamma \geq 0$, thus*

$$S[\gamma \odot \psi(t)] = \gamma \odot S[\psi(t)] \tag{30}$$

Proof. Fuzzy Sumudu transform $\gamma \odot \psi(t)$ is defined as

$$S[\gamma \odot \psi(t)] = \int_0^\infty \gamma \odot \psi(Bt) \odot e^{-t} dt \tag{31}$$

furthermore, we have

$$\int_0^\infty \gamma \odot \psi(Bt) \odot e^{-t} dt = \gamma \odot \int_0^\infty \psi(Bt) \odot e^{-t} dt \tag{32}$$

therefore,

$$S[\gamma \odot \psi(t)] = \gamma \odot S[\psi(t)] \tag{33}$$

□

Lemma 2. *Assume that the $\psi(t)$ is a continuous fuzzy value function, and $\vartheta(t) \geq 0$. Furthermore, if we suppose that the $(\psi(t) \odot \vartheta(t)) \odot e^{-t}$ is improper fuzzy Reiman integrable on $[0, \infty)$, then*

$$
\begin{array}{l}
\int_0^\infty (\psi(Bt) \odot \vartheta(Bt)) \odot e^{-t} dt \\
= (\int_0^\infty \vartheta(Bt)\underline{\psi}(Bt,r)e^{-t}dt, \int_0^\infty \vartheta(Bt)\overline{\psi}(Bt,r)e^{-t}dt)
\end{array}
\tag{34}
$$

Theorem 5. *Suppose $\psi(t)$ is a continuous fuzzy value function, also $S[\psi(t)] = D(B)$, therefore,*

$$S[e^{a_1 t} \odot \psi(t)] = \frac{1}{1 - a_1 B} D\left(\frac{B}{1 - a_1 B}\right) \tag{35}$$

where $e^{a_1 t}$ is considered to be a real value function, also $1 - a_1 B > 0$.

Proof. We have the following relation

$$
\begin{array}{l}
S[e^{a_1 t} \odot \psi(t)] = \int_0^\infty e^{a_1 Bt} e^{-t} \psi(Bt) dt \\
= (\int_0^\infty e^{-(1-a_1 B)t} \underline{\psi}(Bt,r) dt, \int_0^\infty e^{-(1-a_1 B)t} \overline{\psi}(Bt,r) dt)
\end{array}
\tag{36}
$$

Let us consider $z = 1 - a_1 Bt$, then

$$
\begin{array}{l}
S[e^{a_1 t} \odot \psi(t)] = \frac{1}{1-a_1 B} (\int_0^\infty \underline{\psi}(\frac{Bz}{1-a_1 B}, r) e^{-z} dz, \int_0^\infty \overline{\psi}(\frac{Bz}{1-a_1 B}, r) e^{-z} dz) \\
= \{\frac{1}{1-a_1 B} \underline{D}(\frac{B}{1-a_1 B}), \frac{1}{1-a_1 B} \overline{D}(\frac{B}{1-a_1 B})\} = \frac{1}{1-a_1 B} D(\frac{B}{1-a_1 B})
\end{array}
\tag{37}
$$

□

4. Solving Fuzzy Initial Value Problem with Fuzzy Sumudu Transform Method

Consider the following fuzzy initial value problem

$$
\begin{cases}
\phi'(t) = \psi(t, \phi(t)), \\
\phi(0) = (\underline{\phi}(0,r), \overline{\phi}(0,r)), \quad 0 < r \leq 1
\end{cases}
\tag{38}
$$

where $\psi(t, \phi(t))$ is a fuzzy function. The fuzzy function $\psi(t, \phi(t))$ is the mapping of $\psi : R \times E \to E$. By utilizing FST method, we obtain

$$S[\phi'(t)] = S[\psi(t, \phi(t))] \tag{39}$$

The resolving process of Equation (39) is based on the following cases.

Case 1: Assume that the $\phi'(t)$ is (i)-differentiable. Based on the Theorem 4 we extract

$$\phi'(t) = (\underline{\phi}'(t,r), \overline{\phi}'(t,r)) \tag{40}$$

$$S[\phi'(t)] = (\frac{1}{B} \odot S[\phi(t)]) \ominus \frac{1}{B}\phi(0) \tag{41}$$

Equation (41) can be displayed as following relation

$$\begin{cases} S[\underline{\psi}(t,\phi(t),r)] = \frac{1}{B}S[\underline{\phi}(t,r)] - \frac{1}{B}\phi(0,\alpha) \\ S[\overline{\overline{\psi}}(t,\phi(t),r)] = \frac{1}{B}S[\overline{\phi}(t,r)] - \frac{1}{B}\overline{\overline{\phi}}(0,r) \end{cases} \tag{42}$$

where

$$\begin{cases} \underline{\psi}(t,\phi(t),r) = min\{\psi(t,B)|B \in (\underline{\phi}(t,r),\overline{\phi}(t,r))\} \\ \overline{\overline{\psi}}(t,\phi(t),r) = max\{\psi(t,B)|B \in (\underline{\phi}(t,r),\overline{\phi}(t,r))\} \end{cases} \tag{43}$$

Accordingly, Equation (43) can be resolved on the basis of the following assumptions

$$S[\underline{\phi}(t,r)] = U_1(B,r) \tag{44}$$

$$S[\overline{\phi}(t,r)] = U_2(B,r) \tag{45}$$

where $U_1(B,r)$, as well as $U_2(B,r)$ are the solutions of the Equation (43). By applying inverse Sumudu transform, $\underline{\phi}(t,r)$ as well as $\overline{\phi}(t,r)$ are computed as

$$\underline{\phi}(t,r) = S^{-1}[U_1(B,r)] \tag{46}$$

$$\overline{\phi}(t,r) = S^{-1}[U_2(B,r)] \tag{47}$$

Case 2: Assume that the $\phi'(t)$ is (ii)-differentiable. Based on the Theorem 4 we extract

$$\phi'(t) = (\overline{\phi}'(t,r),\underline{\phi}'(t,r)) \tag{48}$$

$$S[\phi'(t)] = (\frac{-1}{B} \odot \phi(0)) \ominus (\frac{-1}{B} \odot S[\phi(t)]) \tag{49}$$

Equation (49) can be displayed as following relation

$$\begin{cases} S[\underline{\psi}(t,\phi(t),r)] = \frac{1}{B}S[\underline{\phi}(t,r)] - \frac{1}{B}\phi(0,r) \\ S[\overline{\psi}(t,\phi(t),r)] = \frac{1}{B}S[\overline{\phi}(t,r)] - \frac{1}{B}\overline{\overline{\phi}}(0,r) \end{cases} \tag{50}$$

where

$$\begin{cases} \underline{\psi}(t,\phi(t),r) = min\{\psi(t,B)|B \in (\underline{\phi}(t,r),\overline{\phi}(t,r))\} \\ \overline{\overline{\psi}}(t,\phi(t),r) = max\{\psi(t,B)|B \in (\underline{\phi}(t,r),\overline{\phi}(t,r))\} \end{cases} \tag{51}$$

Accordingly, Equation (51) can be resolved on the basis of the following assumptions

$$\begin{aligned} S(\underline{\phi}(t,r) = V_1(B,r) \\ S(\overline{\overline{\phi}}(t,r) = V_2(B,r) \end{aligned} \tag{52}$$

where $V_1(B,r)$, and $V_2(B,r)$ are the solutions of the Equation (51). By applying inverse Sumudu transform, $\underline{\phi}(t,r)$ as well as $\overline{\phi}(t,r)$ are computed as

$$\begin{aligned} \underline{\phi}(t,r) = S^{-1}[V_1(B,r)] \\ \overline{\overline{\phi}}(t,r) = S^{-1}[V_2(B,r)] \end{aligned} \tag{53}$$

5. Examples

The following examples have been used to narrate the methodology proposed in this paper.

Example 1. *A tank with a heating system is displayed in Figure 1, where $\tilde{R} = 0.5$, the thermal capacitance is $\tilde{C} = 2$ also the temperature is ψ. The model is formulated as follows* [13,32],

$$\begin{cases} \phi'(t) = -\frac{1}{RC}\phi(t), & 0 \le t \le T \\ \phi(0) = (\underline{\phi}(0,r), \overline{\phi}(0,r)) \end{cases} \tag{54}$$

By utilizing the FST method, we obtain

$$S[\phi'(t)] = S[-\phi(t)] \tag{55}$$

$$S[\phi'(t)] = \int_0^\infty \phi'(Bt) \odot e^{-t} dt \tag{56}$$

where $0 \le B < K$. If $\phi(t)$ is (i)-differentiable and case 1 holds, we extract

$$S[\phi'(t)] = \frac{1}{B} \odot (S[\phi(t)] \ominus \phi(0)) = \frac{1}{B}S[\phi(t)] \ominus \frac{1}{B}\phi(0) \tag{57}$$

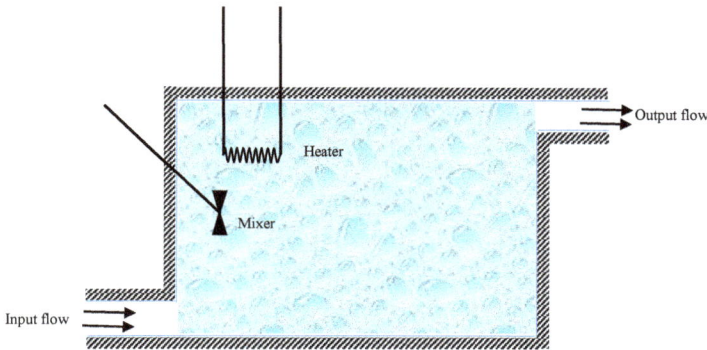

Figure 1. Thermal system.

Therefore

$$-S[\phi(t)] = \frac{1}{B}S[\phi(t)] \ominus \frac{1}{B}\phi(0) \tag{58}$$

Based on the Equation (42), we have

$$\begin{cases} -S[\overline{\phi}(t,r)] = \frac{1}{B}S[\underline{\phi}(t,r)] - \frac{1}{B}\underline{\phi}(0,r) \\ -S[\underline{\phi}(t,r)] = \frac{1}{B}S[\overline{\phi}(t,r)] - \frac{1}{B}\overline{\phi}(0,r) \end{cases} \tag{59}$$

Therefore, the solution of Equation (59) is as follows

$$\begin{cases} S[\overline{\phi}(t,r)] = (\frac{-1}{B^2-1})\overline{\phi}(0,r) + (\frac{B}{B^2-1})\underline{\phi}(0,r) \\ S[\underline{\phi}(t,r)] = (\frac{-1}{B^2-1})\underline{\phi}(t,r) + (\frac{B}{B^2-1})\overline{\phi}(0,r) \end{cases} \tag{60}$$

By utilizing the inverse Sumudu transform we have

$$\begin{cases} S[\overline{\phi}(t,r)] = \overline{\phi}(0,r)S^{-1}(\frac{-1}{B^2-1}) + \underline{\phi}(0,r)S^{-1}(\frac{B}{B^2-1}) \\ S[\underline{\phi}(t,r)] = \underline{\phi}(0,r)S^{-1}(\frac{-1}{B^2-1}) + \overline{\phi}(0,r)S^{-1}(\frac{B}{B^2-1}) \end{cases} \tag{61}$$

where

$$
\begin{cases}
\overline{\phi}(t,r) = e^t \left(\frac{\overline{\phi}(0,r) - \underline{\phi}(0,r)}{2} \right) + e^{-t} \left(\frac{\overline{\phi}(0,r) + \underline{\phi}(0,r)}{2} \right) \\
\underline{\phi}(t,r) = e^t \left(\frac{\underline{\phi}(0,r) - \overline{\phi}(0,r)}{2} \right) + e^{-t} \left(\frac{\underline{\phi}(0,r) + \overline{\phi}(0,r)}{2} \right)
\end{cases}
\tag{62}
$$

Now if $\phi(t)$ be (ii)-differentiable and case 2 holds, we have

$$
S[\phi'(t)] = \left(\frac{-1}{B} S[\phi(t)] \right) \ominus \left(\frac{-1}{B} \phi(0) \right)
\tag{63}
$$

Hence

$$
- S[\phi(t)] = \left(\frac{-1}{B} S[\phi(t)] \right) \ominus \left(\frac{-1}{B} \phi(0) \right)
\tag{64}
$$

Based on the above relations, Equation (54) can be written as follows

$$
\begin{cases}
-S[\underline{\phi}(t,r)] = \frac{1}{B} S[\underline{\phi}(t,r)] - \frac{1}{B} \underline{\phi}(0,r) \\
-S[\overline{\phi}(t,r)] = \frac{1}{B} S[\overline{\phi}(t,r)] - \frac{1}{B} \overline{\phi}(0,r)
\end{cases}
\tag{65}
$$

So, the solution of Equation (65) is displayed as

$$
\begin{cases}
S[\underline{\phi}(t,r)] = \underline{\phi}(0,r) \left(\frac{1}{B+1} \right) \\
S[\overline{\phi}(t,r)] = \overline{\phi}(t,r) \left(\frac{1}{B+1} \right)
\end{cases}
\tag{66}
$$

By utilizing the inverse Sumudu transform, we have

$$
\begin{cases}
\underline{\phi}(t,r) = \underline{\phi}(0,r) S^{-1} \left(\frac{1}{B+1} \right) \\
\overline{\phi}(t,r) = \overline{\phi}(0,r) S^{-1} \left(\frac{1}{B+1} \right)
\end{cases}
\tag{67}
$$

where

$$
\begin{cases}
\underline{\phi}(t,r) = e^{-t} \underline{\phi}(0,r) \\
\overline{\phi}(t,r) = e^{-t} \overline{\phi}(0,r)
\end{cases}
\tag{68}
$$

If the initial condition be a symmetric triangular fuzzy number as $\phi(0) = (-a(1-r), a(1-r))$, then the following cases will hold

Case 1:

$$
\begin{cases}
\underline{\phi}(t,r) = e^t (-a(1-r)) \\
\overline{\phi}(t,r) = e^t (a(1-r))
\end{cases}
\tag{69}
$$

Case 2:

$$
\begin{cases}
\underline{\phi}(t,r) = e^{-t} (-a(1-r)) \\
\overline{\phi}(t,r) = e^{-t} (a(1-r))
\end{cases}
\tag{70}
$$

Corresponding solution plots are displayed in Figures 2 and 3. Corresponding error plots are shown in Figure 4. These errors are the differences of the exact and the approximation solutions for two different methods: FST and Average Euler method [33]. FST is more accurate than the Average Euler method.

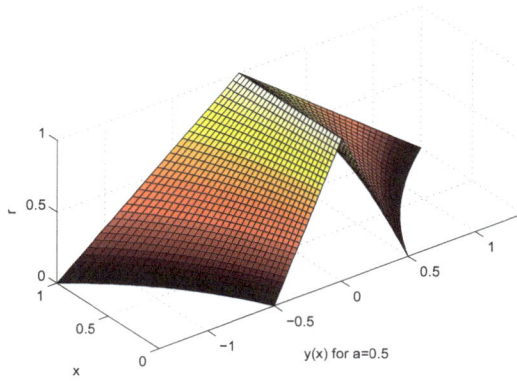

Figure 2. The solution of fuzzy differential equations (FDE) under case 1 consideration.

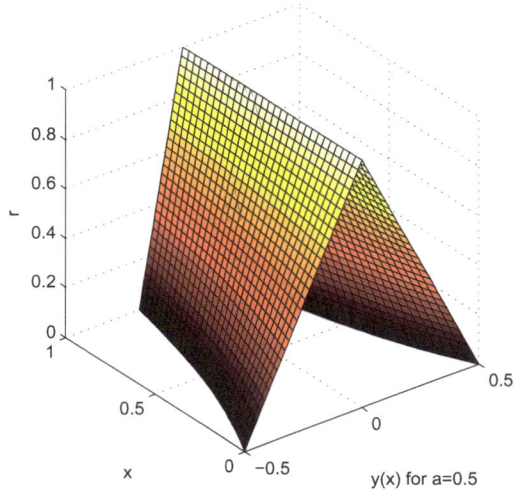

Figure 3. The solution of FDE under case 2 consideration.

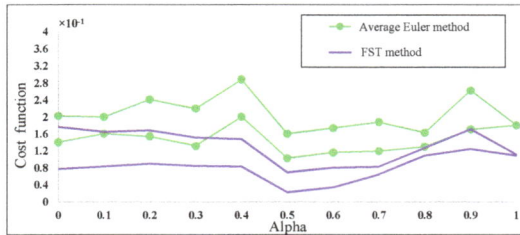

Figure 4. The lower and upper bounds of absolute errors.

Example 2. *A tank system is displayed in Figure 5. Suppose $I = t + 1$ is inflow disturbances of the tank that generates vibration in liquid level ϕ, also $H = 1$ is the flow obstruction, which can be curbed utilizing the valve. $Q = 1$ is the cross section of the tank. The liquid level is illustrated as following relation [34],*

$$\begin{cases} \phi'(t) = -\frac{1}{QH}\phi(t) + \frac{I}{Q}, & 0 \le t \le T \\ \phi(0) = (\underline{\phi}(0,r), \overline{\phi}(0,r)) \end{cases} \tag{71}$$

By utilizing the FST method we obtain

$$-S[\phi(t)] = (\frac{1}{B} \odot S[\phi(t)]) \ominus (\frac{1}{B} S[\phi(0)]) \tag{72}$$

$$S[\phi'(t)](\int_{a_1}^{\infty} \phi'(Bt)e^{-t}dt \tag{73}$$

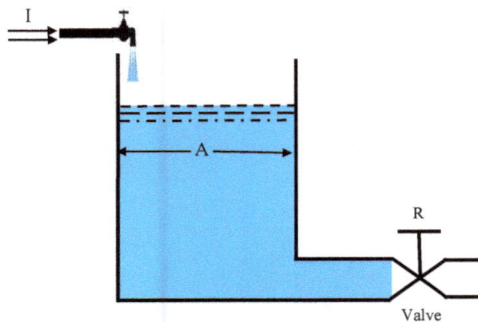

Figure 5. Liquid tank system.

The following relation is extracted by taking into consideration case 2

$$S[\phi'(t)] = (\frac{-1}{B} \odot S[\phi(t)]) \ominus (\frac{-1}{B} S[\phi(0)]) \tag{74}$$

So

$$-S[\phi(t)] + S[t] + S[1] = (\frac{-1}{B} \odot S[\phi(t)]) \ominus (\frac{-1}{B} S[\phi(0)]) \tag{75}$$

Based on the Equation (42), we have

$$\begin{cases} -S[\underline{\phi}(t,r)] + S[t] + S[1] = \frac{1}{B}S[\underline{\phi}(t,r)] - \frac{1}{B}\underline{\phi}(0,r) \\ -S[\overline{\phi}(t,r)] + S[t] + S[1] = \frac{1}{B}S[\overline{\phi}(t,r)] - \frac{1}{B}\overline{\phi}(0,r) \end{cases} \tag{76}$$

Therefore, the solution of Equation (76) is extracted as

$$\begin{cases} S[\underline{\phi}(t,r)] = S[t] + S[1] + \frac{-1}{B}S[\underline{\phi}(t,r)] - \frac{1}{B}\underline{\phi}(0,r) \\ S[\overline{\phi}(t,r)] = S[t] + S[1] + \frac{1}{B}S[\overline{\phi}(t,r)] - \frac{1}{B}\overline{\phi}(0,r) \end{cases} \tag{77}$$

hence,

$$\begin{cases} S[\underline{\phi}(t,r)] = (\frac{1}{B+1})\underline{\phi}(0,r) + B \\ S[\overline{\phi}(t,r)] = (\frac{1}{B+1})\overline{\phi}(t,r) + B \end{cases} \tag{78}$$

By utilizing the inverse Sumudu transform, we obtain

$$\begin{cases} \underline{\phi}(t,r) = \underline{\phi}(0,r)S^{-1}(\frac{1}{B+1}) + S^{-1}(B) \\ \overline{\phi}(t,r) = \overline{\phi}(0,r)S^{-1}(\frac{1}{B+1}) + S^{-1}(B) \end{cases} \tag{79}$$

where

$$\begin{cases} \underline{\phi}(t,r) = e^{-t}\underline{\phi}(0,r) + t \\ \overline{\phi}(t,r) = e^{-t}\overline{\phi}(0,r) + t \end{cases} \tag{80}$$

If the initial condition is taken to be a symmetric triangular fuzzy number as $\phi(0) = (-a(1-r), a(1-r))$, so

$$\begin{cases} \underline{\phi}(t,r) = e^{-t}(-a(1-r)) + t \\ \overline{\phi}(t,r) = e^{-t}(a(1-r)) + t \end{cases} \tag{81}$$

Corresponding solution plot is displayed in Figure 6.

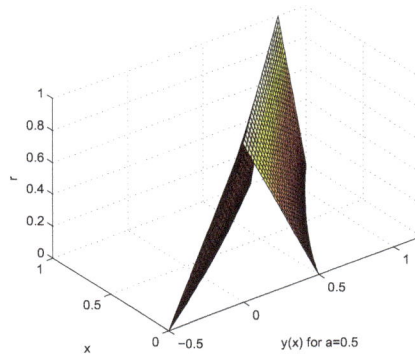

Figure 6. The solution of FDE under case 2 consideration.

Example 3. *The nuclear decay equation can be described as [30],*

$$\begin{cases} N'(t) = -\lambda N(t) \\ N(0) = (\underline{N}(0,r), \overline{N}(0,r)) \end{cases} \tag{82}$$

where $N(t)$ is considered to be the number of radionuclides present, λ is state as the decay constant, also N_0 is taken to be the initial number of radionuclides. Let N_0 be a fuzzy number. By utilizing the FST method the following outcomes can be demonstrated

$$S[N'(t)] = S[-\lambda N(t)] = -\lambda S[N(t)] \tag{83}$$

$$S[N'(t)] = \int_a^\infty N'(st)e^{-t}dt \tag{84}$$

If $N(t)$ is (ii)-differentiable and case 2 holds, we obtain

$$S[N'(t)] = (\frac{-1}{B}S[N(t)]) \ominus (\frac{-1}{B}N(0)) \tag{85}$$

Therefore

$$- \lambda S[N(t)] = (\frac{-1}{B} S[N(t)]) \ominus (\frac{-1}{B} N(0)) \tag{86}$$

According to Equation (42), we will have the below mentioned relation

$$\begin{cases} -\lambda S[\underline{N}(t,r)] = \frac{1}{B} S[\underline{N}(t,r)] - \frac{1}{B} \underline{N}(0,r) \\ -\lambda S[\overline{N}(t,r)] = \frac{1}{B} S[\overline{N}(t,r)] - \frac{1}{B} \overline{N}(0,r) \end{cases} \tag{87}$$

Hence, the solution of Equation (87) is as follows:

$$\begin{cases} S[\underline{N}(t,r)](\lambda - \frac{1}{B}) = \frac{1}{B} \underline{N}(0,r) \\ S[\overline{N}(t,r)](\lambda - \frac{1}{B}) = \frac{1}{B} \overline{N}(t,r) \end{cases} \tag{88}$$

Thus we extract

$$\begin{cases} S[\underline{N}(t,r)] = \frac{1}{-\lambda B+1} \underline{N}(0,r) \\ S[\overline{N}(t,r)] = \frac{1}{-\lambda B+1} \overline{N}(t,r) \end{cases} \tag{89}$$

So, by utilizing the inverse Sumudu transform the following outcomes can be observed

$$\begin{cases} \underline{N}(t,r) = \underline{N}(0,r) S^{-1}(\frac{1}{-\lambda B+1}) \\ \overline{N}(t,r) = \overline{N}(0,r) S^{-1}(\frac{1}{-\lambda B+1}) \end{cases} \tag{90}$$

where

$$\begin{cases} \underline{N}(t,r) = e^{-\lambda t} \underline{N}(0,r) \\ \overline{N}(t,r) = e^{-\lambda t} \overline{N}(0,r) \end{cases} \tag{91}$$

Let $\lambda = 1$ and $N_0 = (1,2,5)$, then

$$\begin{cases} \underline{N}(0,r) = (1+r) \\ \overline{N}(0,r) = (5-3r) \end{cases} \tag{92}$$

So

$$\begin{cases} \underline{N}(t,r) = e^{-t}(1+r) \\ \overline{N}(t,r) = e^{-t}(5-3r) \end{cases} \tag{93}$$

Corresponding solution plot is displayed in Figure 7.

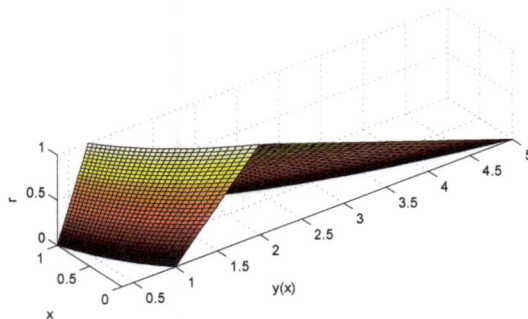

Figure 7. The solution of the nuclear decay equation under case 2 consideration.

6. Conclusions

In this paper, the utilization of FST results in the solution of the first order FDEs in such a manner that it is clarified by using the notion of strongly generalized differentiability. By implementing the methodology of FST, the FDE reduces to an algebraic problem. Some theorems are given to illustrate the properties of the FST. The novel method is validated by three real examples. Numerical experiments along with comparisons demonstrate the excellent behavior of the proposed method. This work makes a significant contribution in initializing a superior starting point for such extensions. Future work involves studying the application of this method in solving FDEs where the uncertainties are in the form of Z-numbers.

Author Contributions: All authors contributed equally to this work. All authors read and approve the final manuscript.

Conflicts of Interest: The authors declare no conflict of interest.

References

1. Zwillinger, D. *Handbook of Differential Equations*; Gulf Professional Publishing: Houston, TX, USA, 1998.
2. Khalili Golmankhaneh, A.; Porghoveh, N.; Baleanu, D. Mean square solutions of second-order random differential equations by using Homotopy analysis method. *Rom. Rep. Phys.* **2013**, *65*, 350–361.
3. Buckley, J.J.; Eslami, E.; Feuring, T. *Fuzzy Mathematics in Economics and Engineering*; Physica-Verlag: Heidelberg, Germany, 2002.
4. Diamond, P.; Kloeden, P. *Metric Spaces of Fuzzy Sets: Theory and Applications*; World Scientific: Singapore, 1994.
5. Kloeden, P. Remarks on Peano-like theorems for fuzzy differential equations. *Fuzzy Set Syst.* **1991**, *44*, 161–164.
6. Jafari, R.; Yu, W. Fuzzy Control for Uncertainty Nonlinear Systems with Dual Fuzzy Equations. *J. Intell. Fuzzy Syst.* **2015**, *29*, 1229–1240.
7. Jafari, R.; Yu, W. Fuzzy Differential Equation for Nonlinear System Modeling with Bernstein Neural Networks. *IEEE Access* **2016**, doi:10.1109/ACCESS.2017.2647920.
8. Jafari, R.; Yu, W. Uncertainty Nonlinear Systems Modeling with Fuzzy Equations. *Math. Probl. Eng.* **2017**, doi :10.1155/2017/8594738.
9. Jafarian, A.; Jafari, R. Approximate solutions of dual fuzzy polynomials by feed-back neural networks. *J. Soft Comput. Appl.* **2012**, doi:10.5899/2012/jsca-00005.
10. Jafari, R.; Yu, W.; Li, X.; Razvarz, S. Numerical Solution of Fuzzy Differential Equations with Z-numbers Using Bernstein Neural Networks. *Int. J. Comput. Int. Sys.* **2017**, *10*, 1226–1237.
11. Jafari, R.; Yu, W.; Li, X. Numerical solution of fuzzy equations with Z-numbers using neural networks. *Intell. Autom. Soft Comput.* **2017**, 1–7, doi:10.1080/10798587.2017.1327154.
12. Friedman, M.; Ming, M.; Kandel, A. Numerical solution of fuzzy differential and integral equations. *Fuzzy Set Syst.* **1999**, *106*, 35–48.
13. Allahviranloo, T.; Ahmadi, M.B. Fuzzy Laplace Transform. *Soft Comput.* **2010**, *14*, 235–243.
14. Ding, Z.; Ma, M.; Kandel, A. Existence of the solutions of fuzzy differential equations with parameters. *Inf. Sci.* **1997**, *99*, 205–217.
15. Truong, V.A.; Ngo, V.H.; Nguyen, D.P. Global existence of solutions for interval-valued integro-differential equations under generalized H-differentiability. *Adv. Differ. Equ.* **2013**, *1*, 217–233.
16. Asiru, M.A. Further properties of the Sumudu transform and its applications. *Int. J. Math. Educ. Sci. Tech.* **2002**, *33*, 441–449.
17. Belgacem, F.B.M.; Karaballi, A.A.; Kalla, S.L. Analytical investigations of the Sumudu transform and applications to integral production equations. *Math. Probl. Eng.* **2003**, *103*, 103–118.
18. Deakin, M.A.B. The Sumudu transform and the Laplace transform. *Int. J. Math. Educ. Sci. Technol.* **1997**, *28*, 159–160.
19. Eltayeb, H.; Kilicman, A. A note on the Sumudu transforms and differential equations. *Appl. Math. Sci.* **2010**, *4*, 1089–1098.
20. Srivastava, H.M.; Khalili Golmankhaneh, A.; Baleanu, D.; Yang, X.J. Local Fractional Sumudu Transform with Application to IVPs on Cantor Sets. *Abstr. Appl. Anal.* **2014**, doi:10.1155/2014/620529.

21. Abdul Rahman, N.A.; Ahmad, M.Z. Fuzzy Sumudu transform for solving fuzzy partial differential equations. *J. Nonlinear Sci. Appl.* **2016**, *9*, 3226–3239.

22. Belgacem, F.B.M.; Karaballi, A.A. Sumudu transform fundamental properties investigations and applications. *J. Appl. Math. Stoch. Anal.* **2006**, doi:10.1155/JAMSA/2006/91083.

23. Liu, Y.; Chen, W. A new iterative method for ordinary equations using Sumudu transform. *Adv. Anal.* **2016**, doi:10.22606/aan.2016.12004.

24. Jafari, R.; Razvarz, S. Solution of Fuzzy Differential Equations using Fuzzy Sumudu Transforms. *IEEE Int. Conf. Innov. Intell. Syst. Appl.* **2017**, doi:10.1109/INISTA.2017.8001137.

25. Bede, B.; Stefanini, L. Generalized differentiability of fuzzy-valued functions. *Fuzzy Set Syst.* **2013**, *230*, 119–141.

26. Seikkala, S. On the fuzzy initial value problem. *Fuzzy Set Syst.* **1987**, *24*, 319–330.

27. Puri, M.L.; Ralescu, D. Fuzzy random variables. *J. Math. Anal. Appl.* **1986**, *114*, 409–422.

28. Wu, H.-C. The improper fuzzy Riemann integral and its numerical integration. *Inform. Sci.* **1999**, *111*, 109–137.

29. Bede, B.; Gal, S.G. Generalizations of the differentiability of fuzzy-number-valued functions with applications to fuzzy differential equations. *Fuzzy Set Syst.* **2005**, *151*, 581–599.

30. Bede, B.; Rudas, I.J.; Bencsik, A.L. First order linear fuzzy differential equations under generalized differentiability. *Inf. Sci.* **2006**, *177*, 1648–1662.

31. Chalco-Cano, Y.; Roman-Flores, H. On new solutions of fuzzy differential equations. *Chaos Solitons Fractals* **2006**, *38*, 112–119.

32. Pletcher, R.H.; Tannehill, J.C.; Anderson, D. *Computational Fluid Mechanics and Heat Transfer*; Taylor and Francis: Abingdon, UK, 1997.

33. Tapaswini, S.; Chakraverty, S. Euler-based new solution method for fuzzy initial value problems. *Int. J. Artif. Intell. Soft Comput.* **2014**, *4*, 58–79.

34. Streeter, V.L.; Wylie, E.B.; Bedford, K.W. *Fluid Mechanics*; McGraw Hill: New York, NY, USA, 1998.

Mathematical and Computational Applications

MDPI

Article

A Simple Spectral Observer

Lizeth Torres [1,*] [iD], **Javier Jiménez-Cabas** [2], **José Francisco Gómez-Aguilar** [3] [iD]
and Pablo Pérez-Alcazar [4] [iD]

1   Cátedras CONACYT, Instituto de Ingeniería, Universidad Nacional Autónoma de México,
    04510 Ciudad de México, México
2   Departamento de Ciencias de la Informática y Electrónica, Universidad de la Costa,
    Barranquilla, Colombia; jjimenez41@cuc.edu.co
3   Cátedras CONACYT, Centro Nacional de Investigación y Desarrollo Tecnológico,
    Tecnológico Nacional de México, 62490 Cuernavaca, México; jgomez@cenidet.edu.mx
4   Facultad de Ingeniería, Universidad Nacional Autónoma de México, 04510 Ciudad de México, México;
    pperezalcazar@fi-b.unam.mx
*   Correspondence: ftorreso@iingen.unam.mx

Received: 3 March 2018; Accepted: 10 April 2018; Published: 1 May 2018

Abstract: The principal aim of a spectral observer is twofold: the reconstruction of a signal of time via state estimation and the decomposition of such a signal into the frequencies that make it up. A spectral observer can be catalogued as an online algorithm for time-frequency analysis because is a method that can compute on the fly the Fourier Transform (FT) of a signal, without having the entire signal available from the start. In this regard, this paper presents a novel spectral observer with an adjustable constant gain for reconstructing a given signal by means of the recursive identification of the coefficients of a Fourier series. The reconstruction or estimation of a signal in the context of this work means to find the coefficients of a linear combination of sines a cosines that fits a signal such that it can be reproduced. The design procedure of the spectral observer is presented along with the following applications: (1) the reconstruction of a simple periodical signal, (2) the approximation of both a square and a triangular signal, (3) the edge detection in signals by using the Fourier coefficients, (4) the fitting of the historical Bitcoin market data from 1 December 2014 to 8 January 2018 and (5) the estimation of a input force acting upon a Duffing oscillator. To round out this paper, we present a detailed discussion about the results of the applications as well as a comparative analysis of the proposed spectral observer vis-à-vis the Short Time Fourier Transform (STFT), which is a well-known method for time-frequency analysis.

Keywords: signal processing; Fourier series; state observer; Short Time Fourier Transform; time-frequency analysis

1. Introduction

The term spectral observer was proposed by Hostetter in his pioneering work [1] to name the algorithm that permits the recursive calculation of the Fourier Transform (FT) of a band-limited signal via state estimation. Since the presentation of such a work, several designs of spectral observers with improved features have been proposed either to deal with noise [2], disturbances, lack of data [3] or to estimate other parameters such as frequency [4]. The main goals of a spectral observer are both the estimation of a given signal and the transformation of such a signal to the frequency domain by means of the recursive identification of the coefficients of a Fourier series [5]. The estimation of a signal in the context of this work means to find the coefficients of a linear combination of functions—sines a cosines functions in our case—that approximates a signal of interest such that it can be reconstructed [6]. Spectral observers are useful in a wide number of applications, e.g., for determining the source of harmonic pollution in power systems [7], for the simulation of the sea surface [8], for fault diagnosis in motors [9,10] or in vibrating structures, such as aerospace and mechanical structures, marine structures, buildings, bridges and offshore platforms.

A spectral observer can be catalogued as an online algorithm to compute the Fourier Transform (FT) during a time window which slides along the signal, i.e., an algorithm to compute the Short Time Fourier Transform (STFT). Therefore, a spectral observer can be used for the time-frequency analysis of frequency variant signals.

The observer that we propose in this contribution is designed from a dynamical system which is constructed from the N derivatives of a n-th order Fourier series.

To perform the estimation, the observer solely requires: (1) The measurement of the signal to be approximated, $s(t)$, which actually is used to compute the observation error $e(t) = s(t) - \hat{y}(t)$, where $\hat{y}(t)$ is the observer output, (2) A frequency step $\omega = 2\pi/T$, where T is a predefined period. The estimation provided by the observer are both the reconstruction of the original signal and the Fourier coefficients to compute the signal frequency components.

This paper is organized as follows: Section 2 presents the core of the proposed method which is the formulation of the spectral observer from the Fourier series. Section 3 presents some examples with test results of the proposed method utilized in different applications. In Section 4 the main results are discussed. Finally, in Section 5 some concluding thoughts are given.

2. The Proposed Method

To construct the proposed observer, we formulate a dynamical synthetic system in state space representation by considering, firstly, that a given signal expressed as $s(t)$ can be approximated by a Fourier series, and secondly, that the Fourier series is the first state of the system and the rest of the states are the N first-order derivatives of the Fourier series expressed by Equation (1), where n is the series order.

$$y(t) = \frac{a_0}{2} + \sum_{k=1}^{n} \left[a_k \cos(k\omega t) + b_k \sin(k\omega t) \right], \tag{1}$$

where $a_0, a_1, b_1, ..., a_n, b_n$ are the Fourier coefficients and ω is the fundamental angular frequency of the signal to be estimated.

In this work, we assume that the signal $s(t)$ to be approximated has not a constant component (offset) or that this offset is removed by an online algorithm prior to be processed by the spectral observer. For this reason, we remove the term a_0 from Equation (1) such that the series for approximating the time function $s(t)$ can be expressed as follows

$$y(t) = \sum_{k=1}^{n} \left[a_k \cos(k\omega t) + b_k \sin(k\omega t) \right]. \tag{2}$$

If the order of the Fourier series is $n = 1$, we need to formulate a dynamical system with $N = 2$ states, each one to recover each coefficient (a_1 and b_1). Thus, the two first states are the Fourier series and its first derivative.

$$v_1(t) = y(t) = a_1 \cos(\omega t) + b_1 \sin(\omega t), \quad v_2(t) = \dot{y}(t) = -\omega a_1 \sin(\omega t) + \omega b_1 \cos(\omega t). \tag{3}$$

where v_i are the states of the synthetic system. Consequently, the dynamical system that results from the change of coordinates, gives:

$$\dot{v}_1(t) = v_2(t), \quad \dot{v}_2(t) = -\omega^2 v_1(t), \tag{4}$$

which basically is the dynamical model of a harmonic oscillator. Now, what happens if the order of the Fourier series increases? If the order increases to $n = 2$, then $N = 4$, since we need to recover four coefficients.

$$
\begin{aligned}
v_1(t) &= y(t) = a_1 \cos(\omega t) + b_1 \sin(\omega t) + a_2 \cos(2\omega t) + b_2 \sin(2\omega t), \\
v_2(t) &= \dot{y}(t) = -\omega a_1 \sin(\omega t) + \omega b_1 \cos(\omega t) - 2\omega a_2 \sin(2\omega t) + 2\omega b_2 \cos(2\omega t), \\
v_3(t) &= \ddot{y}(t) = -\omega^2 a_1 \cos(\omega t) - \omega^2 b_1 \sin(\omega t) - 4\omega^2 a_2 \cos(2\omega t) - 4\omega^2 b_2 \sin(2\omega t), \\
v_4(t) &= y^{(3)}(t) = \omega^3 a_1 \sin(\omega t) - \omega^3 b_1 \cos(\omega t) + 8\omega^3 a_2 \sin(2\omega t) - 8\omega^3 b_2 \cos(2\omega t), \\
\dot{v}_4(t) &= y^{(4)}(t) = \omega^4 a_1 \cos(\omega t) + \omega^4 b_1 \sin(\omega t) + 16\omega^4 a_2 \cos(2\omega t) + 16\omega^4 b_2 \sin(2\omega t).
\end{aligned}
\tag{5}
$$

The dynamical system is then formulated as in Equation (5) which, after some algebraic manipulations, it becomes

$$
\dot{v}_1(t) = v_2(t); \quad \dot{v}_2(t) = v_3(t), \dot{v}_3(t) = v_4(t), \dot{v}_4(t) = -4\omega^4 v_1(t) - 5\omega^2 v_3(t),
\tag{6}
$$

in v-coordinates.

By generalizing Equation (5) for order n, we obtain the following dynamical system:

$$
\begin{aligned}
\dot{v}_1(t) &= v_2(t), \\
\dot{v}_2(t) &= v_3(t), \\
\vdots &= \vdots, \\
\dot{v}_N(t) &= (-1)^{n(\mathrm{mod}\ 2)} \omega^{2n} \begin{bmatrix} 1 & 2^{2n} & \cdots & n^{2n} \end{bmatrix} A_k^{-1} A_\omega^{-1} v(t),
\end{aligned}
\tag{7}
$$

where $N = 2n$, A_ω and A_k are expressed by Equations (8) and (9), respectively.

$$
A_\omega \doteq
\begin{bmatrix}
1 & 0 & 0 & 0 & \cdots & 0 & \cdots & 0 \\
0 & \omega & 0 & 0 & \cdots & 0 & \cdots & 0 \\
0 & 0 & -\omega^2 & 0 & \cdots & 0 & \cdots & 0 \\
0 & 0 & 0 & -\omega^3 & \cdots & 0 & \cdots & 0 \\
\vdots & \vdots & \vdots & \vdots & \ddots & \vdots & \ddots & \vdots \\
0 & 0 & 0 & 0 & \cdots & (-1)^{(m(\mathrm{mod}\ 4) - m(\mathrm{mod}\ 2))/2}\, \omega^m & \cdots & 0 \\
\vdots & \vdots & \vdots & \vdots & \ddots & \vdots & \ddots & \vdots \\
0 & 0 & 0 & 0 & \cdots & 0 & \cdots & (-1)^{(2n(\mathrm{mod}\ 4) - 2)/2}\, \omega^{2n-1}
\end{bmatrix},
\tag{8}
$$

$$
A_k \doteq
\begin{bmatrix}
1 & 0 & 1 & 0 & \cdots & 1 & 0 \\
0 & 1 & 0 & 2 & \cdots & 0 & n \\
1 & 0 & 4 & 0 & \cdots & n^2 & 0 \\
0 & 1 & 0 & 8 & \cdots & 0 & n^3 \\
\vdots & \vdots & \vdots & \vdots & \ddots & \vdots & \vdots \\
1 & 0 & 2^{2n-2} & 0 & \cdots & n^{2n-2} & 0 \\
0 & 1 & 0 & 2^{2n-1} & \cdots & 0 & n^{2n-1}
\end{bmatrix}.
\tag{9}
$$

Before presenting the state observer for system (7), it is necessary to analyze its observability conditions. A dynamical system is said to be observable if it is possible to determine its initial state by knowledge of the input and output over a finite time interval. In this way, a state observer or state estimator is a system that estimates the internal states of a system from the measurement of its inputs and outputs. To verify is a linear systems is observable, the observability rank condition can be used, which is defined in the net lines.

Observability rank condition. A system

$$\begin{aligned}
\dot{v}(t) &= A(t)v(t) \\
y(t) &= Cv(t)
\end{aligned} \tag{10}$$

is said to satisfy the observability rank condition if $\forall v(t), \mathrm{rank}\,(O(v(t))) = N$, where N is the state dimension of (10) and $O(v(t))$ is the observability matrix defined as

$$O(v(t)) = (C\ CA\ CA^2\ ...\ CA^{(N-1)})^T. \tag{11}$$

Thus, according to the definition, the observability matrix for system (7), which in fact can be set as system (10) with

$$A = \begin{bmatrix}
0 & 1 & \cdots & 0 & 0 \\
0 & 0 & 1 & \cdots & 0 \\
 & & \ddots & & \vdots \\
\vdots & & & & 1 \\
\gamma_1(\omega) & & \cdots & & \gamma_n(\omega)
\end{bmatrix}, \tag{12}$$

and $C = [1, 0, ..., 0]$, has full rank. Therefore, since system (7) is observable, a spectral observer can be designed as follows:

$$\dot{\hat{v}}(t) = A\hat{v}(t) + K(y(t) - C\hat{v}(t)), \quad \hat{y}(t) = C\hat{v} = \hat{v}_1. \tag{13}$$

where " $\hat{}$ " means estimation, $\hat{y}(t)$ is the estimated signals and A is given by (12), with constant coefficients expressed by $\gamma_i(\omega)$. The gain of the state observer K, involved in the correction term of Equation (13), can be calculated as $K = S^{-1}C^T$, where S is the unique solution of the following algebraic Lyapunov equation:

$$- \lambda S - A^T S - SA + C^T C = 0 \tag{14}$$

and λ is a parameter that can be used to tune the convergence rate of the observer. A numerical solution for solving Equation (14) for a particular but common case is provided in [11].

The Fourier coefficients can be recovered from the new coordinates by the relation $\hat{c}_k = \Omega(t)\hat{v}(t)$, where $\hat{c}_k = [\hat{a}_1\ \hat{b}_1\ ...\ \hat{a}_n\ \hat{b}_n]^T$ and

$$\Omega(t) = \begin{pmatrix}
\blacktriangle(\omega t) & \blacktriangledown(\omega t) & \blacktriangle(2\omega t) & \blacktriangledown(2\omega t) & \cdots & \blacktriangle(n\omega t) & \blacktriangledown(n\omega t) \\
-\omega\blacktriangledown(\omega t) & \omega\blacktriangle(\omega t) & -2\omega\blacktriangledown(2\omega t) & 2\omega\blacktriangle(2\omega t) & \cdots & -n\omega\blacktriangledown(n\omega t) & n\omega\blacktriangle(n\omega t) \\
-\omega^2\blacktriangle(\omega t) & -\omega^2\blacktriangledown(\omega t) & -4\omega^2\blacktriangle(2\omega t) & -4\omega^2\blacktriangledown(2\omega t) & \cdots & -n^2\omega\blacktriangle(n\omega t) & -n^2\omega\blacktriangledown(n\omega t) \\
\vdots & & & & & & \vdots \\
\omega^\Gamma\blacktriangledown(\omega t) & -\omega^\Gamma\blacktriangle(\omega t) & 2^\Gamma\omega^\Gamma\blacktriangledown(2\omega t) & -2^\Gamma\omega^\Gamma\blacktriangle(2\omega t) & \cdots & n^\Gamma\omega^\Gamma\blacktriangledown(n\omega t) & -n^\Gamma\omega^\Gamma\blacktriangle(n\omega t)
\end{pmatrix}$$

Notice that $\blacktriangle \triangleq \cos$, $\blacktriangledown \triangleq \sin$ and $\Gamma \triangleq n - 1$. Check Figure 1 to see a schema of the estimation.

Before presenting some possible applications of the spectral observer it is important to highlight an important point. Notice that matrix A of the spectral observer depends on the fundamental frequency ω. This means that this variable must be known. In case we want to approximate a periodic signal with a known fundamental frequency, we just need to use it in matrix A. In case we want to fit a periodic signal with unknown fundamental frequency or a non-periodic signal, we must assume, as in the Fourier Transform deduction from the Fourier series, that the period of the signal tends to infinity, which means that the fundamental frequency tends to zero. As a consequence of this assumption, ω should be chosen sufficiently small or according to the desired precision in the recovery of the frequency components. In other words ω is the frequency step that determines the resolution of the discretized frequency domain, such that we have to choose ω thinking how close we want the frequency components.

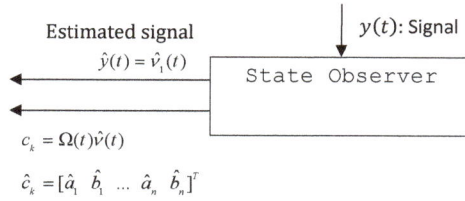

Figure 1. Schema of the reconstruction of a signal by using the spectral observer.

3. Application Examples

This section presents five examples of possible applications of the spectral observer, which were conceived such that the reader can be able to reproduce them.

3.1. Example 1: A Simple Example

Let be the signal $s(t) = 4\cos(t) + \sin(t) + 2\cos(2t) + 5\sin(2t)$. It is obvious that the series order to reproduce the signal is $n = 2$, then the order of system (7) for the conception of the observer should be $N = 4$. Figure 2 shows the estimation of the coefficients that was performed by the observer with a gain $\lambda = 8$, which actually was initialized with $\hat{v}(0) = 0$. The step time to perform the estimation in Simulink was $\Delta t = 0.005$ [s] and the used solver was ODE3.

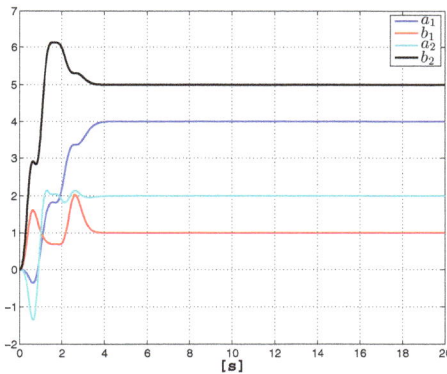

Figure 2. Example 1. Estimated coefficients.

3.2. Example 2: Reconstruction of Basic Signals

This example aims to show the estimation of the coefficients for basic signals such as square and sawtooth waves. The first signal to be estimated is a square wave with angular frequency $\omega = 1$ [rad/s]. The observer was tuned with $\lambda = 15$. The order of the series was set $n = 2$, i.e., $N = 4$. The frequency step was set $\omega = 5$ [rad/s]. The step time to perform the estimation in Simulink was set $\Delta t = 0.01$ [s] and the used solver was ODE3. Figure 3 shows the signal reconstruction performed by the spectral observer and the estimated coefficients. Firstly, notice that the coefficients do not converge towards a constant value; the reason for this is the number of coefficients used to approximate the signal, which is not enough to represent each harmonic that composes it. Even though the coefficients are not constant, the signal is estimated. Notice too that all the coefficients change abruptly at each discontinuity. This feature can be used for edge detection as will be seen in the next example.

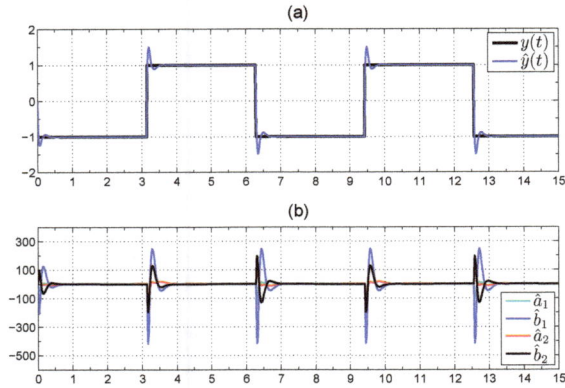

Figure 3. Example 2. (**a**) Square wave reconstruction and (**b**) Estimated coefficients.

Both the observer and conditions that were used to reconstruct the square wave were used to reconstruct the sawtooth signal shown in Figure 4. Notice that the convergence time is less than one second and the coefficients become greater at the discontinuities.

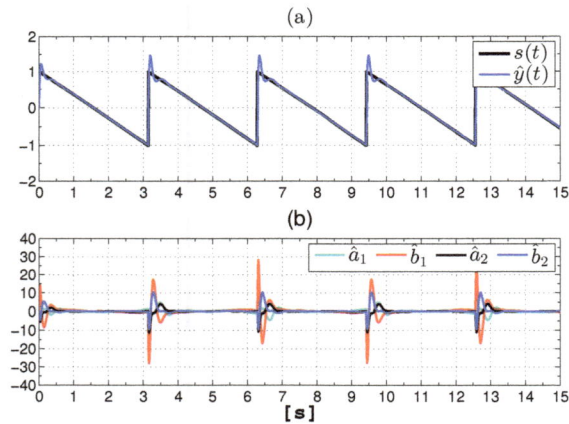

Figure 4. Example 2. (**a**) Triangular wave reconstruction and (**b**) Estimated coefficients.

To end this example, we made several simulations to show how the parameter λ determines the convergence period of the estimation. Figure 5 shows the estimation of the square signal with different values of λ. Notice that the bigger its value, the faster is the convergence.

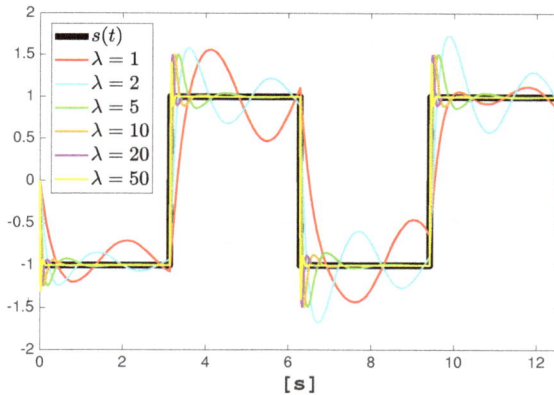

Figure 5. Variations of parameter λ.

3.3. Example 3: Edge Detection by Using the Fourier Coefficients

Edge detection and the detection of discontinuities are important in many fields. In image processing, for example, one often needs to determine the boundaries of the items of which a picture is composed, [12] or in applications that utilize time-domain reflectometry (TDR), which is a measurement technique used to determine the characteristics of transmission lines by observing reflected waveforms. TDR analysis begins with the propagation of a step or impulse of energy into a system and the subsequent observation of the energy reflected by the system. By analyzing the magnitude, duration and shape of the reflected waveform, the nature of the transmission system can be determined. TDR is a common method used to localize faults in transmission lines—such a leaks in pipelines or faults with small impedance in wires—because faults in transmission lines cause discontinuities in the reflected waveforms. For this reason, methodologies to detect discontinuities are required in order to localize the nature and position of the faults.

In order to shows how the spectral observer (13) can be used to detect discontinuities in a function, we present the following example: Let us consider $s(t) = \text{sign}(\sin(0.5t)) - 0.05\text{sign}(\sin(2t))$, which is plot in Figure 6a. The aim of this test is to detect the discontinuities in the principal signal with period $T = 2$ [s]. To identify the discontinuities, the coefficients provided by the spectral observer are used to calculate the following indicator function:

$$r_k = \ln\left(\sqrt{\hat{a}_k^2 + \hat{b}_k^2}\right) \tag{15}$$

The observer to perform the estimation was tuned with $\lambda = 15$. The order of the series was set $n = 2$, i.e., $N = 4$. The frequency step was set $\omega = 1$ [rad/s]. The step time to perform the estimation in Simulink was set $\Delta t = 0.01$ [s] and the used solver was ODE3.

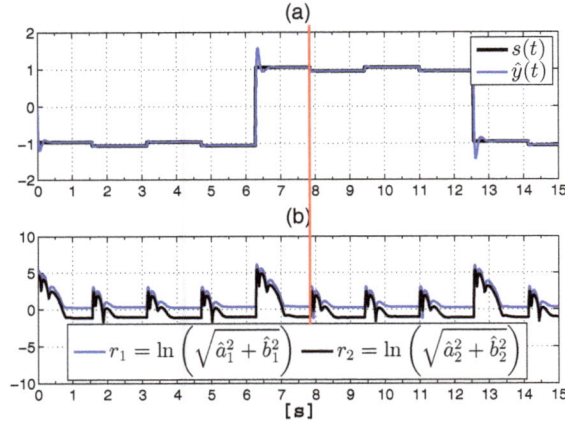

Figure 6. Example 3. (**a**) Signal reconstruction and (**b**) Estimated coefficients.

Figure 6a shows $s(t)$ and its reconstruction $\hat{y}(t)$. Figure 6b shows the index $r_1(t)$ and $r_2(t)$ that becomes greater at the discontinuities indicating where they are.

3.4. Example 4: Fitting Complex Signal: The Bitcoin Price

Bitcoin is the longest running and best known cryptocurrency in the world. It was released as open source in 2009 by the anonymous Satoshi Nakamoto. Bitcoin serves as a decentralized medium of digital exchange, with transactions verified and recorded in a public distributed ledger (the blockchain) without the need for a trusted record keeping authority or central intermediary. Hereafter, we will use the proposed spectral observer for fitting the historical Bitcoin market close data every 1000 [min]. The records were downloaded from the website: https://www.kaggle.com/neelneelpurk/bitcoin/data.

The observer to perform the estimation was tuned with $\lambda = 1$. The order of the series was set $n = 20$, i.e., $N = 40$. The frequency step was set $\omega = 10$ [rad/s]. The step time to perform the estimation in Simulink was set $\Delta t = 0.01$ [s] and the used solver was ODE8.

In Figure 7, the Bitcoin fitting performed by the spectral observer is shown. Figure 8 shows the estimated coefficients which are not constant and look as if they were enveloped by exponential functions. In order to have a model that represents the behavior of the Bitcoin in the specified interval, we can fit each coefficient by means of polynomials after calculating the natural logarithm of each one. In Figure 9, $\ln(|a_1|)$ is plotted versus a cubic polynomial calculated to interpolate it.

$$\ln(|a_1|) = 1.2 \times 10^{-9} t^3 + 1.5 \times 10^{-6} t^2 - 0.0021t - 1.4, \tag{16}$$

We can perform the same procedure for each coefficient to obtain a series with the following form:

$$
\begin{aligned}
\hat{y}(t) = & \left(e^{|(\alpha_{c1}t^3 + \beta_{c1}t^2 + \gamma_{c1}t + \delta_{c1})|} \right) \cos(\omega t) + \left(e^{|(\alpha_{s1}t^3 + \beta_{s1}t^2 + \gamma_{s1}t + \delta_{s1})|} \right) \sin(\omega t) \\
& + \left(e^{|(\alpha_{c2}t^3 + \beta_{c2}t^2 + \gamma_{c2}t + \delta_{c2})|} \right) \cos(2\omega t) + \left(e^{|(\alpha_{s2}t^3 + \beta_{s2}t^2 + \gamma_{s2}t + \delta_{s2})|} \right) \sin(2\omega t) \\
& + \ldots + \left(e^{|(\alpha_{cn}t^3 + \beta_{cn}t^2 + \gamma_{cn}t + \delta_{cn})|} \right) \cos(n\omega t) + \left(e^{|(\alpha_{sn}t^3 + \beta_{sn}t^2 + \gamma_{sn}t + \delta_{sn})|} \right) \sin(n\omega t)
\end{aligned}
\tag{17}
$$

where $\alpha_{ck}, \beta_{ck}, \gamma_{ck}, \delta_{ck}, \alpha_{sk}, \beta_{sk}, \gamma_{sk}, \delta_{sk}$ are the coefficients of the polynomial that approximates the natural logarithms of the coefficients.

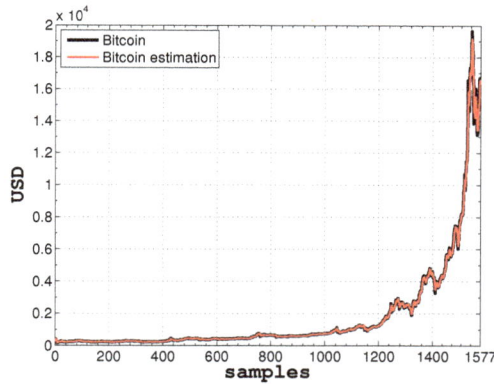

Figure 7. Example 4. Bitcoin fitting.

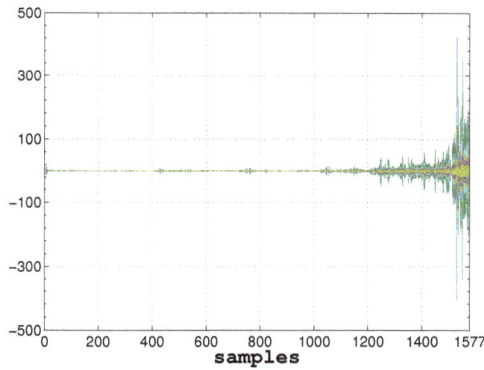

Figure 8. Example 4. Estimated coefficients.

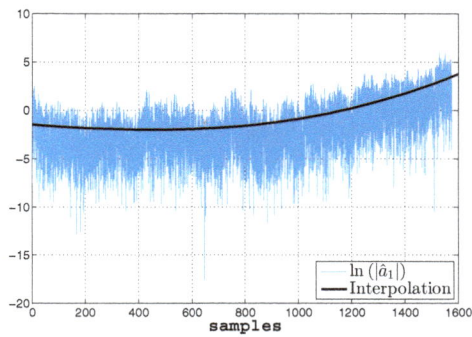

Figure 9. Example 4. $\log |a_1|$ vs. interpolation.

3.5. Example 5: Estimation of the Input Force on a Duffing Oscillator

One of the advantages of the proposed spectral observer is its structure, which is a chain of integrators expressed in state-space representation. Such a structure permits to couple the observer to dynamic models, also expressed in state variables, that represent physical systems for control or estimation purposes. With this in mind, we present an example to show how the spectral observer can be coupled to the model of a given system, even with nonlinear structure such as the Duffing oscillator, in order to estimate an exogenous input affecting its behavior.

The Duffing oscillator is expressed by the following equation:

$$\ddot{x}(t) + \delta\dot{x}(t) + \alpha x(t) + \beta x^3(t) = u(t) \tag{18}$$

where $x(t)$ is the displacement, which is assumed as available in this example, $\dot{x}(t)$ is the velocity and $\ddot{x}(t)$ is the acceleration. In addition, δ, α and β are parameters, which in this example are assumed to be known. Finally, $u(t)$ is an external force, which is unknown and can be estimated by using our proposed observer. To achieve this goal, the following steps need to be executed.

Step 1. Equation (18) must be set in state-space representation. To execute this step, we define $x_1(t) = x(t)$ and $x_2(t) = \dot{x}(t)$ as the state variables, such that we obtain the following equation system:

$$\begin{aligned} \dot{x}_1(t) &= x_2(t), \\ \dot{x}_2(t) &= -\delta x_2(t) - \alpha x(t) - \beta x^3(t) + u(t), \\ y(t) &= x_1(t). \end{aligned} \tag{19}$$

If the Liénard transform [13] is applied to system (18) in order to set it in a more appropriate form for estimation purposes [14], it becomes

$$\begin{aligned} \dot{x}_1(t) &= x_2(t) - \delta x_1(t), \\ \dot{x}_2(t) &= -\alpha x_1(t) - \beta x_1^3(t) + u(t), \end{aligned} \tag{20}$$

Step 2. Since $u(t)$ is unknown and needs to be estimated, we propose its estimation by using a spectral observer with $n = 1$. Therefore, we coupled in cascade equation system (18) with equation system (4) as follows:

$$\begin{aligned} \dot{x}_1(t) &= x_2(t) - \delta x_1(t), \\ \dot{x}_2(t) &= -\alpha x_1(t) - \beta x_1^3(t) + v_1(t), \\ \dot{v}_1(t) &= v_2(t), \\ \dot{v}_2(t) &= -\omega^2 v_1(t), \\ y(t) &= x_1(t), \end{aligned} \tag{21}$$

where $v_1(t) = u(t)$, i.e., it is the force to be estimated.

System (21) can be set in the following form:

$$\underbrace{\begin{pmatrix} \dot{x}_1(t) \\ \dot{x}_2(t) \\ \dot{v}_1(t) \\ \dot{v}_2(t) \end{pmatrix}}_{\dot{\xi}(t)} = \underbrace{\begin{pmatrix} 0 & 1 & 0 & 0 \\ 0 & 0 & 1 & 0 \\ 0 & 0 & 0 & 1 \\ 0 & 0 & -\omega^2 & 0 \end{pmatrix}}_{A} \underbrace{\begin{pmatrix} x_1(t) \\ x_2(t) \\ v_1(t) \\ v_2(t) \end{pmatrix}}_{\xi(t)} + \underbrace{\begin{pmatrix} -\delta y(t) \\ -\alpha y(t) - \beta y^3(t) \\ 0 \\ 0 \end{pmatrix}}_{\varphi(\xi(t))}, \tag{22}$$

$$y(t) = [1\ 0\ 0\ 0]\xi(t) = C\xi(t) = x_1(t), \tag{23}$$

which according to [15] is uniformly observable. Therefore, a state observer expressed as $\dot{\hat{\xi}}(t) = A\hat{\xi}(t) + \varphi(\hat{\xi}(t)) + K(x_1(t) - \hat{\xi}_1(t))$ can be designed for system (22), where K can be calculated by means of Equation (14).

For the simulation, the paramaters of the Duffing oscillator were set: $\alpha = 1$, $\beta = 1$ and $\delta = 0.3$; and their initial conditions were set $x(0) = [-1\ -5]$. The force applied to the oscillator was

$u(t) = 2\cos(2t) + 5\sin(2t) + 4\sin(0.5t)$. The observer was tuned with $\lambda = 10$, $\omega = 1$ [rad/s] and their initial conditions were fixed as $\hat{\xi}(0) = [0\ 0\ 0\ 0]$. Finally, the used solver was ODE 3 with a step time $\Delta t = 0.01$ [s]. The results of the estimation are shown in Figure 10, which particularly presents a comparison between the force and its estimation. Notice that the estimation converges to the force in 1 [s].

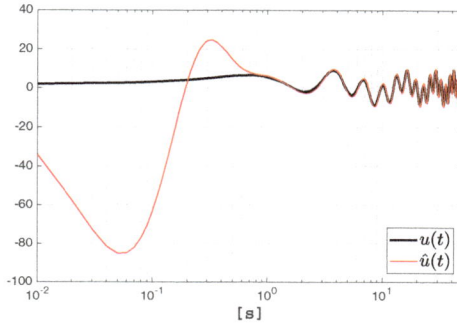

Figure 10. Example 5. Force estimation.

4. Comparative Analysis Vis-à-Vis the STFT

A spectral observer can be used to determine the frequency content of local sections of a signal as it changes over time. The classic technique for performing this task is the Short Time Fourier Transform, which is the Fourier Transform with a suitable chosen windowing function. Ensuing, we present an example to compare the results of using the STFT with the results provided by the spectral observer. For this purpose, we used the MATLAB$^{\copyright}$ codes created by Hristo Zhivomirov to compute STFT and its inverse [16]. The signal analyzed was

$$s(t) = \begin{cases} 0 & 0 \leq t < 100[s] \\ \sin(10\pi t) & 0 \leq t < 300[s] \\ 0 & 300 \leq t < 400[s] \\ \sin(4\pi t) & 400 \leq t < 600[s] \\ 0 & 600 \leq t < 700[s] \end{cases} \tag{24}$$

sampled at 1000 [Hz]. To compute the STFT by using the code of Zhivomirov, the following parameters were set: $\tau_w = 2^8$ [s] as the window length, $h = \tau_w/4$ [s] as the hop size and $n_{fft} = 2^{10}$ as the number of FFT points. The tuning of the spectral observer was done by setting $n = 10$, $\lambda = 1$, $\omega = \pi$ [rad/s] and $\hat{v}(0) = 0$. The solver used for the numerical solution was ODE4 (Runge-Kutta) with a fixed step size $\Delta t = 0.01$ [s]. The spectrograms that were produced by the STFT and the spectral observer, respectively, are presented in Figure 11. To construct the observer spectrogram, we computed de magnitude of each harmonic by means of the following equation:

$$|A_k| = \sqrt{a_k + b_k}. \tag{25}$$

On the one hand, since $n = 10$, the resulting vector containing the magnitude of each harmonic was $A = [A_1\ A_2\ A_3\ A_4\ A_5\ A_6\ A_7\ A_8\ A_9\ A_{10}]$. On the other hand, since the angular frequency step was chosen $\omega = \pi$ [rad/s], the resulting frequency vector was $f = [0.5\ 1\ 1.5\ 2\ 2.5\ 3\ 3.5\ 4\ 4.5\ 5]$ [Hz]. Then, the spectrogram resulted of plotting f versus A.

Notice that the spectrogram generated by using the spectral observer presents a better frequency resolution with respect to the spectral observer. This fact can be better appreciated in Figure 12.

However, this does not mean that the observer's performance is superior, since it is well known that the frequency resolution can be improved by widening the time window length of the STFT, even if this widening implies a decreasing of the time resolution. In the case of the spectral observer, the frequency resolution is adjusted by manipulating the parameter λ.

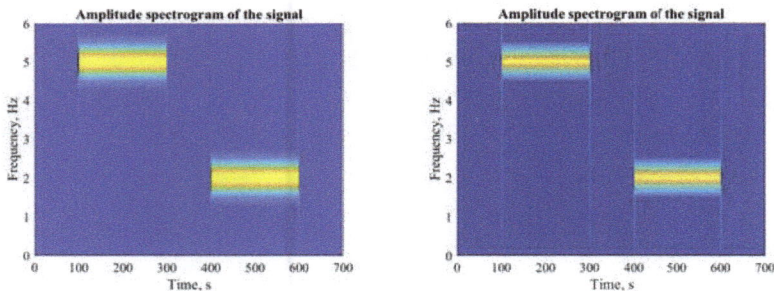

Figure 11. (**a**) STFT spectrogram (**b**) Spectral observer spectrogram.

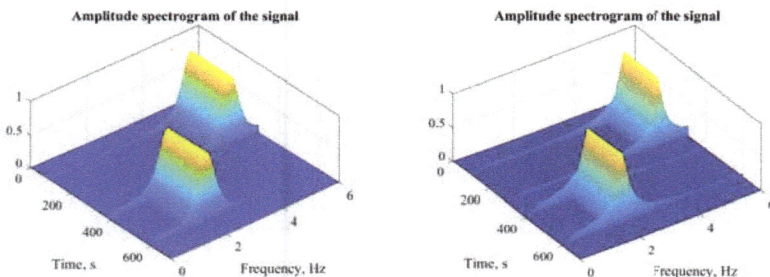

Figure 12. (**a**) STFT spectrogram (**b**) Spectral observer spectrogram.

It is necessary to point out here that both the frequency resolution and the time resolution do not only depend on the parameters τ and λ, also the parameters h y n_{fft} used in the STFT algorithm and ω and n used in the spectral observer, play an important factor; nevertheless, the adjustment of these parameters directly affects in the computational burden and the amount of data to be processed.

5. Results and Discussion

We have introduced an algorithm to reconstruct signals at the same time that their frequency components are calculated: a new spectral observer. In order to show its applicability, we have presented some examples, which in addition, have allowed us to glimpse some advantages and disadvantages of its use. Firstly, we found the following benefits: (1) The structure of the observer, as a chain of integrators, is very adequate for control and parameter estimation purposes; (2) The signal is progressively incorporated at each iteration; (3) The operations required for the observer implementation are with real numbers, which simplifies its programming in single-board computers; (4) The gain of the observer can be easily computed by means of a simple numerical algorithm; (5) The convergence of the observer is exponential, this means that the convergence period can be adjusted by means of a unique parameter λ, which is a clear advantage with respect to other well-known algorithms such as the proposed in [17], where the convergence period cannot be manipulated by a unique parameter. However, there are some drawbacks that we have found for the proposed observer.

(1) The computational cost can be high for a small frequency resolution; (2) The algorithm must be complemented with a methodology to choose ω and λ in order to obtain the best estimation.

To conclude the discussion, it is necessary to emphasize that the spectral observer is an algorithm that, like the STFT, can be used to perform a frequency-time analysis of a frequency varying signal by computing the Fourier Transform (FT) during time intervals. However, there are some differences to remark. (a) The STFT requires operations with complex numbers, the spectral observer does not; (b) The spectral observer computes the FT and its inverse at the same time, which is a clear bonus, because in case of using a recursive STFT we only get the FT, if we want to recover the reconstructed signal, we must compute the Inverse Short Fourier Transform.

6. Conclusions

In this paper, we presented the design of a novel spectral observer, which can be used to approximate periodical and non-periodical signals via state estimation. To design the spectral observer, we constructed a synthetic system in state space representation from the Fourier series. We presented some application examples to reconstruct periodical signals but also a well-know non-periodical one such as the price of the Bitcoin from its genesis. Some important aspects were not discussed in this article that require a deeper analysis, such as a comparison between the computational burden of the spectral observer and that of the Fourier Transform or an analysis of the spectral observer vis-à-vis perturbations and noise. These aspects must be treated in a continuation of this research work.

Author Contributions: Lizeth Torres conceived the spectral observer presented in this article. Javier Jiménez-Cabas and José Francisco Gómez-Aguilar conceived, designed and performed the simulation tests. Pablo Pérez-Alcazar advised the rest of the authors. All the authors wrote the paper.

Conflicts of Interest: The authors declare no conflict of interest.

Appendix A. MATLAB CODES

Appendix A.1. Symbolic Computation of Matrix A_ω

```
syms  w t
n=10; %Order of the Fourier Series
for k=1:2*n;
Aw(k,k)=((-1)^((mod(k-1,4)-mod(k-1,2))/2))*w^(k-1);
end
```

Appendix A.2. Symbolic Computation of Matrix A_ω

```
syms  w t
n=10; %Order of the Fourier Series
for k=1:n
for m=1:n
Ak(2*m-1,2*k-1)=k^(2*m-2);
Ak(2*m,2*k)=k^(2*m-1);
end
end
```

Appendix A.3. Symbolic Computation of Matrix Ω

```
syms  w t
n=10; %Order of the Fourier Series
for k=1:n
for m=2:2*n
O(1,(2*k)-1)=cos(k*w*t);
O(1,(2*k))=sin(k*w*t);
```

```
O(m,(2*k)-1)=diff(O(m-1,2*k-1),'t');
O(m,(2*k))=diff(O(m-1,2*k),'t');
end
end
```

References

1. Hostetter, G.H. Recursive discrete Fourier transformation. *IEEE Trans. Acoust. Speech Signal Proc.* **1980**, *28*, 184–190. [CrossRef]
2. Bitmead, R.R.; Tsoi, A.C.; Parker, P.J. A Kalman filtering approach to short-time Fourier analysis. *IEEE Trans. Acoust. Speech Signal Proc.* **1986**, *34*, 1493–1501. [CrossRef]
3. Orosz, G.; Sujbert, L.; Peceli, G. Spectral observer with reduced information demand. In Proceedings of the 2008 IEEE Instrumentation and Measurement Technology Conference, Victoria, BC, Canada, 12–15 May 2008; pp. 2155–2160.
4. Na, J.; Yang, J.; Wu, X.; Guo, Y. Robust adaptive parameter estimation of sinusoidal signals. *Automatica* **2015**, *53*, 376–384. [CrossRef]
5. Bitmead, R.R. On recursive discrete Fourier transformation. *IEEE Trans. Acoust. Speech Signal Proc.* **1982**, *30*, 319–322. [CrossRef]
6. Torres, L.; Gómez-Aguilar, J.; Jiménez, J.; Mendoza, E.; López-Estrada, F.; Escobar-Jiménez, R. Parameter identification of periodical signals: Application to measurement and analysis of ocean wave forces. *Digit. Signal Proc.* **2017**, *69*, 59–59. [CrossRef]
7. Dash, P.K.; Khincha, H. New algorithms for computer relaying for power transmission lines. *Electr. Mach. Power Syst.* **1988**, *14*, 163–178. [CrossRef]
8. Houmb, O.; Overvik, T. Some applications of maximum entropy spectral estimation to ocean waves and linear systems response in waves. *Appl. Ocean Res.* **1981**, *3*, 154–162. [CrossRef]
9. Blödt, M.; Chabert, M.; Regnier, J.; Faucher, J. Mechanical load fault detection in induction motors by stator current time-frequency analysis. *IEEE Trans. Ind. Appl.* **2006**, *42*, 1454–1463. [CrossRef]
10. Benbouzid, M.E.H.; Vieira, M.; Theys, C. Induction motors' faults detection and localization using stator current advanced signal processing techniques. *IEEE Trans. Power Electron.* **1999**, *14*, 14–22. [CrossRef]
11. Busawon, K.; Farza, M.; Hammouri, H. A simple observer for a class of nonlinear systems. *Appl. Math. Lett.* **1998**, *11*, 27–31. [CrossRef]
12. Engelberg, S. Edge detection using Fourier coefficients. *Am. Math. Mon.* **2008**, *115*, 499–513. [CrossRef]
13. Liénard, A. Étude des oscillations entretenues. *Revue Générale de l'Électricité* **1928**, *23*, 901–912.
14. Torres, L.; Verde, C.; Vázquez-Hernández, O. Parameter identification of marine risers using Kalman-like observers. *Ocean Eng.* **2015**, *93*, 84–97. [CrossRef]
15. Hammouri, H. Uniform observability and observer synthesis. In *Nonlinear Observers and Applications*; Besançon, G., Ed.; Springer: Berlin, Germany, 2007; pp. 35–70.
16. Zhivomirov, H. Short-Time Fourier Transformation (STFT) with Matlab Implementation. Available online: https://www.mathworks.com/matlabcentral/fileexchange/45197-short-time-fourier-transformation--stft--with-matlab-implementation (accessed on 27 April 2018).
17. Kušljević, M.D.; Tomić, J.J. Multiple-resonator-based power system Taylor-Fourier harmonic analysis. *IEEE Trans. Instrum. Meas.* **2015**, *64*, 554–563. [CrossRef]

Mathematical and Computational Applications

MDPI

Article

Differential Evolution Algorithm for Multilevel Assignment Problem: A Case Study in Chicken Transportation

Sasitorn Kaewman [1,*], Tassin Srivarapongse [2], Chalermchat Theeraviriya [3] and Ganokgarn Jirasirilerd [4]

[1] Faculty of Informatics, Mahasarakham University, Maha Sarakham 44150, Thailand
[2] Department of Economics, Rajamangala University of Technology Thanyaburi, Patumthani 12110, Thailand; tassin66@hotmail.com
[3] Faculty of Engineering, Nakhon Phanom University, Nakhon Phanom 48000, Thailand; chalermchat.t@npu.ac.th
[4] Faculty of Liberal Arts and Sciences, Sisaket Rajabhat University, Sisaket 33000, Thailand; ganokgarn.kung@gmail.com
*   Correspondence: sasitorn.k@msu.ac.th; Tel.: +66-89-861-6143

Received: 12 September 2018; Accepted: 30 September 2018; Published: 2 October 2018

Abstract: This study aims to solve the real-world multistage assignment problem. The proposed problem is composed of two stages of assignment: (1) different types of trucks are assigned to chicken farms to transport young chickens to egg farms, and (2) chicken farms are assigned to egg farms. Assigning different trucks to the egg farms and different egg farms to the chicken farms generates different costs and consumes different resources. The distance and the idle space in the truck have to be minimized, while constraints such as the minimum number of chickens needed for all egg farms and the longest time that chickens can be in the truck remain. This makes the problem a special case of the multistage assignment (S-MSA) problem. A mathematical model representing the problem was developed and solved to optimality using Lingo v.11 optimization software. Lingo v.11 can solve to optimality only small- and medium-sized test instances. To solve large-sized test instances, the differential evolution (DE) algorithm was designed. An excellent decoding method was developed to increase the search performance of DE. The proposed algorithm was tested with three randomly generated datasets (small, medium, and large test instances) and one real case study. Each dataset is composed of 12 problems, therefore we tested with 37 instances, including the case study. The results show that for small- and medium-sized test instances, DE has 0.03% and 0.05% higher cost than Lingo v.11. For large test instances, DE has 3.52% lower cost than Lingo v.11. Lingo v.11 uses an average computation time of 5.8, 103, and 4320 s for small, medium and large test instances, while DE uses 0.86, 1.68, and 8.79 s, which is, at most, 491 times less than Lingo v.11. Therefore, the proposed heuristics are an effective algorithm that can find a good solution while using less computation time.

Keywords: assignment problem; chicken transportation; differential evolution algorithm; mathematical model

---

1. Introduction

Thailand has long been known as a farming country due to the influence of Southeast Asian monsoons, which make the landscape, resources, environment, and climate conducive to agriculture. Most of the population works in agriculture or is involved in it in some manner. Although there have been efforts to develop Thailand into an industrialized country, it still largely depends on agriculture. The evolution and development of Thai agriculture has changed over time, reflecting the worldwide flow of changes. Hens are considered to be economic animals, as farmers can generate revenue from them throughout the year. In Thailand, there is a nationwide demand for eggs, which are popular among consumers. Breeding hens, therefore, are important to the economic balance and well-being of the Thai people. However, the problem that most farmers face is that they have high operating costs and generate little profit, thus they struggle to afford the costs incurred for labor, raw materials, transportation, etc.

In recent years, the logistics cost in Thailand is 14% of gross domestic product (GDP), which is valued at more than 1912.9 million baht. Many Thai businesses face the problem of high logistics cost, especially in agricultural businesses. In this study, we focus on reducing the cost of chicken transportation, which is one of the most important business groups of the Thai agricultural industry. Chicken transportation starts from delivering young chickens to egg farms in many different areas, which affects the resource usage of trucks (road conditions). It is not possible to travel on some Thai roads with some types of trucks, and even with some types of trucks the fuel consumption is different from the average speed limit. A suitable assignment approach needs to be decided on so that the total cost is minimized. Moreover, suitable trucks should be assigned to suitable chicken farms. Suitable farms and trucks means the truck does not have much idle space during delivery. Too much idle space means the use of bigger trucks, and bigger trucks always consume more fuel than smaller trucks because the weight of the truck affects the fuel consumption rate.

In transporting chickens from farmers to buyer farms, a number of factors must be taken into account—such as the mode of transportation, time to transport, temperature, etc.—as well as making sure that chickens from different farms are not mixed during transport. Therefore, it essential to find an appropriate vehicle that meets the needs of chicken farmers to avoid mixing chickens during transport to customers. This would help to minimize assignment or production costs and would be beneficial to the chicken farms, resulting in lower production costs and higher quality chickens for the egg farms. Thus, the overall condition of the chicken industry would be improved. Furthermore, chicken farms and egg farms can use the money saved to accomplish other farm activities, such as feeding, vaccinating, or researching—that is, to further develop their farms.

This study investigated a solution to the problem of chicken transport, which is a problem of multistate assignment. This was a case study regarding appropriate vehicle assignment for the transportation of chickens directly from chicken farms to egg farms by using the lowest cost of assignment. The differential evolution was developed to solve the problem because it is effective and uses short computation time. This paper has four contributions:

(1) The special case of the multistage assignment problem is proposed. The attribute that makes it a special case is that the experience of the workers and the type of shipping instrument will be considered in the assignment of trucks to farms. These affect the time to ship chickens in trucks and will affect the total time that chickens can be in the trucks during transport.

(2) The idle space in the truck is considered as the objective function and it is converted into lost opportunity due to higher fuel consumption of bigger trucks, therefore using unsuitable trucks will generate more cost.

(3) The new decoding method of the DE algorithm is presented so that the proposed problem can be solved.

(4) The mathematical model of the proposed problem is presented.

The paper is organized as follows. Section 2 presents the related research. Section 3 presents the problem statement and mathematical model. Section 4 shows the proposed heuristics (differential evolution algorithm). The computational result and a conclusion are presented in Sections 5 and 6, respectively.

2. Literature Review

Assignment problem (AP): The AP evolved from the transportation problem, which is a form of proper task assignment to an employee or machine considering cost effectiveness or profit maximization in some cases. The assignment problem is a type of combinatorial problem. It has been addressed as a transportation problem, where transportation affects other jobs. The aim is to minimize the total cost of transportation. Therefore, this problem could be considered a task assignment problem [1]. The key condition of this problem is that the assignments must be on a one-on-one basis; that is, once a task has been assigned to an employee, it cannot be assigned to another employee as well.

Previously, task assignment problems were resolved by bipartite matching. This matching method was proposed by Frobenius and Konig [2,3]. Later, Dantzig [4] presented the assignment problem in the form of a linear programming problem and used the simplex method to solve it. However, there might be limitations on the size of the problem, the number of variables, and the number of constraint equations. This means that if the limitations were too great or the computer was not sufficiently capable, the simplex method would not work. Kuhn [5] subsequently presented a solution to assigning tasks through the Hungarian method, which is a quick way to solve problems compared to simplex.

Generalized assignment problem (GAP): GAP is more flexible than AP. Unlike AP, where assignments are made on a one-on-one basis, with GAP, one task can be assigned to multiple employees or to the same employee. However, it must not exceed the capacity of the employee. Thus, GAP is more comprehensive and more closely resembles the actual situation than AP. GAP was first proposed by Ross and Soland [6] and was found to be an NP-hard problem [7]. This is often corrected by the exact method with small problems. The problem of 200 tasks and 20 employees was considered to be the biggest problem that could be solved by the exact method [8]. Therefore, the heuristics method has been used to solve GAP.

The proposed problem is the multistage GAP. The GAP has no restrictions that the case study has, such as (1) the longest traveling time; (2) the effect of the shipping instrument and the experience of the workers, which affects the operating cost; (3) the idle space in the truck is considered as the objective function; (4) at least half of the demand of the egg farm has to be considered; and (5) the multistage GAP was never found in the literature.

We will move on to review the methodology to solve the special multistage assignment (S-MSA) problem. Many worldwide methods have been used to solve GAP. Both exact methods—such as branch and bound, branch and price, etc.—and heuristics methods have been used to solve the problem. Metaheuristics is one of the most commonly used methods to solve GAP. Metaheuristics is an approximation method where it is not guaranteed that the solution optioned from the method is the optimal solution. The advantage is that it uses much less computation time than the exact method, which makes it possible to solve real-world problems that cannot be solved by exact methods.

The well-known metaheuristics are the krill herd (KH) algorithm [9–11], the cuckoo algorithm [12,13], the monarch butterfly optimization (MBO) [14,15], the hybridizing harmony search [16], the Lévy-flight krill herd algorithm [17], the bat algorithm [18], elephant herding optimization (EHO) [19], and the earthworm optimization algorithm (EWA) [20].

Enhancing the quality of the heuristics method can be achieved in many ways, such as adjusting suitable parameters for the proposed method, increasing the search area by introducing ways to move from local optimal, and applying local search to increase the intensive search of the method [21–25]. The most recent successful method was proposed by Wang [26]. The method is called the krill herd (KH) algorithm. Many papers have proposed improving the solution quality of the KH, such as adding new attributes to the algorithm [22], using a hybrid KH with other methods [23–26], exchanging information between top krill during the motion calculation process [27], using the best parameters [28], and adding local searches to improve search ability [29]. Aside from being applicable to many types of problems, KH is valid for function optimization [30] as well. An excellent review of the KH method has been proposed by Wang et al. [31].

Metaheuristics has been applied to many combinatorial optimization and real-world problems, such as the parallel hurricane optimization algorithm [32], firefly-inspired krill herd (FKH) [33], the moth search (MS) algorithm [34], monarch butterfly optimization [35–37], across neighborhood search (ANS) [38], chaotic particle-swarm krill herd (CPKH) [39], chaotic cuckoo search (CCS) [40], self-adaptive probabilistic neural network [41], and the differential evolution (DE) algorithm [42].

Differential evolution is a frequently used heuristics method. It is a way to apply the principles of evolution, and the steps are as follows: (1) initial solution, (2) mutation, (3) recombination, and (4) selection. Differential evolution was first used by Storn and Price [43] and has since been used since to solve many problems, such as production scheduling [44] and manufacturing problems [45]. Liao et al. [46] proposed two hybrid DEs to obtain truck sequences for cross-docking operations. Later, Liao et al. [47] proposed six metaheuristic algorithms for sequencing inbound trucks for multi-door cross-docking operations under a fixed schedule of outbound truck departure. Hou [48] proposed discrete DE (DDE) by modifying a mutation operator for the vehicle routing problem (VRP) with simultaneous pickups and deliveries (VRPPD). Recently, Dechampai et al. [49] proposed DE to solve the capacitated VRP with the flexibility of mixing pickup and delivery services and maximum duration of a route in the poultry industry. Sethanan and Pitakaso [50] improved DE by adding two more steps, reincarnation and survival process, to improve its intensification search [51]. It has been proven that using different pairs of mutation and recombination processes gives different solution qualities, such as in Pitakaso and Sethanan [50] and Boon et al. [52]. Sethanan and Pitakaso [50,51] suggested that improving the DE algorithm, such as by adding more steps to the original DE, inserting local search, and adding more attributes, can improve the solution quality, but the design of the decoding method must make sure that the most important rules in solution quality are obtained. In this paper, the excellent design of a decoding method from real numbers obtained from the DE mechanism to find the solution of the proposed problem is reported. The decoding method not only makes it possible to obtain the solution, but also local search has been added to the decoding method routine. Therefore, excellent results can be gained from the proposed method.

3. Problem Statement and Mathematical Modeling

This section explains the problem statement and the mathematical model to represent the problem so that the reader has more understanding of the proposed problem.

3.1. Problem Statement

Figures 1 and 2 represent the unsolved problem and the solved problem, respectively. Chicken farms have different amounts of young chickens. The chicken farms transport the chickens to egg farms, which grow the chickens and collect the eggs to sell to end customers. The egg farms have different capacities and demands for young chickens from chicken farms. The transportation of chickens needs to use trucks. The chicken farms have workers and loading/unloading instruments such as forklifts, carts, etc. This can make for different delivery times for the chicken farm in different types of trucks. When the truck arrives at the egg farm, the egg farm also has instruments for unloading chickens from the truck, which causes different shipping times. The objective function is not only to minimize the total distance of assignment under many constraints, which will be explained later, but also to minimize the idle space of the truck when transporting chickens. The idle space is assumed to be a lost opportunity to use the suitable truck for the suitable route. Moreover, it is possible that some trucks are not used. Therefore, the problem comprises the following components:

(1) Assign the right truck to the right egg farm so that free space is minimized. Different farms have different levels of experience with different trucks, which can affect the loading of chickens onto the truck, which can affect the total time that the chickens are in transport.
(2) Assign the right egg farm to the right chicken farm to minimize so travel distance. The workers at the chicken farm and the egg farm need to be coordinated so that shipping chickens out on the trucks uses as little time as possible, therefore the quality of the chickens remains unchanged.

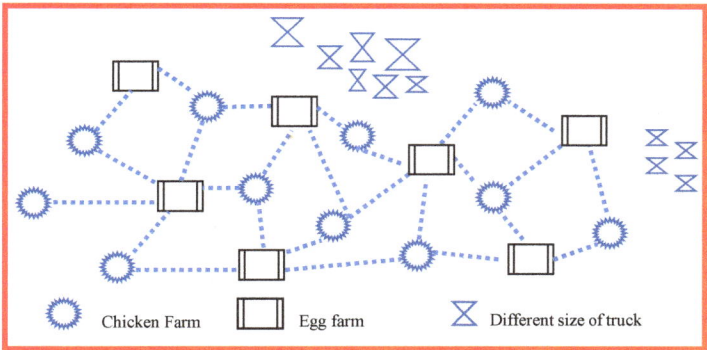

Figure 1. Unsolved problem.

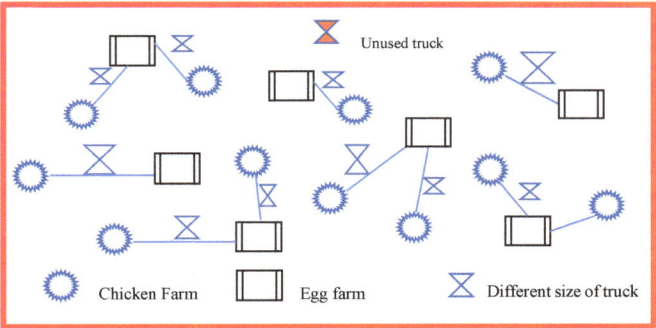

Figure 2. Solved problem.

The problem is a special case of the multistage assignment problem (S-MAP). The objective function is to minimize total distance and minimize free space in the truck. This objective needs to be optimized while keeping the following limitations or constraints:

(1) Transportation of chickens from chicken farms to egg farms is done only by direct transport. There is no chicken picking from other farms and no eggs are sent to other farms. This is because egg farms need to control the quality and breed of the chickens and prevent communicable diseases.
(2) A chicken farm can send chickens to an egg farm more than once.
(3) A chicken farm can sell all of its chickens.
(4) Egg farms may receive chickens from more than one farm, but they may not exceed the capacity of such farms.
(5) Egg farms will receive chickens for at least 50% of the demand. However, some egg farms may not be able to do this. Transportation time should not exceed 8 h, which includes the time spent loading the chickens onto the vehicle, transfer, and removal.
(6) The vehicles used for transport are sufficient for the needs.
(7) Chicken farms may use more than one type of vehicle and each type can be used for more than one round.

There are 40 chicken farms in the case study, and the capacity of a chicken farm is from 5000 to 20,000 chickens per planning period. There are four types of truck available, which have a capacity of 2000 to 12,000 chickens per round of travel. There are 60 egg farms in the system, for which chicken farms need to deliver at least 50% of the demand.

3.2. Mathematical Modelling

The S_MAP can be modeled as follows:

Indices

$i = 1, 2, 3, 4$ (4 truck types)
$j = 1, 2, 3, \ldots , J$ (J number of chicken farms; 40 farms)
$k = 1, 2, 3, 4$ (4 rounds of transportation)
$l = 1, 2, 3, \ldots , L$ (L number of egg farms; 60 farms)

Decision Variables

x_{ijkl}	Decision variable
$x_{ijkl} = 1$	Truck (i) is assigned to transport chickens from chicken farm (j) on the round of transporting (k) to egg farm (l)
$x_{ijkl} = 0$	Truck (i) is assigned to transport chickens from chicken farm (j) on the round of transporting (k) to egg farm (l)
q_{ijkl}	Number of chickens transported by truck (i) from chicken farm (j) on round (k) to egg farm (l)

Parameters

c_{ijl}	Cost of truck type assignment (i) to transport chickens from farm (j) to egg farm (l)
t_i	Capacity of truck (i) to transport chickens (unit: chicken)
oc_{ijl}	Cost of lost opportunity due to truck not completely loaded (i) from chicken farm (j) to egg farm (l) (unit: baht per chicken)
d_l	Demand for chickens by egg farm (l)
s_j	Capacity of chicken breeding of farm (j)
tup_{ij}	Time spent loading chickens onto truck (i) at chicken farm (j)
zup_{ij}	Performance of employees at chicken farm (j) to carry chickens into truck (i)
tdn_{il}	Time spent removing chickens from truck (i) at egg farm (l)
zdn_{il}	Performance of employees at egg farm (l) to remove chickens from truck (i)
ttr_{ijkl}	Time spent traveling from the truck station to chicken farm (j) then to egg farm (l), and then from egg farm (l) back to the truck station
td_{jl}	Time spent for truck (i) to go from the station to chicken farm (j) to convey chickens on round (k) to egg farm (l)
$time_{jkl}$	Time spent for an assignment of each round, not to exceed 8 h, consisting of the time to load chickens at chicken farm (i) $\left(tup_{ijkl}\right)$, time to transport from farm (i) to egg farm (l) $\left(ttr_{ijkl}\right)$, and time to remove chickens from the truck at egg farm (l) $\left(tdn_{ijkl}\right)$, or the overall time to use the truck
$ttruck_i$	Usage time of truck type

Remark 1. *The time for ttr_{ijkl} and td_{ijkl} does not include the usage time to load and remove chickens from the truck.*

Objective Function

$$min\sum_{i}^{I}\sum_{j}^{J}\sum_{k}^{K}\sum_{l}^{L}\left(c_{ijl} \times x_{ijkl}\right) + \underbrace{\sum_{i}^{I}\sum_{j}^{J}\sum_{k}^{K}\sum_{l}^{L}[\left(\left(x_{ijkl} \times t_i\right) - q_{ijkl}\right) \times oc_{ijl}]}_{Part2} \tag{1}$$

$$x_{ijkl} \in \{0,1\} \quad for \quad \forall ijkl \tag{2}$$

$$q_{ijkl} \in \{0,1\} \quad for \quad \forall ijkl \tag{3}$$

$$\sum_l^L x_{ijkl} \leq 1 \quad for \quad \forall ijk \tag{4}$$

$$\sum_i^I \sum_j^J \sum_k^K q_{ijkl} \leq d_l \quad for \quad \forall l \tag{5}$$

$$\sum_i^I \sum_j^J \sum_k^K q_{ijkl} \geq 0.5 \times d_l \quad for \quad \forall l \tag{6}$$

$$\sum_i^I \sum_k^K \sum_l^L q_{ijkl} = s_j \quad for \quad \forall j \tag{7}$$

$$q_{ijkl} \leq M \times x_{ijkl} \quad for \quad \forall ijkl \tag{8}$$

$$q_{ijkl} \leq t_i \quad for \quad \forall ijkl \tag{9}$$

$$tup_{ij} = zup_{ij} \times q_{ijkl} \quad for \quad \forall ijkl \tag{10}$$

$$tdn_{il} = zdn_{il} \times q_{ijkl} \quad for \quad \forall ijkl \tag{11}$$

$$ttr_{ijkl} = td_{jl} \times x_{ijkl} \quad for \quad \forall ijkl \tag{12}$$

$$time_{ijkl} = tup_{ijkl} + tdn_{ijkl} + ttr_{ijkl} \quad for \quad \forall ijkl \tag{13}$$

$$time_{ijkl} \leq 8 \quad for \quad \forall ijkl \tag{14}$$

$$\sum_j^J \sum_k^K \sum_l^L time_{ijkl} \leq ttruck(i) \quad for \quad \forall i \tag{15}$$

$$x_{ijkl} \geq x_{ij(k+1)l} \quad for \quad \forall ijkl \tag{16}$$

This mathematical model was created to solve multistage assignment problems. The objective function aims to investigate how to assign the task in order to gain the lowest cost of the assignment, for which distance is the key factor (part1 of Equation (1)). The cost of opportunity loss is due to the truck not being completely loaded. This means that there is free space on the truck for each round of transporting (part2 of Equation (1)).

The relevant limitations begin from the decision variables $\left(x_{ijkl} \right)$, which are the counting numbers and equal to 0 or 1 only (Equation (2)). The number of chickens prepared for transport $\left(q_{ijkl} \right)$ must be equal in the positive integer only (Equation (3)). Egg farm (i) can receive chickens from farm (j) by using truck (i) for transportation round (k) only one time (Equation (4)). Epidemics may be prevented because chickens are transported from different places, and an egg farm (l) may not receive as many chickens as it needs (Equation (5)). Each egg farm (i) will receive at least 50% of the chickens it needs (Equation (6)). Each chicken farm (j) will be able to deliver all chickens to the egg farm until there are no more chickens at the farm (Equation (7)). If there is no assignment $\left(x_{ijkl} = 0 \right)$, the number to transport is equivalent to 0 $\left(q_{ijkl} = 0 \right)$ as well (Equation (8)).

The quantity to be transported (q) in each round shall not exceed the capacity of the truck (Equation (9)). The time spent loading chickens onto the truck is represented by Equation (10). The quantity of chickens to be transported (q), the performance of the workers in loading chickens (zup), and the time taken to remove chickens from the truck are represented by Equation (11). The quantity of chickens to be transported (q), the performance of the workers in removing chickens from the truck (zdn), and the time spent on transportation $\left(ttr_{ijkl} \right)$ from the station to the chicken farm, and then from the chicken farm to the egg farm, and finally getting back to the station, which does not include time for loading and removing chickens, depends on distance and driving speed (Equation (12)). Equation (13) is the overall time for an assignment. The time spent loading chickens (tup), time spent traveling (*ttrans*), and time spent removing chickens from the truck (*tdn*) must not exceed 8 h (Equation (14)). The usage time of each type of truck may not exceed the working hours that have been set (Equation (15)). Equation (16) fixes a round of assignments starting at the first round.

4. Proposed Algorithm

The proposed method that is used to solve the S-MSA is the differential evolution (DE) algorithm. DE comprises four steps: (1) initialize the target vectors, (2) perform the mutation process, (3) perform the recombination process, and (4) perform the selection process. The proposed method can be explained as follows.

4.1. Initial Vector (Initial Population)

This step requires defining the number of the population and the random number of the population. The population must be at least four, because the DE process uses three vectors to determine the direction of the searching. If the population is small, it will not spread to find the answer. When the initial population is determined, it will be entered into the encoding process as the next DE method. The end product of this step is the first target vector.

The initialization needs to generate the vector representing the problem solution. We call it encoding. The encoding is the coding for three factors, such as the type of truck, chicken farm, and egg farm. In this case study, there are five vectors and each vector has four positions. This process starts by specifying a random number for each position in each vector, for which the random numbers are equivalent between 0 and 1, as shown in Tables 1–3 respectively.

Table 1. Initial target vectors of truck types.

Position	Vectors of Truck Types				
	1	2	3	4	5
1	0.853	0.956	0.886	0.521	0.545
2	0.335	0.397	0.639	0.863	0.471
3	0.757	0.391	0.519	0.098	0.443
4	0.967	0.293	0.063	0.595	0.824

Table 2. Initial target vectors of chicken farms.

Position	Vectors of Chicken Farms				
	1	2	3	4	5
1	0.234	0.713	0.396	0.417	0.806
2	0.257	0.082	0.696	0.829	0.091
3	0.512	0.979	0.186	0.475	0.502
4	0.164	0.780	0.873	0.522	0.671

Table 3. Initial target vectors of egg farms.

Position	Vectors of Egg Farms				
	1	2	3	4	5
1	0.544	0.766	0.067	0.171	0.634
2	0.004	0.045	0.313	0.300	0.042
3	0.564	0.436	0.051	0.498	0.485
4	0.499	0.250	0.263	0.178	0.396

From the random number of the position in each initial vector, the initial vectors of truck type, chicken farm, and egg farm are sorted by random numbers. The lowest of the random numbers is set to be in the first position. The higher random numbers will be set in the last row of each column, as shown in Tables 4–6.

Table 4. Positions of initial vectors of truck types.

Position	Vectors of Truck Types				
	1	**2**	**3**	**4**	**5**
1	3	4	4	2	3
2	1	3	3	4	2
3	2	2	2	1	1
4	4	1	1	3	4

Table 5. Positions of initial vectors of chicken farms.

Position	Vectors of Chicken Farms				
	1	**2**	**3**	**4**	**5**
1	2	2	2	1	4
2	3	1	3	4	1
3	4	4	1	2	2
4	1	3	4	3	3

Table 6. Positions of initial vectors of egg farms.

Position	Vectors of Egg Farms				
	1	**2**	**3**	**4**	**5**
1	3	4	2	1	4
2	1	1	4	3	1
3	4	3	1	4	3
4	2	2	3	2	2

For this study, first, vectors of truck type, chicken farm, and egg farm were assigned. The lowest random number, or position no. 1, was selected as the first order, as shown in Table 7.

Table 7. Sequence of position in vectors of truck type, chicken farm, and egg farm.

Position	Truck Type	Chicken Farm	Egg Farm
1	2	1	1
2	4	4	3
3	1	2	4
4	3	3	2

The vectors shown in Tables 1–4 need to be decoded to get the proposed problem's solution. This step is called decoding. Decoding is the sequential ordering in the assignment table, as shown in Table 8, to encode. For the operation of decoding, all constraints must be accomplished at the same time. For example, each type of truck cannot be overloaded and must not exceed the specified working hours. The working hours start from the time spent loading chickens at chicken farms. Travel time includes the time to travel from the station to the chicken farm and from the chicken farm to the egg farm, including the time to get back to the station and the time to remove chickens from the truck at the egg farm. The chicken farm can deliver all of the chickens without the rest of the chickens at the farm. The first assignment must be completed before another assignment can be given to the chicken farm. The egg farm will receive at least 50% of the demand in the beginning, after which more chickens will be added. The chicken farm in the first sequence usually receives the amount of chickens it demands. However, the farm in the latter sequence may not get the number of chickens it wants. All factors involved must be considered simultaneously. An important factor in task assignment is the quantity assigned each time, which will be affected by the type of truck and the number of rounds, which includes transportation costs. The amount of chickens to be transported ($AssignQ$) is determined by the number of chickens from one farm, including the demand from the egg farm to assign up to 50% of its demand before providing the rest. The transported chickens are counted from the quantity of chickens at the farm and the demands of the egg farm. If any quantity is less, it will be transported using such a quantity. The assignment is shown in Equation (17)

$$AssignQ = \begin{cases} Q_d^A \ if \ Q_d \leq Q_s \\ Q_s \ otherwise \end{cases} \tag{17}$$

$AssignQ$ means quantity to be transported, Q_d^A means the initial chicken demands of the egg farm equivalent to 50% of the entire demand, and Q_s means productivity of the chicken farm.

Table 8. Results of the decoding method.

Time	Truck		Round	Chicken Farm			Egg Farm		
	Type	Empty		Farm	Assignment	Remaining	Farm	Assignment	Remaining
1	2	3000	1	2	5000	0	1	5000	0
2	3	0	1	3	4000	6000	2	4000	0
3	3	1500	1	3	2500	3500	4	2500	0
4	3	500	1	3	3500	0	3	3500	1500
5	4	500	1	1	1500	8500	3	1500	0
6	2	3000	1	1	5000	3500	1	5000	0
7	3	500	1	1	3500	0	2	3500	500
8	4	1500	1	4	500	4500	2	500	0
9	3	1500	1	4	2500	2000	4	2500	0
10	4	0	1	4	2000	0	3	2000	3000

Another factor is the quantity to carry, which has a great influence on the type of truck selected. Choosing the proper vehicle will save transportation costs. If a truck is chosen that has less capacity than the amount it is required to carry, it will require several trips to complete the transport. This results in fuel consumption and increases transportation cost. If a truck is chosen that is larger than the amount to transport, there is free space, which raises the cost. Therefore, selecting the truck type should be based on a capacity that is larger than—but similar to—the quantity to be delivered, to avoid several rounds of transportation, which could cost more than a single trip with a large truck, as shown in Equation (18)

$$min \ TruckQ_k \geq AssignQ \tag{18}$$

$TruckQ_k$ means the capacity of truck type k; $k = 1, 2, 3$, and 4.

The assignment of chicken farms will be carried out in the order shown in Table 10, starting from the initial order farm since the first farm has been completely assigned. Then, another farm can be assigned when the all farms have been assigned and no chickens remain. The assignment for the egg farms is different, as they require at least 50% of their demands. After that, additional chickens will be provided. The chicken farm in the first sequence usually receives the amount of chickens it demands. However, the farm in the latter sequence may not get the chickens it wants. Once the assigned quantity has been determined, the type of vehicle will be assigned. Also, the second round of transport will take place when the same vehicle is assigned from the same chicken farm to the same egg farm until the requirement is met. With the initial decoding for an assignment to provide 50% of chickens to egg farms as the limitation, the first assignment is shown in Table 8, in which chicken farm no. 2 has been assigned as the first and has an order of 5000 chickens, while the first egg farm needs 10,000 chickens. Therefore, this egg farm will receive at least 50% of its demand (5000 chickens) from chicken farm no. 2. To avoid several rounds of transportation, trucks that have more capacity than the assigned quantity are used.

However, choosing a more capable truck will result in costs of lost opportunity. In order to avoid this cost, trucks should be selected with a capacity closest to the quantity to be commissioned. In the case of 5000 chickens, this will be assigned to a truck that has capacity for 8000 chickens, which also results in 3000 empty spaces and lost opportunity cost.

4.2. Perform Mutation Process

Mutation is a method of modifying the values in the vector position, which is a step to extend the scope for finding the answers. It starts from three random vectors from the initial population in the same group (truck type, chicken farm, and egg farm), combining the first vector with the difference from the other two to form a new vector. This principle is unique to differential evolution and can be expressed as

$$v_{i,j,G+1} = x_{r_1,j,G} + F(x_{r_2,j,G} - x_{r_3,j,G})$$ (19)

$x_{r_1,j,G}$ = target vector (r_1) of chicken farm group (j) acquired from random G population, for which there are three vectors, and the next vectors are $x_{r_2,j,G}$ and $x_{r_3,j,G}$.

$v_{i,j,G+1}$ = mutant vector, or the vector from the steps of modifying the values of the vector in population position (i) of vector group (j) for the new population $(G + 1)$.

F = mutant factor; F is equal to 0.8 (Qin et al., 2009), acquired from the experiment to find the optimal value.

$i = 1, 2, 3, \ldots, N$ (N means the number of population).

$j = 1, 2, 3$ means the vector from type of truck, chicken farm, and egg farm, respectively.

The target vectors of truck type, chicken farm, and egg farm, as shown in Tables 1–3 respectively, are taken into the modification of the vector position to obtain the mutant vector of the truck type (Table 9), of the chicken farm (Table 10), and of the egg farm (Table 11). Evaluating the mutant vector, $F = 0.8$ [53], this is a good starting point to randomly select three vectors from $X_{i,G}$ and then substitute them into Equation (19), which produces 0.853 + 0.8 (0.757–0.335), which is equal to 1.191. Then, it leads to the recombination of the vectors.

Table 9. Mutant vectors of truck type.

Position	Mutant Vectors of Truck Type				
	1	2	3	4	5
1	1.191	0.961	0.982	0.735	0.850
2	0.412	0.849	−0.019	0.525	0.553
3	0.848	0.308	0.321	0.157	0.384
4	0.629	0.745	0.159	1.207	0.906

Table 10. Mutant vectors of chicken farm.

Position	Mutant Vectors of Chicken Farm				
	1	2	3	4	5
1	0.512	0.872	−0.154	0.663	0.477
2	0.035	0.028	0.146	0.875	−0.017
3	0.438	0.474	−0.054	0.229	1.074
4	0.386	1.498	0.705	0.192	0.428

Table 11. Mutant vectors of egg farm.

Position	Mutant Vectors of Egg Farm				
	1	2	3	4	5
1	0.940	0.915	0.277	0.073	0.988
2	−0.012	−0.104	0.326	0.038	−0.029
3	0.168	−0.141	0.091	0.596	0.202
4	0.947	0.563	0.053	0.336	0.870

4.3. Perform the Recombination Process

The recombination process uses the mutant vector $(v_{i,j,G})$ and the target vector $(x_{i,j,G})$ to select once again to be a trial vector $(u_{i,j,G})$ by choosing only one vector. The process must not choose the same vector. The method can be expressed as

$$u_{i,j,G} = \begin{cases} v_{i,j,G} & when \quad CR \leq rand \\ x_{i,j,G} & when \quad CR > rand \end{cases} \qquad (20)$$

The recombination method aims to use the target vector from the first step to find the difference of vectors in the mutant vector of position but still allow the target vector to return to the process by setting the crossover rate (CR) to 0.2. If the random number for each vector is greater than or equal to a CR of 0.2, the mutant vector will be selected. If the random number is less than the CR, the target vector will be selected, as shown in Table 12.

Table 12. Random number values for each vector.

Type of Vector	Random Number of Vectors				
	1	2	3	4	5
Truck type	0.614	0.130	0.963	0.601	0.074
Chicken farm	0.141	0.229	0.764	0.147	0.622
Egg farm	0.697	0.603	0.158	0.744	0.796

Applying the random number obtained from Table 12 to the mutant vector demonstrates that most of the vectors taken into consideration in the vector selection procedure were the ones that already passed the mutant vector process. The selected vector of the truck type is shown in Table 13, that of the chicken farm is shown in Table 14, and that of the egg farm is shown in Table 15. Then, the vector selection procedure continues.

Table 13. Vectors after mutation of vector positions of truck type.

Position	Vectors of Truck Type				
	1	2	3	4	5
1	1.191	**0.956**	0.982	0.735	**0.545**
2	0.412	**0.397**	−0.019	0.525	**0.471**
3	0.848	**0.391**	0.321	0.157	**0.443**
4	0.629	**0.293**	0.159	1.207	**0.824**

Note: Target vectors are in bold.

Table 14. Vectors after mutation of vector positions of chicken farm.

Position	Vectors of Chicken Farm				
	1	2	3	4	5
1	**0.234**	0.872	−0.154	**0.417**	0.477
2	**0.257**	0.028	0.146	**0.829**	−0.017
3	**0.512**	0.474	−0.054	**0.475**	1.074
4	**0.164**	1.498	0.705	**0.522**	0.428

Note: Target vectors are in bold.

Table 15. Vectors after mutation of vector positions of egg farm.

Position	Vectors of Egg Farm				
	1	2	3	4	5
1	0.94	0.915	**0.067**	0.073	0.988
2	−0.012	−0.104	**0.313**	0.038	−0.029
3	0.168	−0.141	**0.051**	0.596	0.202
4	0.947	0.563	**0.263**	0.336	0.87

Note: Target vectors are in bold.

4.4. Perform the Selection Process

The selection process is the vector selection step in DE that compares the cost of assignment (fitness function) with the cost of the target and trial vectors from the mutation process.

$$x_{i,j,G+1} = \begin{cases} u_{i,j,G+1} & if f\left(u_{i,j,G+1}\right) \leq f\left(x_{i,j,G}\right) \\ x_{i,j,G} & otherwise \end{cases}$$

If the cost of the assignment from the trial vector is less than or equal to the cost of the assignment from the target vector, then the trial vector will be selected and collected for the next population. On the other hand, if the cost of the assignment given by the trial vector is greater than the target vector, then the target vector will be collected from this population to be the vector for further population. Repeat steps 2–6 until the best answer is acquired.

From the explanation in Sections 4.1–4.4, the proposed heuristics procedure is shown in Algorithm 1.

Algorithm 1. Pseudocode of the proposed heuristics.

Set NP, CR, F, NP (size of vector)
Generate initial solution
Begin
For G = 1 to G_{max}, where G = iterations and G_{max} = maximum iterations
For N = 1 to NP
Generate random target vector $X_{r_1,j,G}$ and RV and update BV
Produce mutant vector N (mutation process) (Equation (19))

$$v_{i,j,G+1} = x_{r_1,j,G} + F\left(x_{r_2,j,G} - x_{r_3,j,G}\right)$$

Produce trial vector N (recombination process)
- Using Equation (20):

$$u_{i,j,G} = \begin{cases} v_{i,j,G} & when \ CR \leq rand \\ x_{i,j,G} & when \ CR > rand \end{cases}$$

Produce new target vector (selection\process)

$$X_{i,j,G+1} = \begin{cases} U_{i,j,G} & if f\left(U_{i,j,G}\right) \leq f\left(X_{i,j,G}\right) \\ X_{i,j,G} & otherwise \end{cases}$$

End

---

5. Computational Framework and Result

The proposed heuristics were executed and compared with the solution generated by Lingo v.11. We reprogrammed the proposed heuristics in C++ and simulated it on a computer with Intel (R) Core i7-3520M CPU @ 2.90 GHz Ram 8.00 GB. We tested our algorithms with three groups of test instances, small, medium, and large. The simulation was executed five times until the best solution was selected, as shown in the table. Details of the test instances are shown in Table 16.

Table 16. Details of the test instances.

Group Test Instance	Number of Test Instances	Number of Chicken Farms	Number of Egg Farms	Number of Trucks	Compare Method	St
Small	12	5	5	5–10	Exact	it
Medium	12	10	10	10–20	Exact	it
Large	12	20	20	20–30	Lower bound	it
Case study	1	40	60	54	Lower bound	it

ST, stopping criteria; it, number of iterations; compare method, method that the proposed heuristic will be compared with.

From Table 16, we test our 37 tested instances, composed of 12 small, medium, and large instances and 1 case study. For small and medium test instances, the proposed method was compared with the exact method. The exact method used here is Lingo v.11. For the large instances and the case study, the proposed method was compared with the lower bound generated by Lingo v.11 within 72 h.

The first experiment was executed with the small and medium test instances. The stopping criterion for Lingo v.11 was when it found the optimal solution. The best solution and computation time were collected. The stopping criterion for DE was when it found the optimal solution (the same as Lingo v.11) or when it reached 500 iterations. The results are shown in Tables 17 and 18 for small and medium randomly generated datasets, respectively. The simulation was executed in 12 test instances, each of which had a size of 5 × 5 (number of egg farms × number of chicken farms). The best solutions out of five runs are shown in Tables 17 and 18.

Table 17. Results of small samples (5 × 5) showing cost and time of assignment.

Dataset	Lingo v.11		Differential Evolution		%Diff.
	Cost (Baht) (a)	Time (s)	Cost (Baht) (b)	Time (s)	
1	9723	6.9	9723	0.5	0.00
2	8753	2	8753	0.2	0.00
3	5330	10.7	5330	0.2	0.00
4	7056	4.8	7059	0.2	0.04
5	7317	4.7	7317	0.2	0.00
6	6098	8.9	6107	0.7	0.15
7	7004	9.6	7004	0.3	0.00
8	7649	9.4	7649	0.8	0.00
9	7761	4.8	7761	1.7	0.00
10	7894	12.8	7894	1.8	0.00
11	7566	13.7	7575	2.9	0.12
12	7683	20.8	7683	0.8	0.00
Average	**7486.17**	**9.09**	**7487.92**	**0.86**	**0.03**

Note: %diff. $= \frac{b-a}{a} \times 100\%$.

Table 18. Experimental results of the medium sample (10 × 10) showing cost of assignment and processing time.

Dataset	Lingo v.11		Differential Evolution		%Diff.
	Cost (Baht) (a)	Time (s)	Cost (Baht) (b)	Time (s)	
1	11,989	175	11,989	0.6	0.00
2	11,397	79	11,401	0.6	0.04
3	11,572	84	11,572	0.5	0.00
4	12,898	91	12,898	2.8	0.00
5	12,184	92	12,184	1.4	0.00
6	11,315	98	11,315	0.9	0.00
7	14,508	105	14,508	1.8	0.00
8	12,613	93	12,613	2.9	0.00
9	10,902	92	10,921	1.6	0.17
10	11,870	108	11,890	1.8	0.17
11	15,817	114	15,849	1.5	0.20
12	12,114	79	12,114	3.7	0.00
Average	**12431.58**	**100.83**	**12437.83**	**1.68**	**0.05**

Notes: % diff. $= \frac{b-a}{a} \times 100\%$.

Table 17 shows a small group of problem instances. In one out of five instances, the proposed heuristic could not find the optimal solution. The average gap (%diff.) of DE from the solution generated by Lingo v.11 was 0.03% and it used 10.57 times (9.09/0.86) less computation time. The simulation was executed for 12 test instances, each with a size of 10 × 10 (number of egg farms × number of chicken farms). The best solutions out of five runs are shown in Table 18.

Table 18 shows a medium-sized test, for which the sample size is 10 × 10. Lingo v.11 took 100.83 s on average to find a 0.05% better solution than DE, but DE used only 1.68 s computation time on average to find that solution.

The next experiment was executed with a large size of test instances. The stopping criterion of Lingo v.11 was 72 h or 4320 min. The best solution found within that time was collected to compare with the result generated by DE. The stopping criterion of DE was set at 1000 iterations. The solution is shown in Table 19. This group includes the case study (40 × 60).

Table 19. Experimental results for large samples (20 × 20) and the case study showing the cost of assignment and processing time.

Dataset	Lingo v.11		DE		%Diff.
	Cost (Baht) (a)	Time (min)	Cost (Baht) (b)	Time (min)	
1	33,249	4320	32,716	5.1	1.63
2	29,943	4320	29,094	5.1	2.92
3	37,672	4320	36,128	3.6	4.27
4	38,891	4320	37,781	4.8	2.94
5	39,781	4320	37,895	5.9	4.98
6	31,480	4320	30,084	11.2	4.64
7	58,984	4320	57,738	14.5	2.16
8	89,872	4320	87,573	15.8	2.63
9	90,164	4320	89,079	19.1	1.22
10	35,878	4320	34,871	11.5	2.89
11	29,095	4320	28,049	3.7	3.73
12	29,892	4320	29,152	3.7	2.54
Case study	125,593	4320	114,932	10.3	9.28
Average	**51,576.46**	**4320**	**49,622.46**	**8.79**	**3.52**

Note: %diff. $= \frac{a-b}{b} \times 100\%$.

Table 19 shows a large-scale problem test, for which the sample size was 20 × 20. The result generated by Lingo v.11 within 72 h had an average cost of 51,576.46 baht, while the result generated by DE was 49,622.46 baht, or 4.52% less, using 491 times less computation time. The comparison of Lingo v.11 and DE shown in Tables 17–19 was statistically tested using Wilcoxon signed-rank test with 95% confidence interval. The statistical test results are shown in Table 20.

Table 20. Statistical test results.

Problem Size	Significance Level		Critical Value		Result
	p-Value	W	*p*-Value	W	
Small	0.89812	17.5	0.05	8	Lingo V.11 = DE
Medium	0.65994	10	0.05	6	Lingo V.11 = DE
Large	0.00148	0	0.05	17	Lingo V.11 ≥ DE

From Table 20, we can see that in the small and medium groups, the performance of DE and Lingo v11 was not significant different, and for the large size, DE had significantly lower cost than Lingo v.11.

6. Conclusions and Suggestions

The purpose of resolving the multistage assignment problem is to minimize the cost of assignment. The case study in this research consisted of the main cost of transportation, which relies on the distance to transport as well as the cost of opportunity loss related to truck incapacity.

The resolution started from the development of a mathematical model that consisted of the cost of chicken transportation and opportunity loss. The model had to comply with the conditions as well. Then, the mathematical model was applied to find the best answer using Lingo v.11. Unfortunately, when the problem size is large, Lingo v.11 is not able to solve the problem into optimality, therefore the metaheuristics have to be further developed to get the solution for the case study and the large problem size.

Later, DE was developed to solve the multistage assignment problem, and the results of the efficiency of DE vs. Lingo v.11 were compared. In the case that Lingo v.11 can find the optimal solution, we compared the proposed heuristic (DE) with the optimal solution. The time Lingo v.11 used to find the optimal solution was recorded for all test instances. In this case, DE will use number of iterations (set to 500 from the preliminary test). The computation results show that in small- and medium-sized test instances, DE uses much less computation time than Lingo v.11 while obtaining less than a 1% cost difference (0.03% and 0.05%, respectively) from the optimal solution. In the large-sized problem instances, DE found a 3.52% better solution while using 491 times less computation time than Lingo v.11. Thus, we can see that the performance of DE is better when the problem size is larger. DE obtains better solutions than Lingo v.11 when it uses much more computation time. DE is suitable to solve big problems that an exact method like Lingo v.11 cannot solve.

From the computation results shown in Tables 18–20, we can see that when the problem size is small, Lingo v.11 can always find the optimal solution and DE sometimes has worse solution quality than Lingo v.11. This is the weak point of the proposed heuristics: in small- and medium-sized test instances, it cannot find the optimal solution even when we increase the iterations to 1000 or 1500. This means that, in small-sized test instances, DE converts fast and sticks on the local optimal. When there is a large size of test instances, DE can find a better solution than Lingo v.11, because when the problem size is large, it is hard for the exact method to solve to optimality, and when the computation time is set at 72 h, Lingo v.11 is not yet finished with the search while DE, the metaheuristic, can finish the search activity.

Future research should study more complicated assignment problems as well as the current problems or other metaheuristic methods to enhance solutions through hybrid methodologies. Algorithm designers need to add a search mechanism that allows the proposed solution to escape from the local optimal to the general mechanism of DE so that the ability to escape from the local optimal will be increased.

Author Contributions: S.S. designed the algorithm; T.S. and C.T. gathered data and programed the algorithm; G.J. made the summary and conclusion.

Funding: We would like to express our thanks to Mahasarakham University, Rajamangala University of Technology Thanyaburi, Ubon Ratchathani University, Nakhon Phanom University, and Sisaket Rajabhat University for funding this project.

Conflicts of Interest: The authors declare no conflict of interest.

References

1. Monge, G. Sur le CalculIntégraldes Équations Aux Differences Partielles. Available online: http://verbit.ru/MATH/TALKS/India/History-MA.pdf (accessed on 10 May 2016).
2. Frobenius, F.G. *Ferdinand Georg Frobenius; Gesammelte Abhandlungen, Band III*; Serre, J.-P., Ed.; Springer: Berlin, Germany, 1968.
3. Konig, D. Vonalrendszerek és determinánsok. *Mathenatikaies Termeszettudomanyi Ertesito* **1915**, *33*, 221–229.
4. Dantzig, G.B. Application of the simplex method to a transportation problem. In *Activity Analysis of Production and Allocation, Proceedings of Linear Programming, Chicago, Illinois, 1949*; Wiley: New York, NY, USA, 1951; pp. 359–373.
5. Kuhn, H.W. The Hungarian method for the assignment problem. *Nav. Res. Logist. Q.* **1956**, *2*, 83–97. [CrossRef]
6. Ross, G.T.; Soland, R.M. A branch and bound algorithm for the generalized Assignment problem. *Math. Program.* **1975**, *8*, 91–103. [CrossRef]
7. Fisher, M.L.; Jaikumar, R. A generalized assignment heuristic for vehicle routing. *Networks* **1981**, *11*, 109–124. [CrossRef]
8. Liu, L.; Mu, H.; Song, Y.; Luo, H.; Li, X.; Wu, F. The equilibrium generalized Assignment problem and genetic algorithm. *Appl. Math. Comput.* **2012**, *218*, 6526–6535. [CrossRef]
9. Wang, G.; Tan, Y. Improving Metaheuristic Algorithms with Information Feedback Models. *IEEE Trans. Cybern.* **2017**, *99*, 1–14. [CrossRef] [PubMed]
10. Wang, G.; Guo, L.; Gandomi, A.H.; Hao, G.; Wang, H. Chaotic Krill Herd algorithm. *Inf. Sci.* **2014**, *274*, 17–34. [CrossRef]
11. Wang, G.; Gandomi, A.H.; Alavi, A.H. An effective krill herd algorithm with migration operator in biogeography-based optimization. *Appl. Math. Model.* **2014**, *38*, 2454–2462. [CrossRef]
12. Cui, Z.; Sun, B.; Wang, G.; Xue, Y.; Chen, J. A novel oriented cuckoo search algorithm to improve DV-Hop performance for cyber-physical systems. *J. Parallel Distrib. Comput.* **2016**, *103*. [CrossRef]
13. Wang, G.; Alavi, A.H.; Zhao, X.; Hai, C.C. Hybridizing harmony search algorithm with cuckoo search for global numerical optimization. *Soft Comput.* **2016**, *20*, 273–285. [CrossRef]
14. Feng, Y.; Wang, G. Binary Moth Search Algorithm for Discounted {0-1} Knapsack Problem. *IEEE Access* **2018**. [CrossRef]
15. Wang, G.; Deb, S.; Cui, Z. Monarch Butterfly Optimization. *Neural Comput. Appl.* **2015**. [CrossRef]
16. Wang, G.; Guo, L.; Gandomi, A.H.; Cao, L.; Alavi, A.H.; Duan, H.; Li, J. Lévy-Flight Krill Herd Algorithm. *Math. Probl. Eng.* **2013**. [CrossRef]
17. Wang, G.; Guo, L.; Duan, H.; Wang, H.; Liu, L.; Shao, M. Hybridizing Harmony Search with Biogeography Based Optimization for Global Numerical Optimization. *J. Comput. Theor. Nanosci.* **2013**, *10*, 2312–2322. [CrossRef]
18. Wei, Z.J.; Wang, G. Image Matching Using a Bat Algorithm with Mutation. *Appl. Mech. Mater.* **2012**, *203*, 88–93. [CrossRef]
19. Wang, G.; Coelho, L.; Gao, X.Z.; Deb, S. A new metaheuristic optimisation algorithm motivated by elephant herding behavior. *Int. J. Bio-Inspir. Comput.* **2016**, *8*, 394. [CrossRef]
20. Wang, G.; Deb, S.; Coelho, L.D.S. Earthworm optimization algorithm: A bio-inspired metaheuristic algorithm for global optimization problems. *Int. J. Bio-Inspir. Comput.* **2018**, *12*, 1–12. [CrossRef]

21. Wang, G.; Guo, L.; Wang, H.; Duan, H.; Liu, L.; Li, J. Incorporating mutation scheme into krill herd algorithm for global numerical optimization. *Neural Comput. Appl.* **2014**, *24*, 853–871. [CrossRef]

22. Wang, G.; Gandomi, A.H.; Alavi, A.H. Stud krill herd algorithm. *Neurocomputing* **2014**, *128*, 363–370. [CrossRef]

23. Wang, G.; Gandomi, A.H.; Yang, X.; Alavi, A.H. A new hybrid method based on krill herd and cuckoo search for global optimisation tasks. *Int. J. Bio-Inspir. Comput.* **2016**, *8*, 286–299. [CrossRef]

24. Wang, H.; Yi, J.H. An improved optimization method based on krill herd and artificial bee colony with information exchange. *Memet. Comput.* **2017**, *10*, 177–198. [CrossRef]

25. Wang, G.; Gandomi, A.H.; Alavi, A.H.; Deb, S. A hybrid method based on krill herd and quantum-behaved particle swarm optimization. *Neural Comput. Appl.* **2016**, *27*, 989–1006. [CrossRef]

26. Wang, G.; Guo, L.; Gandomi, A.H.; Alavi, A.H.; Duan, H. Simulated Annealing-Based Krill Herd Algorithm for Global Optimization. *Abstr. Appl. Anal.* **2013**, *2013*, 213853. [CrossRef]

27. Guo, L.; Wang, G.; Gandomi, A.H.; Alavi, A.H.; Duan, H. A new improved krill herd algorithm for global numerical optimization. *Neurocomputing* **2014**, *138*, 392–402. [CrossRef]

28. Wang, G.; Gandomi, A.H.; Alavi, A.H. Study of Lagrangian and Evolutionary Parameters in Krill Herd Algorithm. In *Adaptation and Hybridization in Computational Intelligence. Adaptation, Learning, and Optimization*; Fister, I., Fister, I., Jr., Eds.; Springer: Cham, Switzerland, 2015.

29. Wang, G.; Gandomi, A.H.; Alavi, A.H.; Deb, S. A Multi-Stage Krill Herd Algorithm for Global Numerical Optimization. *Int. J. Artif. Intell. Tools* **2016**, *25*. [CrossRef]

30. Wang, G.; Deb, S.; Gandomi, A.H.; Alavi, A.H. A Opposition-based krill herd algorithm with Cauchy mutation and position clamping. *Neurocomputing* **2016**, *177*, 147–157. [CrossRef]

31. Wang, G.; Gandomi, A.H.; Alavi, A.H.; Gong, D. A comprehensive review of krill herd algorithm: Variants, hybrids and applications. *Artif. Intell. Rev.* **2017**. [CrossRef]

32. Rizk, M.; Rizk, A.; Ragab, A.; El-Sehiemy, R.A.; Wang, G. A novel parallel hurricane optimization algorithm for secure emission/economic load dispatch solution. *Appl. Soft Comput.* **2018**, *63*, 206–222. [CrossRef]

33. Wang, G.; Gandomi, A.H.; Alavi, A.H.; Dong, Y. A Hybrid Meta-Heuristic Method Based on Firefly Algorithm and Krill Herd. In *Handbook of Research on Advanced Computational Techniques for Simulation-Based Engineering*; IGI: Hershey, PA, USA, 2016; pp. 521–540. [CrossRef]

34. Wang, G. Moth search algorithm: A bio-inspired metaheuristic algorithm for global optimization problems. *Memet. Comput.* **2018**, *10*, 151–164. [CrossRef]

35. Feng, Y.; Wang, G.; Deb, S.; Lu, M.; Zhao, X. Solving 0–1 knapsack problem by a novel binary monarch butterfly optimization. *Neural Comput. Appl.* **2017**, *28*, 1619–1634. [CrossRef]

36. Feng, Y. Solving 0–1 knapsack problems by chaotic monarch butterfly optimization algorithm with Gaussian mutation. *Memet. Comput.* **2016**. [CrossRef]

37. Wang, G.; Deb, S.; Zhao, X.; Cui, Z. A new monarch butterfly optimization with an improved crossover operator. *Oper. Res.* **2016**, *3*, 731–755. [CrossRef]

38. Wu, G. Across neighborhood search for numerical optimization. *Inf. Sci.* **2016**, *329*, 597–618. [CrossRef]

39. Wang, G.; Gandomi, A.H.; Alavi, A.H. A chaotic particle-swarm krill herd algorithm for global numerical optimization. *Kybernetes* **2013**, *42*, 962–978. [CrossRef]

40. Wang, G.; Deb, S.; Gandomi, A.H.; Zhang, Z.; Alavi, A.H. A Fusion of Foundations, Methodologies and Applications. *Soft Comput.* **2016**, *20*, 3349–3362. [CrossRef]

41. Yi, J.; Wang, J.; Wang, G. Improved probabilistic neural networks with self-adaptive strategies for transformer fault diagnosis problem. *Adv. Mech. Eng.* **2016**, *8*. [CrossRef]

42. Storn, R.; Price, K. Differential evolution—A simple and efficient heuristic for global Optimization over continuous spaces. *J. Glob. Optim.* **1977**, *11*, 341–359. [CrossRef]

43. Pitakaso, R. Differential evolution algorithm for simple assembly line balancing type 1 (SALBP-1). *J. Ind. Prod. Eng.* **2015**, *32*, 104–114. [CrossRef]

44. Pitakaso, R.; Sethanan, K. Modified differential evolution algorithm for simple assembly line balancing with a limit on the number of machine types. *Eng. Optim.* **2015**, *48*, 253–271. [CrossRef]

45. López Cruz, I.L.; Van Willigenburg, L.G.; Van Straten, G. Optimal control of nitrate in lettuce by a hybrid approach: Differential evolution and adjustable control weight gradient algorithms. *Comput. Electron. Agric.* **2003**, *40*, 179–197. [CrossRef]

46. Liao, T.W.; Egbelu, P.J.; Chang, P.C. Two hybrid differential evolution algorithms for optimal inbound and outbound truck sequencing in cross docking operations. *Appl. Soft Comput.* **2012**, *12*, 3683–3697. [CrossRef]
47. Liao, T.W.; Egbelua, P.J.; Chang, P.C. Simultaneous dock assignment and sequencing of inbound trucks under a fixed outbound truck schedule in multi-door cross docking operations. *Int. J. Prod. Econ.* **2013**, *141*, 212–229. [CrossRef]
48. Hou, L.; Zhou, H.; Zhao, J. A novel discrete differential evolution algorithm for stochastic VRPSPD. *J. Comput. Inf. Syst.* **2010**, *6*, 2483–2491.
49. Dechampai, D.; Tanwanichkul, L.; Sethanan, K.; Pitakaso, R. A differential evolution algorithm for the capacitated VRP with flexibility of mixing pickup and delivery services and the maximum duration of a route in poultry industry. *J. Intell. Manuf.* **2015**, *28*, 1357–1376. [CrossRef]
50. Sethanan, K.; Pitakaso, R. Differential evolution algorithms for scheduling raw milk transportation. *Comput. Electron. Agric.* **2016**, *121*, 245–259. [CrossRef]
51. Sethanan, K.; Pitakaso, R. Improved differential evolution algorithms for solving generalized assignment problem. *Expert Syst. Appl.* **2016**, *45*, 450–459. [CrossRef]
52. Boon, E.T.; Ponnambalam, S.G.; Kanagara, G. Differential evolution algorithm with local search for capacitated vehicle routing problem. *Int. J. Bio-Inspir. Comput.* **2013**, *7*. [CrossRef]
53. Qin, A.K.; Huang, V.L.; Suganthan, P.N. Differential Evolution Algorithm with Strategy Adaptation for Global Numerical Optimization. *IEEE Trans. Evol. Comput.* **2009**, *13*, 398–417. [CrossRef]

Mathematical and Computational Applications

MDPI

Article

Modeling and Simulation of a Hydraulic Network for Leak Diagnosis

José-Roberto Bermúdez [1,2], Francisco-Ronay López-Estrada [1,*] ⓘ, Gildas Besançon [2],
Guillermo Valencia-Palomo [3,*] ⓘ, Lizeth Torres [4] ⓘ and Héctor-Ricardo Hernández [1]

[1] TURIX-Dynamics Diagnosis and Control Group, Tecnológico Nacional de México, Instituto Tecnológico de
    Tuxtla Gutiérrez, Tuxtla Gutiérrez 29050, Mexico; bermudez_r10@hotmail.com (J.-R.B.);
    hhernandezd@ittg.edu.mx (H.-R.H.)
[2] GIPSA-lab, CNRS, Grenoble INP, Université Grenoble Alpes, 38000 Grenoble, France;
    Gildas.Besancon@gipsa-lab.grenoble-inp.fr
[3] Tecnológico Nacional de México, Instituto Tecnológico de Hermosillo, Hermosillo 83170, Mexico
[4] Cátedras CONACYT, Instituto de Ingeniería, Universidad Nacional Autónoma de México,
    Ciudad de México 04510, Mexico; ftorreso@iingen.unam.mx
* Correspondence: frlopez@ittg.edu.mx (F.-R.L.-E.); gvalencia@ith.mx (G.V.-P.)

Received: 28 September 2018; Accepted: 4 November 2018; Published: 6 November 2018

Abstract: This work presents the modeling and simulation of a hydraulic network with four nodes
and two branches that form a two-level water distribution system. It also proposes a distribution of
hydraulic valves that allows emulating a leak using a valve and different network configurations,
e.g., simple ducts, closed networks and branched networks. The network is modeled in the steady
state considering turbulent flow. Numerical experiments are performed, and the results show that
the proposed network is useful for the design of leakage diagnosis and control algorithms in different
configurations and leakage scenarios.

Keywords: hydraulic networks; leaks in pipes; pipe model; leak diagnosis

1. Introduction

Safety and proper functioning of hydraulic networks are very important, since these systems are
used in the industrial and governmental sectors for the transportation of different types of fluids such
as gases, hydrocarbons and water. In this sense, one of the main problems in the distribution networks
of liquids are hydraulic leaks, where a hydraulic leak is defined as the uncontrolled output of fluid that
occurs in any section of the network. A leak can occur when there is corrosion on the inside or outside
of the pipe, or it can be caused by blows, by the theft of transported product, etc. [1]. Leaks have serious
consequences such as environmental pollution, economic losses and human deaths from hydrocarbon
explosions. Leaks occur more frequently in worn pipes, pipes with low maintenance, pipe joints or in
some accessories such as elbows and valves, among others [2].

Due to the strong environmental and social impact caused by leaks in water distribution systems,
it is necessary to reduce this problem [3], taking into account that the water demands increase as the
population multiplies, and water losses in highly populated cities increase in a high percentage as
well. The Organization for Economic Cooperation and Development (OECD) conducted a survey
in 42 cities with the highest leakage problem in their distribution networks. In the list, one can find
Mexican cities like Tuxtla Gutiérrez with a flow loss of 70%, San Luis Potosí with 50% and Mexico City
with more than 40%. Other important cities around the world are Paris with loses of 10%, Hong Kong
(China) with more than 15% and Liverpool with more than 20% [4]. These percentages of water losses
in distribution networks give evidence of the need to develop solutions in order to solve or at least
mitigate this problem.

Different works have been developed that focus on leak detection and control of hydraulic network systems. For example, Van Pham et al. [5] used a receding horizon optimal control method applied to the flow model of the network, guaranteeing the convergence of pressures in case of transients; Torres et al. [6] designed a state observer through redundant relationships in order to isolate sensor and actuator faults and unknown extractions of fluid; Wang et al. [7] used a robust predictive control algorithm to control pressures and flows in the demanding nodes in a small network. In order to validate these algorithms, it is necessary to test them experimentally in laboratory plants that can reproduce, at scale, the behavior of real hydraulic networks. In this context, different pilot plants have been built for experimentation with emphasis on hydraulic leaks. For example, at the laboratory of the Engineering Institute of UNAM(Mexico), a hydraulic network was designed with a vertical coil arrangement where leak detection and localization algorithms have been developed using model- and data-based techniques using monitoring software in Labview® and MATLAB® for flow and pressure analysis [8,9]. In the hydraulic laboratory at CINVESTAVcampus Guadalajara (Mexico), the pilot plant has a rectangular geometry, and it has also enabled the development of leak detection algorithms with model-based methods [10,11]. GIPSA-lab (France) and the University of Catalonia (Spain) also have facilities enabling the experimentation with hydraulics networks [7,12], and the list can continue. On the other hand, when there are no pilot plants available or the physical extension of the hydraulic network is prohibitive for a laboratory, it is necessary to use specialized software to emulate the behavior of the network. One of these software programs is EPANET®, which is a public domain, water distribution system modeling software package developed by the United States Environmental Protection Agency; it performs extended-period simulation of hydraulic and water-quality behavior within pressurized pipe networks, and it has been also used for leak detection [13]. Flow-Master® is another software program that has been used for the representation of hydraulic networks and the development of leakage detection and localization techniques [14].

In line with the laboratories previously mentioned, as well as their important usefulness in the development of new leak detection and localization techniques, this paper proposes a model of an extended hydraulic network with two lateral connections that is under construction at the Instituto Tecnológico de Tuxtla Gutiérrez. It will allow experimenting with a hydraulic network of a few nodes that emulate some real problems that arise in a real distribution network. Therefore, this work presents the modeling and simulation of a 200 m hydraulic network, with a storage depot of 2500 L, a hydraulic pump of 5 hp, four nodes in its two lateral sockets and valves for the simulation of leaks. SolidWorks® is used to present its geometric arrangement in order to consider the space available at the laboratory where it will be located. The physical layout of the network consists of a base with two lateral branches that connect to a second level. The steady-state model of the hydraulic network obtained in the MATLAB® environment is also described. In this way, a unique prototype that simulates a real hydraulic network is obtained. It will serve as an experimental basis to develop leakage diagnosis algorithms, in addition to optimal control techniques applied to the network to reduce water waste due to hydraulic leaks. The results are presented with the description of the parameters obtained in leak-free and leak conditions.

2. Modeling of the Hydraulic Network

2.1. Hydraulic Concepts

A distribution network comprises a collection of interconnected pipe sections in a specific configuration, each one with a length, a diameter and a roughness according to the material. Sections of the pipeline may contain pumps and accessories, such as elbows and valves. The end points of each pipe section are identified as union nodes or fixed grade nodes. A joint node is a point where two or more sections of pipe join together, and it is also a point of consumption where the flow can enter and exit the system. A node of fixed degree is a point where a constant piezometric height is maintained, such as a connection to a reservoir, an elevated storage tank or any other constant pressure region. In any pipe network, the following equation is fulfilled [15]:

$$N_P = N_J + N_L + N_F - 1, \tag{1}$$

where N_P is the number of pipe sections (also called lines), N_J is the number of union nodes, N_L is the number of closed loops and N_F is the number of fixed degree nodes.

On the other hand, hydraulic pumps can be described in several ways. For some applications, a constant power input is specified. For other applications, a curve is adapted to the actual operating data of the pump. For pumps described for their useful power P_u, the energy added by the pump between the nodes i, j is expressed in terms of the flow rate by:

$$H_j - H_i = \mu\, P_u / Q_{ij}; \tag{2}$$

where H is the pressure head (mwc), μ it is a constant that depends on the units used and Q is the flow (m^3/s). Alternatively, a pump can be described by its flow values at different operating points; these points relate the discharges (output flows) with the differences in piezometric height through the pump. A quadratic polynomial can be adjusted through these points to obtain a characteristic curve that describes the operation of the pump in the form:

$$H_j - H_i = a_0 \eta^2 + b_0 \eta\, Q_{ij} + c_0 Q_{ij}^2, \tag{3}$$

where a_0, b_0 y c_0 are the coefficients of the curve that represent the real operation when the pump works at maximum speed. At least three data points are required to determine the coefficients, where η is the proportion of the rotational speed at any time, with respect to the rotational speed associated with the data used to determine the coefficients [16].

2.2. Energy Losses and Mass Balance in Hydraulic Networks

When there is fluid transport in a hydraulic network, an advance resistance is generated due to the roughness of the material, where the relative roughness is defined by:

$$\epsilon = k_s / d, \tag{4}$$

where k_s is the absolute roughness (mm) and d is the diameter of the pipeline (mm). Another important parameter of the flow in hydraulic networks is the friction factor f, which depends on the geometry of the pipe and the Reynolds number (Re). The friction factor depends on the flow turbulence, which is considered to be laminated when Re < 2000, and it is considered turbulent when Re > 4000 [17]. In a turbulent regime, the friction factor depends not only on the Reynolds number, but also on the relative roughness (ϵ). In the transition region $2000 \le$ Re ≤ 4000, to estimate the friction, a cubic interpolation is made with the border values of both regimes. Taking into account both types of flow (laminar and turbulent), the friction factor is calculated with:

$$f = \begin{cases} 64/\text{Re}, & \text{Re} < 2000; \\ \left(-2 \log_{10} \left(\dfrac{\epsilon}{3.7} + \dfrac{5.74}{\text{Re}^{0.9}} \right) \right)^{-2}, & \text{Re} > 4000. \end{cases} \tag{5}$$

When the system is in a steady state, it is possible to calculate the friction factor with the Darcy–Weisbach equation using measurements obtained from the hydraulic network, to next use the Colebrook equation to calculate the roughness coefficient of the hydraulic network material. However, in practice, an explicit approximation using the Swamee–Jain equation is used [18]:

$$f = 0.25 \left(\log_{10} \left(\frac{\varepsilon}{3.7} + \frac{5.74}{\text{Re}^{0.9}} \right) \right)^{-2}. \tag{6}$$

Therefore, considering the roughness of the material and the friction caused by the flow through the pipeline, the pressure energy losses are defined by the Darcy–Weisbach equation:

$$H_i - H_j = h_{ij} = f L Q_{ij}^2 / 2g A^2 d, \tag{7}$$

where L is the length (m) of the pipeline, h_{ij} is the loss of pressure (mwc) when the flow passes through the pipe, A is the area of the cross-section (m^2) of the pipeline, d is the diameter of the pipeline (mm) and g is the gravity (m/s^2). It is possible to represent some parameters of the Darcy–Weisbach equation in a single coefficient of the form:

$$r_{ij} = f L / \left(2g A^2 d \right), \tag{8}$$

and Equation (7):

$$H_i - H_j = h_{ij} = r_{ij} Q_{ij}^2, \tag{9}$$

where r_{ij} is the coefficient of resistance to flow. It is important to mention that the equivalent length of the accessories in the hydraulic network is included in the variable L of Equation (8).

For the flows that are distributed in hydraulic networks, the principle of the conservation of mass in each node of the hydraulic network is considered, and it is presented in the following equation:

$$\sum_i Q_{ij} = D_j, \tag{10}$$

where the sum considers all the flows arriving at the node j from any adjacent node i and D_j is the demand or consumption in the node j. On the other hand, to represent a leak at any point of the hydraulic network, the Torricelli equation is used [19]:

$$Q_f = \lambda_f \sqrt{H_f}, \tag{11}$$

where Q_f it is the flow of the leak, λ is the leakage coefficient and H_f is the pressure at the point where the leak occurs. The leakage coefficient is calculated according to the following expression:

$$\lambda_f = c_d A_d \sqrt{2g}, \tag{12}$$

where c_d is a discharge coefficient in the leak and A_d is the area of the hole in the pipe where the leak occurs. These equations are considered in the modeling of hydraulic networks.

In this case, the model developed is in the steady state regime in order to guarantee flow in all regions of the hydraulic network even when a leakage flow is present at any point in the pipeline. The leak, in this case, is modeled with Torricelli's equation. However, it is important to mention that in a real hydraulic network system, the flow regime is normally turbulent, which means that a more accurate model is needed to consider the transient effects in order to perform a pressure wave analysis in the case of an abrupt leak.

In the design of the hydraulic network, the Schedule 80 PVC pipe is proposed for its elastic properties and low cost compared to other materials. Other characteristics of the network, such as the inner diameter, the length of the pipe and the coil arrangement of the system were proposed according to the available space in the hydraulic laboratory. The diameter and length are physical parameters that are directly considered in the model. In the case of the coil arrangement and elbows, these are used to redirect the flow, causing more turbulence and more energy losses. However, the energy losses caused by accessories are included indirectly by considering an equivalent straight pipe with that energy loss.

2.3. Configurations of the Proposed Hydraulic Network

The proposed hydraulic network is designed to represent three geometric arrangements; this is possible according to the opening and closing of valves distributed in the system. The first configuration is represented in the instrumentation diagram of Figure 1, where all the valves remain open. The second configuration is achieved by closing valves G1 and G2, to get the shape of a simple pipe, as presented in Figure 2. Finally, the third configuration consist of a duct with two branches, which can be obtained when the valves G3 and G4 are closed, leading to the arrangement shown in Figure 3. This emphasizes that the proposed hydraulic network is reconfigurable, and it will allow the development of leak detection and control algorithms in hydraulic networks with different topologies. The physical layout of the hydraulic network (as it will be at the end of its construction) is shown in Figure 4. The first level of this hydraulic network is already built and instrumented [20] and the construction of the second level is work in process that should be completed by the end of 2018.

The importance of the different configurations of the proposed network is because they represent real cases of hydraulic distribution networks. For example, the configuration of a single pipeline is commonly used in hydrocarbon distribution networks where the fluid is transported by very extensive (kilometers of) single pipelines; the branched network is commonly used to transport fluids in industry, where there is a single source/reservoir of fluid and several demand points at different ends of the network; while the interconnected network (closed network) is used in the transport of drinkable water, allowing, in some cases, the water supply of a complete city.

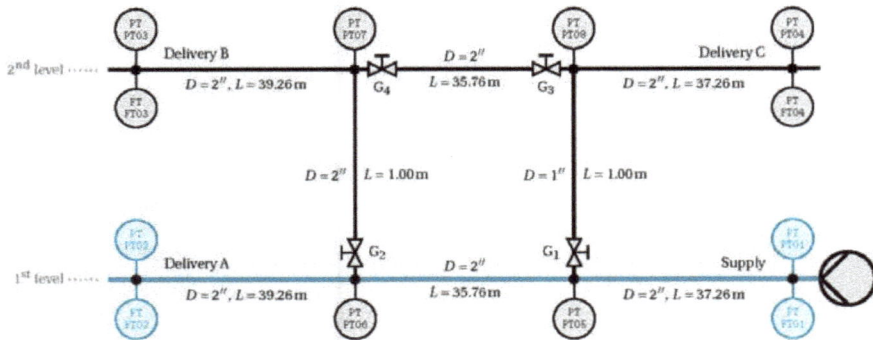

Figure 1. Complete configuration of the hydraulic network.

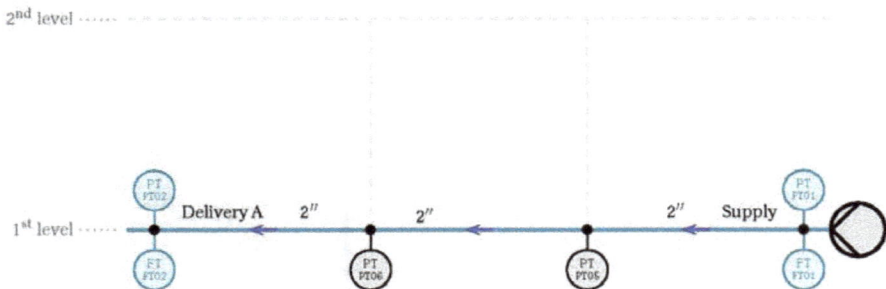

Figure 2. Simple pipe configuration.

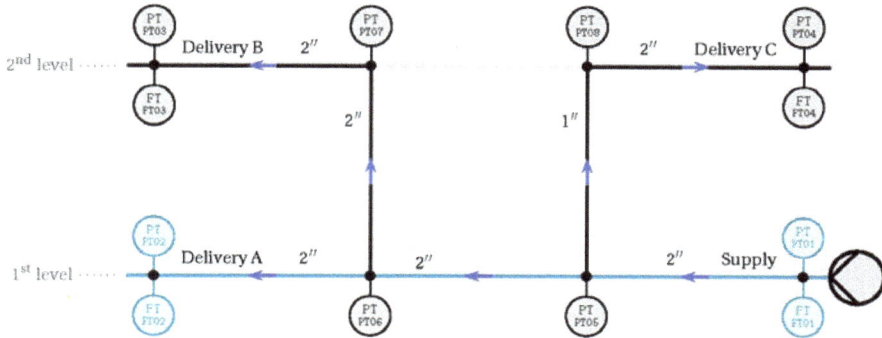

Figure 3. Sectioned hydraulic network configuration.

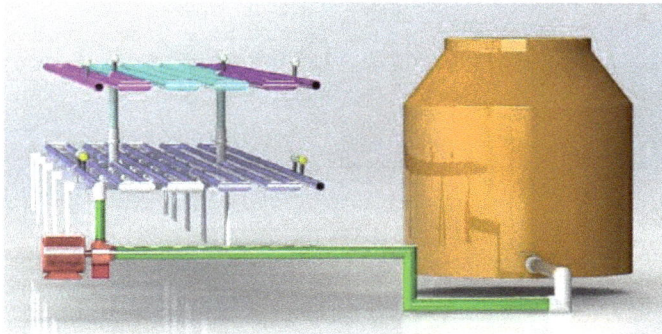

Figure 4. Front view of the hydraulic network in SolidWorks®.

2.4. Modeling of the Proposed Hydraulic Network

Figure 5 shows the pressures on the nodes and the directions of the flows through the hydraulic network, as well as the nomenclature considered to develop the mathematical model of the network.

To model the hydraulic network, it is necessary to apply the energy loss equations in each line or pipe of the system and a mass balance in each node.

On the other hand, as was previously mentioned, the lower level of the network is a section already built and instrumented at the Instituto Tecnológico de Tuxtla Gutiérrez (ITTG) hydraulics laboratory. The piezometric height values are known according to the water level of the deposit (where the flow is stored and recirculated to flow into the network). Furthermore, the pressure provided by the pump is known through sensors at the first level of the network, and it is considered as the boundary parameter that causes the flow in the network. These are known values in the system of equations that models the proposed network and that will be solved simultaneously. The known parameters include the pressures H_0, H_4, H_5, H_8 referring to the nodes of Figure 5, which conform to 1 mwc, and as for Node 1 of the hydraulic pump H_1, the values are according to the characteristic curve generated by a pressure of 5.60 mwc and a flow of 3.83 m^3/s, while the roughness value of the Schedule 80 PVC material is $\epsilon = 0.0235$ mm, calculated with the Swamee–Jain equation with the experimental values obtained from the simple duct configuration that was already built [20]. The proposed lengths and diameters in each region of the network are presented in Table 1. Using Equations (3), (9) and (11), a system of nonlinear equations is obtained that will be solved in MATLAB®, where it is possible to assign different values arbitrarily to the leakage coefficient λ to simulate different scenarios without leakage and with leakage in Node 9, which represents a hole that introduces to the model a virtual node between Nodes 2 and 3. In this way, the pressure and flow values distributed throughout the hydraulic network will be obtained. Therefore, from the system represented in Figure 5, the next set of equations is generated; for pressure drops in the system, the model is presented as follows,

$$H_0 - H_1 = F(Q_{pump}), \tag{13}$$

$$H_1 - H_2 = r_{12} Q_{12}^2, \tag{14}$$

$$H_2 - H_9 = r_{29} Q_{29}^2, \tag{15}$$

$$H_3 - H_4 = r_{34} Q_{34}^2, \tag{16}$$

$$H_2 - H_6 = r_{26} Q_{26}^2, \tag{17}$$

$$H_6 - H_5 = r_{65} Q_{65}^2, \tag{18}$$

$$H_6 - H_7 = r_{67} Q_{67}^2, \tag{19}$$

$$H_7 - H_8 = r_{65} Q_{65}^2, \tag{20}$$

$$H_3 - H_7 = r_{37} Q_{65}^2, \tag{21}$$

$$H_9 - H_3 = r_{93} Q_{93}^2, \tag{22}$$

where the coefficient r_{ij} is described in Equation (8) and F indicates that the pressure difference between Nodes 0 and 1 of the network is a function of the flow produced by the pump.

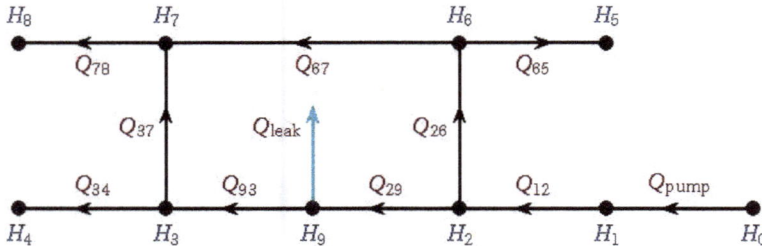

Figure 5. Nomenclature of the system variables.

Table 1. Dimensions of the pipe sections.

Section	Length (m)	Diameter (mm)
$1 \to 2$	37.26	48.6
$2 \to 9$	x (*)	48.6
$9 \to 3$	$35.76 - x$	48.6
$3 \to 4$	39.26	48.6
$2 \to 6$	1.00	24.3
$3 \to 7$	1.00	48.6
$6 \to 5$	37.26	48.6
$6 \to 7$	35.76	48.6
$7 \to 8$	39.26	48.6

(*) Denotes an arbitrary escape position between Nodes 2 and 3.

Considering the mass balance in the nodes of the system, the following set of equations is obtained,

$$Q_{pump} - Q_{12} = 0, \tag{23}$$

$$Q_{12} - Q_{29} = Q_{26}, \tag{24}$$

$$Q_{29} - Q_{leak} = Q_{93}, \tag{25}$$

$$Q_{93} - Q_{34} = Q_{37}, \tag{26}$$

$$Q_{26} - Q_{65} = Q_{67}, \tag{27}$$

$$Q_{37} + Q_{67} = Q_{78}, \tag{28}$$

where the leakage flow is calculated with the Torricelli Equation (11):

$$Q_{leak} = \lambda\sqrt{H_9}. \tag{29}$$

3. Numerical Experiments

The set of Equations (13)–(29) obtained in the previous section is solved in MATLAB® using the nonlinear optimization method called "trust region", whose purpose is to solve a system of equations $f(x) = 0$ in order to find the optimal solution through an iterative process minimizing $\|f(x)\|^2$. The efficiency of this numerical method is due to its rapid convergence [21]; in fact, it only takes 3–5 s to simulate, on a PC with a Core i5 processor, 24 h of running time.

The first numerical experiment is performed in the closed network without leaks. The solution of the model provided the values of flow and pressure in different regions of the hydraulic network. The results are shown in Figure 6 where the pressure values (mwc) are located at the nodes and the flows (m^3/s) in the lines of the figure. The pressure values at the ends of the hydraulic network (Nodes 0, 4, 5, 8) are known values with a hydraulic height of 1 mwc. This height corresponds to the water reservoir, as there will be a single deposit that will be used to supply, store and demand water to/from the network. Therefore, all the network ends connect to this deposit, and the water is recirculated through the system. It also should be noted here that the displayed pressure values are gauge pressure values, i.e., zero-referenced against ambient air pressure, so they are equal to absolute pressure minus atmospheric pressure.

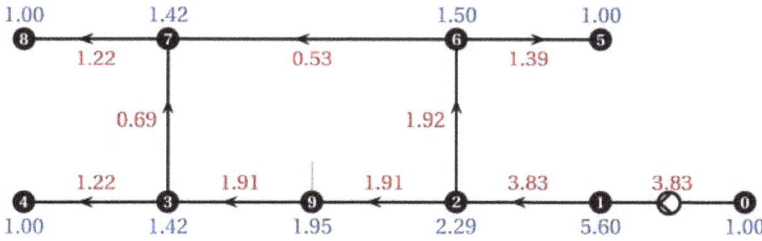

Figure 6. Pressures and flows in the network without leakage.

In the same way, a second numerical experiment is performed considering a leak in Node 9 at a z distance with respect to Node 0; the obtained values are presented in Figure 7. In this case, the results show that, when a leak occurs, the pressures decrease in each node of the network, and there are variations in the flow rates. The leak was proposed in Node 9; however, if the leak is in any other region of the pipe, new values of flows and pressures would be obtained in the network due to the mass and energy conservation, respectively. However, the flow in each section of the hydraulic network is still turbulent considering its speed and its diameters with respect to the Reynolds number.

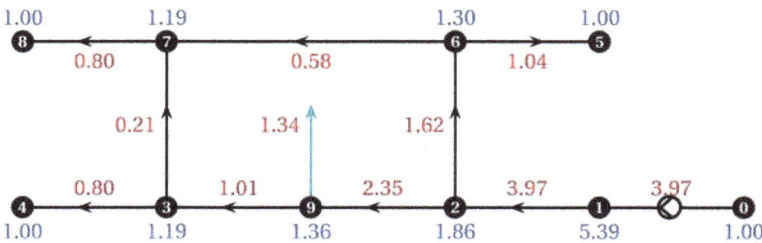

Figure 7. Pressures and flows in the network with a leakage in Node 9.

The numerical experiment whose result is shown in Figure 7 corresponds to a specific and unique leakage configuration. However, using the mathematical model of the network, it is possible to simulate varying conditions over time, e.g., programmed consumptions in the nodes that change during the day or leaks that increase progressively, among others. To execute these numerical experiments—called "extended period" in the specialized literature—from the steady state equations, a sweep is made of the values that the parameters can take over time using "for" cycles, as shown in the following pseudocode:

```
params = value_list;
sols = [];
for param = params
sol = simulateNetwork(param);
sols = [sols,sol];
end
```

With the previous consideration, the network was simulated with a progressive leak in Node 9, and the gradual change in the leak coefficient was modeled using an exponential function:

$$\lambda = 1 \times 10^{-4} \left(1 - \exp(-t \times 3600/3)\right) \qquad [\text{m}^{5/2}\text{s}^{-1}], \tag{30}$$

where t is time in hours. This function was selected considering that its evolution over time represents an increase from zero to a value $\lambda_{max} = 1 \times 10^{-4}$ that would correspond to a leak through a circular hole of about a quarter of an inch in diameter. In the graphs of Figures 8 and 9, the effects of a leak over the network are shown, i.e., the increase in the flow rate and the loss of the pressure head at the outlet of the pump. In Figure 8, it can be seen that the flow of the pump Q_{pump} represented between Nodes 0 and 1 remains constant over time when the system does not leak; however, this flow increases exponentially when a leak is present in the system.

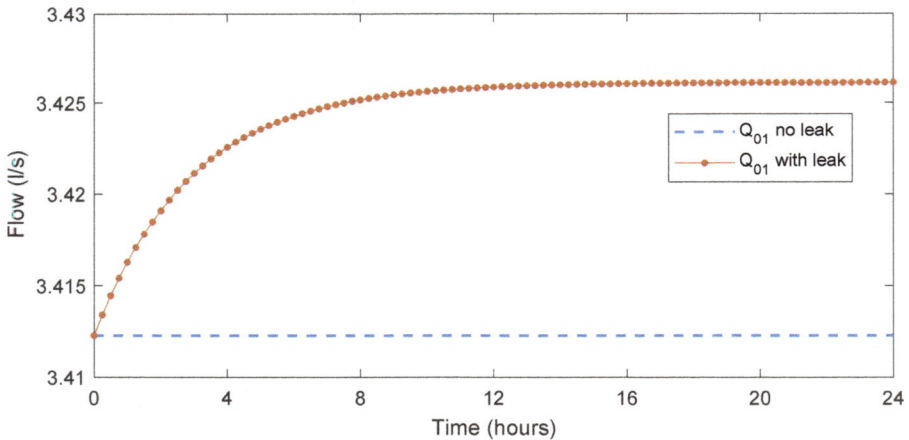

Figure 8. Flow variation in the presence of a leak.

In relation to the pressure variation in the system due to the occurrence of a leak, in Figure 9, it is appreciated how the pressure H_1 remains constant in the absence of leaks in the system, but decreases when the leak appears.

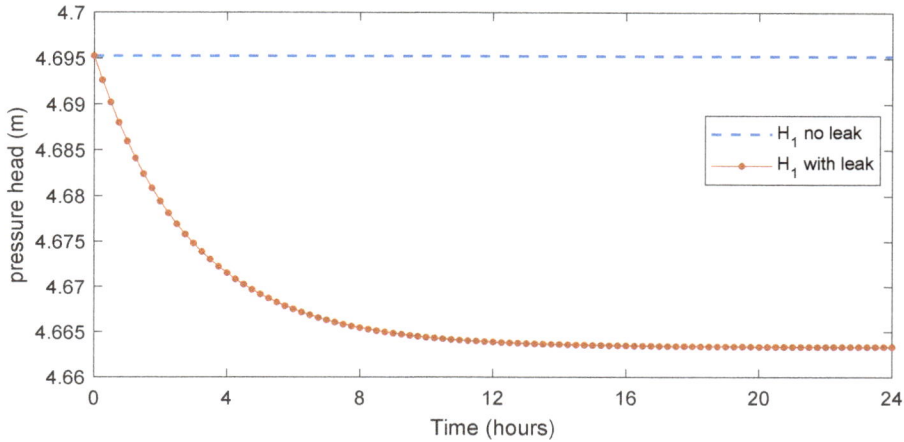

Figure 9. Pressure head loss at the pump outlet caused by a leak.

4. Conclusions

This paper presented the design of a prototype of a hydraulic network that allows the development of leakage control and diagnosis algorithms for three different configurations: single ducts, ducts with branches and closed networks. The proposed hydraulic network was modeled, obtaining its system of equations for the steady state case, which was solved using nonlinear optimization algorithms. Numerical experiments in MATLAB® showed that the proposed network structure guarantees turbulent flow in all its lines for the considered configurations and that its construction and use as a scale model of pressurized networks is feasible. It is important to mention that the developed model is part of a section already built of the hydraulic network, which is the simple pipeline section. The known and experimentally-validated parameters of the already built section of the network were taken into account in the extended model of the network. However, it is necessary to perform a mean square error analysis between the experimental values of flow and pressure and the values obtained numerically to further adjust the parameters of the model. This analysis remains as future work. Finally, the goal of the model was to ensure the description of the flow in all sections of the network, leaving the study of the transitory effects for future work. The inclusion of the transitory effects in the model will allow the extension of the leakage diagnosis and control techniques, as well.

Author Contributions: F.-R.L.-E. and G.V.-P. conceived and designed the experiments; J.-R.B. and H.-R.H. performed the experiments; G.B. and L.T. analyzed the data; J.-R.B. wrote the paper.

Funding: This research was funded by the National Council for Science and Technology (CONACYT) of Mexico, Grant Number PN-2016/3595, and Tecnológico Nacional de México (TecNM) under the program Apoyo a la Investigación Científica y Tecnológica, Grant Number 6358.17-P.

Conflicts of Interest: The authors declare no conflict of interest.

References

1. Solis Luna, N.B. Determinación Remota de Fugas de Gas y Petróleo Por Medio de Cámaras Infrarrojas. Ph.D. Thesis, Instituto Politécnico Nacional, Mexico City, Mexico, 2009.
2. Fuentes-Mariles, O.; Palma-Nava, A.; Rodríguez-Vázquez, K. Estimación y localización de fugas en una red de tuberías de agua potable usando algoritmos genéticos. *Ing. Investig. Techol.* **2011**, *12*, 235–242.
3. Doney, K. Leak Detection in Pipelines Using the Extended Kalman Filter and the Extended Boundary Approach. Ph.D. Thesis, University of Saskatchewan, Saskatoon, SK, Canada, 2007.
4. OECD. Water Governance in Cities. In *OECD Studies on Water*; OECD: Paris, France, 2016.

5.   Van Pham, T.; Georges, D.; Besancon, G. Predictive control with guaranteed stability for water hammer equations. *IEEE Trans. Autom. Control* **2014**, *59*, 465–470. [CrossRef]

6.   Torres, L.; Verde, C.; Carrera, R.; Cayetano, R. Algoritmos de diagnóstico para fallas en ductos. *Techol. Cienc. Agua* **2014**, *5*, 57–78.

7.   Wang, Y.; Blesa, J.; Puig, V. Robust Periodic Economic Predictive Control based on Interval Arithmetic for Water Distribution Networks. *IFAC-PapersOnLine* **2017**, *50*, 5202–5207. [CrossRef]

8.   Verde, C. Multi-leak detection and isolation in fluid pipelines. *Control Eng. Pract.* **2001**, *9*, 673–682. [CrossRef]

9.   Torres, L.; Besancon, G.; Georges, D. A collocation model for water-hammer dynamics with application to leak detection. In Proceedings of the 47th IEEE Conference on Decision and Control, Cancun, Mexico, 9–11 December 2008; pp. 3890–3894.

10.   Navarro, A.; Begovich, O.; Besançon, G.; Dulhoste, J. Real-time leak isolation based on state estimation in a plastic pipeline. In Proceedings of the IEEE International Conference on Control Applications (CCA), Denver, CO, USA, 28–30 September 2011; pp. 953–957.

11.   Delgado-Aguiñaga, J.; Besançon, G.; Begovich, O.; Carvajal, J. Multi-leak diagnosis in pipelines based on Extended Kalman Filter. *Control Eng. Pract.* **2016**, *49*, 139–148. [CrossRef]

12.   Ocampo-Martinez, C.; Barcelli, D.; Puig, V.; Bemporad, A. Hierarchical and decentralised model predictive control of drinking water networks: Application to barcelona case study. *IET Control Theory Appl.* **2012**, *6*, 62–71. [CrossRef]

13.   Scola, I.R.; Besançon, G.; Georges, D. Optimizing Kalman optimal observer for state affine systems by input selection. *Automatica* **2018**, *93*, 224–230.

14.   Soldevila, A.; Fernandez-Canti, R.M.; Blesa, J.; Tornil-Sin, S.; Puig, V. Leak localization in water distribution networks using Bayesian classifiers. *J. Process Control* **2017**, *55*, 1–9. [CrossRef]

15.   Wood, D.J.; Rayes, A.G. Reliability of algorithms for pipe network analysis. *J. Hydraul. Div.* **1981**, *107*, 1145–1161.

16.   Wylie, E.B.; Streeter, V.L. *Hydraulic Transients*; FEB Press: Ann Arbor, MI, USA, 1983.

17.   Crane. *Flujo de Fluidos en Válvulas, Accesorios y Tuberías*; McGraw-Hill: New York, NY, USA, 1989.

18.   Genić, S.; Aranđelović, I.; Kolendić, P.; Jarić, M.; Budimir, N.; Genić, V. A review of explicit approximations of Colebrook's equation. *FME Trans.* **2011**, *39*, 67–71.

19.   Mott, R.L. *Mecánica de Fluidos Aplicada*; Pearson Educación: Turin, Italy, 1996.

20.   Bermúdez, J.R.; Santos-Ruiz, I.; López-Estrada, F.R.; Torres, L.; Puig, V. Diseño y modelado dinámico de una planta piloto para detección de fugas hidráulicas. In Proceedings of the Congreso Nacional de Control Automático CNCA 2017, Mexico City, Mexico, 4–6 October 2017; Asociación Mexicana de Control Automático: Mexico City, Mexico, 2017; Volume 1, pp. 2–7.

21.   Coleman, T.F.; Li, Y. An interior trust region approach for nonlinear minimization subject to bounds. *SIAM J. Optim.* **1996**, *6*, 418–445. [CrossRef]

MDPI

St. Alban-Anlage 66

4052 Basel

Switzerland

Tel. +41 61 683 77 34

Fax +41 61 302 89 18

www.mdpi.com

Mathematical and Computational Applications Editorial Office

E-mail: mca@mdpi.com

www.mdpi.com/journal/mca

www.ingramcontent.com/pod-product-compliance
Lightning Source LLC
Chambersburg PA
CBHW051727210326
41597CB00032B/5631